COHOMOLOGY AND DIFFERENTIAL FORMS

COHOMOLOGY AND DIFFERENTIAL FORMS

Izu Vaisman

Department of Mathematics
University of Haifa
Haifa, Israel

DOVER PUBLICATIONS, INC.
Mineola, New York

About the Author

Izu Vaisman was a member of the Faculty of Mathematics and Mechanics at the University Al. I. Cuza, Iasi, Romania, from 1959 to 1976. He has been Professor of Mathematics at the University of Haifa, Haifa, Israel, from 1976 to the present.

Translation Editor:
Samuel I. Goldberg, *University of Illinois*

Copyright
Copyright © 1973 by Izu Vaisman
All rights reserved.

Bibliographical Note

This Dover edition, first published in 2016, is a slightly corrected republication of the work originally published in the series "Pure and Applied Mathematics: A Series of Monographs and Textbooks" by Marcel Dekker, Inc., New York, in 1973.

Library of Congress Cataloging-in-Publication Data

Names: Vaisman, Izu. | Goldberg, Samuel I.
Title: Cohomology and differential forms / Izu Vaisman, Faculty of Mathematics and Mechanics, University Al I. Cuza, Iasi, Romania [and] Department of Mathematics, University of Haifa, Haifa, Israel.
Description: Dover edition. | Mineola, New York : Dover Publications, 2016. | "Translation editor: Samuel I. Goldberg, University of Illinois"— Galley t.p. verso. | Slightly corrected; originally published: New York : Marcel Dekker, Inc., 1973. | Includes bibliographical references and index.
Identifiers: LCCN 2015051405| ISBN 9780486804835 | ISBN 0486804836
Subjects: LCSH: Geometry, Differential. | Differential forms. | Cohomology theory.
Classification: LCC QA649 .V2813 2016 | DDC 514/.23—dc23 LC record available at http://lccn.loc.gov/2015051405

CONTENTS

Preface   *v*

Chapter 1   CATEGORIES AND FUNCTORS   1
1. Classes and Sets 1
2. Pseudocategories and Categories 4
3. Morphisms, Objects, and Operations 9
4. Abelian Categories 18
5. Functors and Homology 25
6. Atlases 36

Chapter 2   SHEAVES AND COHOMOLOGY   43
1. Presheaves on a Topological Space 43
2. Sheaves of Sets 48
3. Sheaves with Values in a Cantorian Category 54
4. Cohomology with Coefficients in Presheaves 60
5. The Case $\mathscr{C} = A\text{-Mod}$ 68
6. Cohomology with Coefficients in a Sheaf 78

Chapter 3   FIBER AND VECTOR BUNDLES   93
1. Fiber Bundles 93
2. Fiber Bundles with Structure Group 101
3. Vector Bundles 113
4. Operations with Vector Bundles. Characteristic Classes 123

Chapter 4   DIFFERENTIAL GEOMETRY   135
1. Differentiable Manifolds 135
2. Vector and Tensor Fields 146
3. Differential Forms and Integration 155
4. Absolute Differential Calculus 165
5. Riemannian and Foliated Riemannian Manifolds 174
6. Complex and Almost Complex Manifolds 188

Chapter 5   COHOMOLOGY CLASSES AND DIFFERENTIAL FORMS   201
1. The Theorems of de Rham and Allendoerfer–Eells 201
2. Theorems of de Rham Type for Complex and Foliated Manifolds 215
3. Characteristic Classes of Differentiable Vector Bundles 230
4. Elliptic Operators. Elliptic Complexes 247
5. Cohomology and Harmonic Forms 262

References   277

Index   281

PREFACE

There are in differential geometry three important structures with specific sheaves of germs of differentiable functions: the differentiable manifolds with the sheaf of germs of locally constant functions, the foliated manifolds with the sheaf of germs of differentiable functions which are constant on the leaves, and the complex analytic manifolds with the sheaf of germs of holomorphic functions. In each case, there is a corresponding cohomology theory which plays an essential role.

In the present monograph, which is a result of lectures given at the University of Iasi (Romania) and of some personal research, we wish to expose the known method of calculating the cohomology spaces of the manifold with coefficients in the above mentioned sheaves using globally defined differential forms, and to give some applications.

It is assumed that the reader knows enough algebra, analysis, topology, and classical differential geometry, and that he has sufficient training in abstract mathematical reasoning.

In the first part, our principal aim is to expose the Čech cohomology with coefficients in a sheaf of Abelian and non-Abelian groups. For a general setting of the problems and to avoid repetition we begin with a chapter on categories and functors, where we only intend to acquire the corresponding language. The second chapter develops Čech cohomology theory, including the non-Abelian case. The geometric theory of the nonabelian Čech cohomology, i.e., the theory of fiber bundles with structure group is exposed in the third chapter. In the fourth chapter we review the principal subjects of differential geometry, considering the general case of manifolds modeled on Banach spaces. Finally, the last chapter contains the solution of the problems announced, the theorems of de Rham and Dolbeault–Serre and the corresponding theorem which we gave for foliated manifolds. An informative treatment of the theorem of Allendoerfer and Eells is also given. In the same chapter, applications of these theorems to characteristic classes and the general theory of harmonic forms are treated.

We consider it useful for the reader to have in the preliminary chapters a more complete development than would be strictly necessary for our cohomology problems. In the book, there is a basic text where all things are proved, with the exception of technical details, and a supplementary text where we give, without proof, information which is less important here but still worthy of consideration by the reader.

Let us mention some new features (in form or content) to be found in this book: the introduction of pseudocategories and strengthened categories; the derivative functors; the theory of atlases; a new definition of sheaves with values in a category; the study of foliated Riemann manifolds and all things related to them (in the fifth chapter); etc.

The limitations of our work did not permit us to develop other "de Rham theorems," e.g., those arising in the deformation theory of pseudogroup structures, but the reader will find here all that is necessary for an independent study of them.

The literature at the end of the book contains only works which we employed in writing it and the references are not always to those who introduced the notions or obtained the results but rather to those whose papers we employ. Cross references are given in the usual manner, e.g., "Proposition 2.3.7," "Formula 2.4(8)", etc., where in the same chapter (or section), the chapter (section) number is omitted. Notation the meaning of which is clear from the context is not explained.

We hope that this monograph will be useful to both graduate students and research workers in differential geometry and global analysis.

I should like to express here my gratitude to the Mathematical Faculty of the University of Iași for providing me the opportunity to teach courses on the subject of this book, to Prof. R. Miron, Prof. V. Cruceanu, Dr. L. Maxim, and Dr. V. Oproiu for reading the manuscript, and to the Department of Mathematics of the University of Illinois for inviting me to visit, an invitation which led me to write this book in English (I am sorry that, for reasons beyond my control, this visit could not be undertaken).

Preface

It is a pleasure for me to acknowledge my special gratitude to Prof. Samuel I. Goldberg of the University of Illinois for his interest in my mathematical activity, for reviewing this book, and correcting its English, and for his invaluable help in publication.

Last, but not least, I should like to express my gratitude to my wife Silvia for having created for me very favorable conditions during the writing of this book.

I. Vaisman

Chapter 1

CATEGORIES AND FUNCTORS

1 Classes and Sets

The objects with which we usually operate in mathematics are called classes; they arise by abstraction from the intuitive notion of a collection of things. This intuitive image is unsatisfactory for mathematical operations because it gives rise to contradictions, so we must consider classes as defined by a determined system of axioms. We consider here the same system of axioms as in [7].

In this system, we have one primitive notion, the *class*, and one primitive relation, *belongs to*. The classes will be denoted by different letters and the relation A *belongs to* B will be denoted by $A \in B$. The sign $\notin$ will be used to deny the relation $\in$. A class A is called a *set* if there is at least one class B such that $A \in B$, and in this case we shall say that A is an *element* of B. We consider also the *equality* relation of classes, $A = B$ (and *inequality*, $A \neq B$), as logical identity, i.e., $A = B$ if and only if for any property $P(X)$ of classes, $P(A)$ and $P(B)$ are simultaneously true or false.

The first two axioms of the system which we expose for sets and classes are:

A1. *Two classes are equal if and only if they have the same elements.*

A2. *For any property $P(X)$ of classes, there is a class whose elements are just the sets X for which $P(X)$ is true.*

From A1, it follows that the class determined by A2 is unique; it will be denoted by $\{X \mid P(X)\}$. If this class is a set, $P(X)$—the characteristic property of this set—is called a *collectivizing property*. For example, the property $X \notin X$ is not a collectivizing property because, $A = \{X \mid X \notin X\}$ would be a set and we would derive Russell's antinomy: $A \in A$ if and only if $A \notin A$. A is also an example of a class which is not a set.

A class A is called a *subclass* of the class B if any element of A is also an element of B. We shall write then $A \subseteq B$ or $B \supseteq A$. In particular, if A is a set, it will be called a *subset* of B. We give now a third axiom of the system

A3. *Every subclass of a set is itself a set.*

Using axiom A2, we easily get the existence of the following classes

$$\varnothing = \{X \mid X \neq X\}$$
$$\bigcup A = \{X \mid \text{there is a } Y \in A \text{ such that } X \in Y\} \qquad (1)$$
$$\mathfrak{P}(A) = \{X \mid X \subseteq A\},$$

called respectively the *empty class*, the *union* of the elements of A and the *power class* of A. We now give three new axioms:

A4. $\varnothing$ *is a set.*
A5. *If A is a set, $\bigcup A$ and $\mathfrak{P}(A)$ are again sets.*
A6. *If A and B are sets, $\{A, B\} = \{X \mid X = A \text{ or } X = B\}$ is a set.*

As a consequence, if A is a set there is a set $\{A\}$ whose single element is A. This is $\mathfrak{P}(A)$ for $A = \varnothing$ and $\{X \mid X \in \{\varnothing, A\} \text{ and } X \neq \varnothing\}$ for $A \neq \varnothing$.

For two classes A and B we define the *cartesian product*

$$A \times B = \{Z \mid Z = (X, Y), X \in A, Y \in B\}, \qquad (2)$$

where $(X, Y) = \{\{X\}, \{X, Y\}\}$ is the *ordered pair*. A *binary relation* from A to B is a subclass of $A \times B$ and, if the relation is such that to every element of A corresponds only one element of B, it is called a *map* or a *function*. Now, we shall introduce a new axiom:

A7. *If F is a function from a set A to a class B, the class of elements of B which correspond to elements of A is again a set.*

The set given by A7 is the *range* of F and is denoted by $\{X_a\}_{a \in A}$, where X is the element corresponding to a.

2. Classes and Sets

An important notion is that of a *universal set*. This is a set U, satisfying the following conditions:

(a) If $X \in U$, then $X \subseteq U$.
(b) If $X \in U$, then $\mathfrak{P}(X) \in U$.
(c) If $X, Y \in U$, then $\{X, Y\} \in U$.
(d) If $F = \{F_i\}_{i \in I}$ and $F_i \in U$, $I \in U$, then $\bigcup F \in U$.

We add now to our system the axiom

A8. *Every set is an element of some universal set.*

As a consequence the existence of infinite sets is implied.

The axiomatic system of sets and classes must still be completed by two special axioms. The first excludes singular sets satisfying the condition $X \in X$, which are not used in classical mathematics. This axiom is:

A9. *For every nonempty class A there is an $X \in A$ such that there is no set Y for which $Y \in X$ and $Y \in A$.*

Finally, the last axiom assures the existence of sets which cannot be obtained in a constructive manner from the previous axioms. It is the famous *axiom of choice*, which can be given in the following form:

A10. *For every set A, there is a function defined on the set of nonempty subsets of A which associates with every subset one of its elements.*

Using the given system of axioms one can now obtain the usual construction of set theory and the theory of functions and relations. This can be performed as in the so-called naive set theory which we assume known. Hence we shall not go further with the development of the set theory, except for a few comments on partial maps.

Let A, $A' \subseteq A$, and B be three nonempty sets and $f: A' \to B$ a map. Then we call f a *partial map* from A to B and denote it by $f: A \dashrightarrow B$. For this map, A' is the *domain*, $f(A')$ is the *range* or *image*, A is the *source*, and B is the *target*, and they will be denoted respectively by dom f, im f, source f, tar f. Two partial maps will be considered as equal if the respective maps are equal and if they have the same source, domain, and target (and, hence, also the same image).

If $f: A \dashrightarrow B$ and $g: C \dashrightarrow D$, their composition defines a partial map $g \circ f: A \dashrightarrow D$ given by

$$(g|_{\text{im } f \,\cap\, \text{dom } g}) \circ f|_{f^{-1}(\text{im } f \,\cap\, \text{dom } g)},$$

whose domain is $f^{-1}(\text{im } f \cap \text{dom } g)$ and whose image is $\text{im } g|_{\text{im } f \cap \text{dom } g}$. If $\text{im } f \cap \text{dom } g = \emptyset$, $g \circ f$ is not defined.

For partial maps, injectivity and surjectivity have the usual meaning. If $f: A \longrightarrow B$ is injective, there is a partial map $f^{-1}: B \longrightarrow A$ such that $\text{dom } f^{-1} = \text{im } f$ and $\text{im } f^{-1} = \text{dom } f$ and if 1_M is the identity map of a set M we have

$$f^{-1} \circ f = 1_{\text{dom } f}, \qquad f \circ f^{-1} = 1_{\text{im } f}. \tag{3}$$

We note the relations

$$f \circ 1_M = f, \qquad 1_M \circ g = g \tag{4}$$

where $f: M \longrightarrow N$, $g: N \longrightarrow M$, and the fact that the composition of partial maps is associative when defined.

Generally, calculation with partial maps is performed with the rules of usual maps, taking into account the respective domains and images.

2 Pseudocategories and Categories

Usually, the classes studied in mathematics have different structures. In this section we consider a general structure which defines a mathematical language.

1 Definition *A pseudocategory* $\mathfrak{P}$ consists of a class $\mathfrak{O} = \{A, B, \ldots\}$, whose elements are called *objects*, disjoint sets $\mathfrak{M}(A, B)$, one for each ordered pair of objects, and partial maps

$$\circ: \mathfrak{M}(A, B) \times \mathfrak{M}(B, C) \longrightarrow \mathfrak{M}(A, C),$$

one for every triple of objects, such that:

(a) the operation given by $\circ$ is associative when defined;
(b) for every $A \in \mathfrak{O}$ there is an element $1_A \in \mathfrak{M}(A, A)$ such that
$\circ(f, 1_A) = f$, $\circ(1_A, g) = g$, $(f \in \mathfrak{M}(B, A), g \in \mathfrak{M}(A, B))$.

2. Pseudocategories and Categories

If $\mathfrak{P}$ is a pseudocategory we shall give, if necessary, the symbols $\mathfrak{O}$, $\circ$, and $\mathfrak{M}$ the index $\mathfrak{P}$ or some other distinguishing sign. The elements of $\mathfrak{M}(A, B)$ are called *morphisms* of *source* A and *target* B and they are denoted by $u: A \to B$ or $A \xrightarrow{u} B$ instead of $u \in \mathfrak{M}(A, B)$. The operation $\circ$ is called the *composition* of morphisms, and 1_A, which is obviously unique, is called the *identity morphism* of A. Another notation for $\circ(u, v)$ will be $v \circ u$ or vu.

We obtain a typical example of a pseudocategory if we take the class of objects consisting of the nonempty sets and the morphisms being the partial maps of sets. The composition of morphisms is the usual composition of maps and the identity morphisms the identity maps. This pseudocategory will be denoted by Ens^p (from the French word *ensemble*, which means set).

2 Definition If $\mathfrak{P}$ is a pseudocategory such that its objects are sets (possibly with a supplementary structure), the morphisms are maps, and the composition is the composition of maps, $\mathfrak{P}$ will be called a *Cantorian pseudocategory*[†]. If we only have that the objects of $\mathfrak{P}$ are sets, this pseudocategory will be called *quasi-Cantorian*.

Ens^p is an example of a Cantorian pseudocategory. Another important example is obtained by taking as objects nonempty topological spaces and as morphisms partial continuous maps with open domain. This pseudocategory will be denoted by Top^p.

3 Definition A quasi-Cantorian (Cantorian) pseudocategory whose objects are topological spaces (and whose morphisms are continuous maps with the usual composition) is called a *quasitopological* (*topological*) *pseudocategory*.

4 Definition A pseudocategory for which the composition of morphisms $A \xrightarrow{u} B$, $B \xrightarrow{v} C$ is always defined, i.e.,

$$\circ: \mathfrak{M}(A, B) \times \mathfrak{M}(B, C) \to \mathfrak{M}(A, C)$$

is called a *category*.

[†]This name has a different meaning in other texts. Here, we merely wish to distinguish objects that are part of set theory from those that are part of other mathematical theories.

In particular we shall consider Cantorian, quasi-Cantorian, topological, and quasitopological categories.

If the class of objects of a category is a set the category is called *small*.

The structure of a category is that given at the beginning of this section and it is just the structure which we need for a general mathematical language. We remark that it is also possible to consider more general structures which we call *hyperpseudocategories* and *hypercategories* (*great categories* [40]); these are obtained from Definitions 1 and 4 by admitting that the $\mathfrak{M}(A, B)$ can be classes. Then, to avoid antinomies, proper universal sets must be chosen [40].

There are many examples of categories. Thus, if we take the objects to be sets and the morphisms to be maps with the usual composition and identity we get the category of sets which will be denoted by Ens. In the same manner, the topological spaces and continuous maps give the category Top of topological spaces. For both Ens and Top we have to exclude the empty set, but we shall denote also by Ens and Top the previous categories enlarged by the empty set as a new object and by the sets of morphisms $\mathfrak{M}(A, \emptyset) = \emptyset$ and $\mathfrak{M}(\emptyset, A) =$ a set consisting of one element for every object A.

The groups and their homomorphisms define the category Grp of groups.

The Abelian groups and their homomorphisms form the category Ab of Abelian groups. In the same manner, we have the category of rings, of A-modules (where A is a commutative and unitary ring) which is denoted by A-Mod, and also categories corresponding to other algebraic structures.

Another category is obtained if the objects are the *pointed sets*, i.e., sets with a distinguished *base point*, and the morphisms are maps which preserve base points, This category will be denoted by Ens· and we also have an analogous category Top· of pointed topological spaces and continuous maps preserving base points.

Let us give an important example of a quasitopological category, where the morphisms are classes of maps. We recall that two continuous maps between topological spaces, $f, g: X \to Y$, are homotopic if there is a continuous map $F: X \times I \to Y$, $I = [0, 1]$, such that $F(x, 0) = f(x)$ and $F(x, 1) = g(x)$ for $x \in X$. It is easy to verify that homotopy is an equivalence relation on the set Y^X of all continuous maps from X to Y and that it is compatible with the composition of maps [42]. Hence, if we take the objects the topological spaces and the morphisms the homotopy classes of continuous maps, we get a category. This is the so-called *homotopy category* of topological spaces.

2. Pseudocategories and Categories

There are also important categories which are not necessarily cantorian or quasicantorian. Thus, if M is a set and $\leq$ is a partial ordering of it, then M is a small category whose objects are the elements $a, b, \ldots$ of M, the sets $\mathfrak{M}(a, b)$ being empty for $a \not\leq b$ and consisting of one element if $a \leq b$. The composition of morphisms has an obvious definition. We remark that by a partial ordering we understand a binary relation on M which is reflexive and transitive. Clearly, any small category for which the sets of morphisms contain at most one element has a partially ordered set of objects, the relation $a \leq b$ having the meaning $\mathfrak{M}(a, b) \neq \varnothing$. Hence, we can define a partially ordered set to be a small category with the morphism sets as above.

Particular cases of the previous example are the set $\mathfrak{P}(M)$ of subsets of a set M and the set $\mathfrak{D}(T)$ of open subsets of a topological space T. In both cases, the partial ordering is the inclusion of sets and the morphisms can be identified with the inclusion mappings. Hence $\mathfrak{P}(M)$ is cantorian and $\mathfrak{D}(T)$ is topological.

We now define some pseudocategories (categories) associated with given pseudocategories (categories).

If $\mathfrak{P}$ is a pseudocategory, we define the *dual pseudocategory* $\mathfrak{P}^*$ with the same objects as $\mathfrak{P}$ and morphism sets $\mathfrak{M}^*(A, B) = \mathfrak{M}(B, A)$. The composition of morphisms in $\mathfrak{P}^*$ is given by that in $\mathfrak{P}$, i.e., $v \circ^* u = u \circ v$ and the identity morphisms are again those of $\mathfrak{P}$. It is obvious that if $\mathfrak{P}$ is a category its dual $\mathfrak{P}^*$ is also a category.

The existence of a dual category gives a *duality principle* in the theory of categories and we can obtain the dual of any notion or theorem by taking it in the dual category, i.e., by changing the sense of morphisms.

Let $\mathfrak{P}_1$ and $\mathfrak{P}_2$ be two pseudocategories. We define their product $\mathfrak{P}_1 \times \mathfrak{P}_2$, the objects of which are the ordered pairs (A_1, A_2) where A_1 is an object of $\mathfrak{P}_1$ and A_2 an object of $\mathfrak{P}_2$, and the morphism sets

$$\mathfrak{M}_{\mathfrak{P}_1 \times \mathfrak{P}_2}[(A_1, A_2), (B_1, B_2)] = \mathfrak{M}_{\mathfrak{P}_1}(A_1, B_1) \times \mathfrak{M}_{\mathfrak{P}_2}(A_2, B_2).$$

The composition of morphisms in $\mathfrak{P}_1 \times \mathfrak{P}_2$ is componentwise, i.e.,

$$(u_1, u_2) \circ (v_1, v_2) = (u_1 \circ v_1, u_2 \circ v_2),$$

if, of course, the components of the right member exist. The identity morphisms will be of the form $(1_A, 1_B)$.

In the same manner, we can define the product of n pseudocategories. Clearly, the product of categories is also a category.

Consider a pseudocategory $\mathfrak{P}$ and the class of all the morphisms in $\mathfrak{P}$. This can be considered as the class of objects of a new pseudocategory $S(\mathfrak{P})$ if we define the morphisms in $S(\mathfrak{P})$ to be the commutative diagrams of the form

$$\begin{array}{ccc} A & \xrightarrow{u} & B \\ {\scriptstyle a}\downarrow & & \downarrow{\scriptstyle b} \\ A' & \xrightarrow{v} & B' \end{array}, \qquad (1)$$

i.e., the pairs of morphisms in $\mathfrak{P}(A \xrightarrow{a} A', B \xrightarrow{b} B')$ such that

$$v \circ a = b \circ u,$$

where it is assumed that these composites exist.

The composition of two diagrams (1) is defined by the composition of their vertical arrows, when possible. The identity diagrams will have the vertical arrows $(1_A, 1_B)$.

$S(\mathfrak{P})$ is called the *arrow pseudocategory* of $\mathfrak{P}$ and if $\mathfrak{P}$ is a category, so is $S(\mathfrak{P})$.

A generalization is obtained if the arrows are replaced by other diagrams of a given type, which gives general categories of diagrams [40].

5 Definition A pseudocategory $\mathfrak{P}'$ is a *subpseudocategory* of a pseudocategory $\mathfrak{P}$ if it satisfies the following conditions

$$\mathfrak{O}' \subseteq \mathfrak{O}, \quad \mathfrak{M}'(A, B) \subseteq \mathfrak{M}(A, B), \quad A, B \in \mathfrak{O}',$$
$$\circ' = \circ|_{\mathfrak{O}'}, \quad 1'_A = 1_A, \quad A \in \mathfrak{O}'. \qquad (2)$$

The third condition means that the composition of morphisms in $\mathfrak{P}'$ is induced by that in $\mathfrak{P}$. If $\mathfrak{M}'(A, B) = \mathfrak{M}(A, B)$, $\mathfrak{P}'$ is a *full subpseudocategory* of $\mathfrak{P}$. In particular, we shall talk of *subcategories* and *full subcategories*. For example, Ens is a subcategory of Ens^p, and Top is a subcategory of Top^p, but they are not full. Ab is a full subcategory of Grp, the category of Hausdorff spaces and continuous maps is a full subcategory of Top, etc.

Finally, we shall show how to associate with any topological pseudo-category a quasitopological category. Let $\mathfrak{P}$ be a topological pseudo-category and $f: A \to B$ a morphism in $\mathfrak{P}$, i.e., a continuous mapping from an open subset $D \subseteq A$ to B.

3. Morphisms, Objects and Operations

Let x be a point in D. We define the *germ* of f at x, and denote by $[f]_x$, the class of morphisms in $\mathfrak{M}(A, B)$ of $\mathfrak{P}$ which are defined on open neighborhoods of x and are such that any two of them are equal in a corresponding neighborhood. Now, we can introduce the category $\mathfrak{L}(\mathfrak{P})$, whose objects are pairs (A, a), where a is a point in A, and whose morphism sets are

$$\mathfrak{M}_{\mathfrak{L}(\mathfrak{P})}[(A, a), (B, b)] = \{[f]_a \mid f: A \to B \text{ and } f(a) = b\}.$$

The composition of morphisms in $\mathfrak{L}(\mathfrak{P})$ is induced in an obvious manner by that in $\mathfrak{P}$ and the identity morphism is the germ of the identity map. The conditions of Definitions 1 and 4 are easily verified, so $\mathfrak{L}(\mathfrak{P})$ is a quasitopological category.

6 Definition The category $\mathfrak{L}(\mathfrak{P})$ is called the *local category* associated with the pseudocategory $\mathfrak{P}$.

Local categories associated with topological categories play an important role in differential geometry.

3 Morphisms, Objects and Operations

The definitions and results which we give here are considered for categories, but some of them are also valid for pseudocategories. As an exercise the reader should make the proper generalizations. These notions are inspired by set theory and the case of general categories is obtained by looking for definitions of the respective notions which do not make use of the elements of the sets.

Let $\mathfrak{C}$ be a category and $u: A \to B$ a morphism in $\mathfrak{C}$. Then, for any object X in $\mathfrak{C}$ we have the mappings

$$'u: \mathfrak{M}(X, A) \to \mathfrak{M}(X, B), \; u': \mathfrak{M}(B, X) \to \mathfrak{M}(A, X)$$

defined by

$$'u(v) = u \circ v, \; u'(w) = w \circ u,$$

where $v \in \mathfrak{M}(X, A)$, $w \in \mathfrak{M}(B, X)$.

1   Definition   The morphism u is an *injection* (respectively a *surjection*) in $\mathfrak{C}$ if the mapping $'u$ (respectively u') is univalent. A morphism which is both an injection and a surjection is called a *bijection*.

Clearly, injection and surjection are dual notions.

2   Proposition   *If u and v are morphisms in a category $\mathfrak{C}$ and $v \circ u$ exists and is an injection (surjection) then u is an injection (v is a surjection). The identity morphisms are bijections.*

The first part is a consequence of the following immediate relations

$$'(vu) = 'v'u, \qquad (vu)' = u'v',$$

and the second part follows directly from the definition of a bijection.

In Ens, injections, surjections, and bijections given by Definition 1 are the same as the classical ones. If $u: A \to B$ is an injective map between two sets and $v_i: X \to A$ ($i = 1, 2$), $v_1 \neq v_2$, then there is an $x \in X$ with $v_1(x) \neq v_2(x)$, hence $uv_1(x) \neq uv_2(x)$ and it follows that $'u$ is univalent. Conversely, if $'u$ is univalent u is also univalent; for otherwise we would have $a_i \in A$ ($i = 1, 2$) with $u(a_1) = u(a_2)$ and $a_1 \neq a_2$, and by defining $v_1(X) = a_1$, $v_2(X) = a_2$ the univalence of $'u$ is contradicted. Hence, injections in Ens in the sense of Definition 1 and in the classical sense are the same. The proof for surjections and bijections is left to the reader.

The previous results do not hold in any cantorian category. For instance, if $u: A \to B$ is some map of sets, u is an injection in the category with objects $\{A, B\}$ and morphisms $\{1_A, 1_B, u\}$ but it may not be an injection in Ens. In the category of rings, the inclusion $J \subset Q$ of the integers in the rational numbers is obviously a surjection, but this is not a surjection in Ens. In the category of topological Hausdorff spaces and continuous maps, $u: A \to B$ is a surjection if $u(A)$ is dense in B, even if u is not a surjective map. The last two cases provide examples of bijections in cantorian categories which are not bijections in Ens. In the category A-Mod we again have a situation where injections and surjections are injective and surjective mappings, respectively. Hence these notions must be used carefully in the categorical and set theoretical languages. In any case, we have the following obvious result.

3   Proposition   *If $\mathfrak{C}$ is a cantorian or topological category, u is a morphism in $\mathfrak{C}$, and u is an injective (surjective) map, then u is also injective (surjective) in $\mathfrak{C}$.*

3. Morphisms, Objects and Operations

Let $u: A \to B$ be a morphism in a category $\mathfrak{C}$. We shall say that u has a *right inverse* or is a *retraction* if there is a morphism $v: B \to A$ in $\mathfrak{C}$ such that $uv = 1_B$. Analogously, u has a *left inverse* or is a *coretraction* if there is $w: B \to A$ such that $wu = 1_A$; in this case A is a *retract* of B. If u has both right and left inverses they are equal (from $uv = 1_B$ we have $w(uv) = w = (wu)v = 1_A v = v$) and we say that u has an *inverse*.

4 Definition A morphism u which has an inverse is called an *isomorphism*.

Obviously, isomorphisms can be considered in pseudocategories too.

From Proposition 2, it follows immediately that every isomorphism is a bijection. In Ens, the isomorphisms are the bijections and in Top they are the homeomorphisms of topological spaces. There are categories where not every bijection is an isomorphism, e.g., Top, and the category of rings, where the inclusion of the integers in the rational numbers is a bijection but, obviously, it has no inverse, and, hence, is not an isomorphism. If all the bijections of the category $\mathfrak{C}$ are isomorphisms, $\mathfrak{C}$ is called a *balanced category*.

The inverse of an isomorphism u is again an isomorphism and will be denoted by u^{-1}.

5 Proposition *The composit of two injections, surjections, bijections, or isomorphisms is again an injection, surjection, bijection, or isomorphism, respectively.*

For the first three cases this follows from $'(vu) = {'v}{'u}$ and $(vu)' = u'v'$, and in the last case we have $(vu)^{-1} = u^{-1}v^{-1}$.

We introduce some further terminology. The morphisms $u: A \to A$ are also called *endomorphisms* and the isomorphisms $u: A \to A$ are called *automorphisms*. The set of automorphisms of an object A is a group, denoted by Aut A, with respect to the composition of morphisms. Sometimes, especially for the categories of algebraic structures, the name morphism is replaced by *homomorphism* and the notation $\mathfrak{M}(A, B)$ is replaced by Hom(A, B). Also, an injection is called a *monomorphism* and a surjection an *epimorphism*.

6 Definition A small category where every morphism has an inverse is called a *groupoid*.

We now introduce several important kinds of objects in a category.

7   Definition   Let $\mathfrak{C}$ be a category and A an object of $\mathfrak{C}$. If for every X in $\mathfrak{C}$, $\mathfrak{M}(A, X) \neq \phi$, A is called a *total source*. If $\mathfrak{M}(A, X)$ has exactly one element, A is called an *initial object*. If $\mathfrak{M}(X, A) \neq \emptyset$, A is called a *total target*. If $\mathfrak{M}(X, A)$ has exactly one element, A is called a *final object*. An object which is both initial and final is called a *zero object*.

One says that an initial object has the *initial universality property* and a final object has the *final universality property*. Obviously the initial objects of a category are isomorphic (if they exist) and the same is true for final and zero objects.

For instance, in the categories Ens and Top enlarged with the empty set, this set is an initial object. Also, any object consisting of a single point is a final object.

As we have seen, initial, final, and zero objects, if they exist, are determined up to isomorphism. This is a situation which is often met and we shall connect it with the following notion.

8   Definition   Let p be a property of the objects of a category $\mathfrak{C}$. If $\mathfrak{A}$ is a class of objects of $\mathfrak{C}$ such that any object of $\mathfrak{A}$ has the property p, and if for any object of $\mathfrak{C}$ having the property p there is a uniquely determined isomorphic object in $\mathfrak{A}$, we call $\mathfrak{A}$ a *representative class for the property p*.

In particular, we can have initial, final or zero representative objects.

Let $\mathfrak{C}$ be a category and A an object of $\mathfrak{C}$. We shall denote by $\mathfrak{C}_A$ the subcategory of $S(\mathfrak{C})$ whose objects are morphisms $X \xrightarrow{u} A$ of target A and whose morphisms are commutative diagrams of the form

$$\begin{array}{ccc} X & \xrightarrow{u} & A \\ {\scriptstyle t}\downarrow & & \downarrow{\scriptstyle 1_A} \\ Y & \xrightarrow{v} & A \end{array} \quad \text{or} \quad \begin{array}{ccc} X & \xrightarrow{u} & A \\ {\scriptstyle t}\downarrow & \nearrow{\scriptstyle v} & \\ Y & & \end{array} \qquad (1)$$

Dually, let us consider the category $\mathfrak{C}^A$ of morphisms $A \xrightarrow{u} X$ of source A and diagrams of the form

$$\begin{array}{ccc} A & \xrightarrow{u} & X \\ & {\scriptstyle v}\searrow & \downarrow{\scriptstyle t} \\ & & Y \end{array} \qquad (2)$$

3. Morphisms, Objects and Operations

By definition, an object of $\mathfrak{C}_A$ which is an injection in $\mathfrak{C}$ is called a *subobject* of A; a subobject is considered determined up to an isomorphism of the category $\mathfrak{C}_A$. The dual of the subobject is a *quotient object*, which is a surjection of $\mathfrak{C}^A$ determined up to an isomorphism of $\mathfrak{C}^A$. It is to be remarked that this definition of a subobject is not always satisfactory for the classical categories, for instance in Top. In such cases one must introduce supplementary conditions.

The subobjects of a given object A form a full subcategory S_A of $\mathfrak{C}_A$. If now $f: B \to A$ is a morphism in the category $\mathfrak{C}$ we can introduce a full subcategory S'_A of S_A whose objects $X \xrightarrow{u} A$ satisfy the following restrictive condition: there is a morphism $f': B \to X$ such that $f = uf'$. An initial object of the category S'_A is called the *range* of the morphism f and is denoted by range f.

The range is determined up to an isomorphism of S'_A. In Ens, the range of the map $f: B \to A$ is isomorphic with the inclusion of $f(B)$ in A. It follows that in Ens the inclusion of $f(B)$ in A is a representative object for the range of f. More generally, in this category the inclusions form a representative class for subobjects and we shall use only the subobjects of that class. Wherever possible, the same consideration will be given in other Cantorian categories.

The dual notion of the range is the *corange* of a morphism $f: B \to A$, denoted by corange f. This means that corange f is a surjection $v: B \to Y$ for which there is an $f': Y \to A$ with $f = f'v$ and which is such that if $w: B \to Z$ is another surjection with the same property, there is a uniquely determined morphism $t: Z \to Y$ satisfying $v = tw$.

For subobjects we can consider some important operations. If $u_i: B_i \to A (i \in M)$ is a set of subobjects of A, those subobjects $u: B \to A$ which have the property that for every $i \in M$ there is a v_i such that $u = u_i v_i$ form a full subcategory S''_A of S_A and a final object (if it exists) of S''_A is called the *intersection* of the objects B_i and is denoted by $\bigcap_{i \in M} B_i$. The dual notion is that of the *cointersection*.

Consider the diagram

where the vertical arrows are injections. If there is a morphism $f': A' \to B'$ which makes this diagram commutative, we say that the subobject A' of A is sent by f into the subobject B' of B.

Considering again the subobjects $u_i: B_i \to A$ ($i \in M$), we define their *union* by the subobject $C \xrightarrow{c} A$ of A such that $u_i = cv_i$ ($i \in M$) and if $f: A \to B$ and every A_i is sent by f into a subobject C' of B, C is also sent by f into C'. The union (if it exists) is determined up to an isomorphism.

In the category of sets, Ens, there is a representative class for both intersections and unions: it is the class of usual intersections and unions of sets.

The categories $\mathfrak{C}_A$ and $\mathfrak{C}^A$ can be generalized. Let A_i ($i \in M$) be a set of objects in a category $\mathfrak{C}$, M being the set of indices. We define the category $\mathfrak{C}^{A_i}$ whose objects are systems of morphisms $A_i \xrightarrow{u_i} X$ ($i \in M$) of $\mathfrak{C}$ and whose morphisms are the morphisms t with commutative diagrams

$$\begin{array}{c} A_i \xrightarrow{u_i} X \\ {}_{v_i}\searrow \downarrow{}^{t} \\ Y \end{array}, \quad i \in M. \qquad (3)$$

Dually, we shall consider the category $\mathfrak{C}_{A_i}$ whose objects are systems of the form $X \xrightarrow{u_i} A_i$ and whose morphisms are the morphisms t with commutative diagrams

$$\begin{array}{c} X \xrightarrow{u_i} A_i \\ {}^{t}\downarrow \nearrow{}_{v_i} \\ Y \end{array}, \quad i \in M. \qquad (4)$$

9 Definition If there is an initial object in the category $\mathfrak{C}^{A_i}$, it will be denoted by $\vee_i A_i$ or $\oplus_i A_i$ and called the *direct sum* of the objects A_i of $\mathfrak{C}$. Dually, a final object in $\mathfrak{C}_{A_i}$ is denoted by $\times_i A_i$ or $\prod_i A_i$ and is called the *direct product* of the objects A_i in $\mathfrak{C}$.

Direct sums and products do not always exist and, if they exist, they are determined up to an isomorphism in $\mathfrak{C}^{A_i}$ and $\mathfrak{C}_{A_i}$, respectively.

For instance, in Ens the direct sum is the disjoint union of sets and the direct product is the Cartesian product. In Top, the direct sum is the disjoint union with the topology whose open sets are the sets whose intersection with every term of the sum is open, and the direct product is the Cartesian product with the topology whose basis of open sets is given by products of open subsets of the factors such that only a finite number of

3. Morphisms, Objects and Operations

them are proper subsets. In the category of A-modules the direct sums and products are the well known classical operations of linear algebra.

The two previous operations can be generalized. We suppose that the set M of indices of the objects A_i in $\mathfrak{C}$ has a partial ordering relation (reflexive and transitive) $\leqslant$ and that there are given morphisms $f_j^i: A_i \to A_j$ for $i \leqslant j$ which satisfy the following conditions:

(a) $\qquad\qquad\qquad f_i^i = 1_{A_i},$

(b) $\qquad\qquad\qquad f_k^j \circ f_j^i = f_k^i, \quad i \leqslant j \leqslant k.$

Such a system $(A_i, f_j^i)_{i,j \in M}$ is called an *inductive family of objects in* $\mathfrak{C}$.

Now, we shall consider the category $\mathfrak{C}_\leqslant^{A_i}$ whose objects are systems of morphisms $A_i \xrightarrow{u_i} X$ of $\mathfrak{C}$, such that for $i \leqslant j$ we have a commutative diagram

$\qquad\qquad\qquad\qquad\qquad\qquad\qquad\qquad\qquad\qquad\qquad\qquad$ (5)

and whose morphisms are defined as for $\mathfrak{C}^{A_i}$, i.e., they are given by Diagrams (3).

10 Definition If $(A_i, f_j^i)_{i,j \in M}$ is an inductive family in $\mathfrak{C}$, an initial object of the category $\mathfrak{C}_\leqslant^{A_i}$ is called the *inductive* or *direct limit* of the family.

For the inductive limit, we use the notation $\varinjlim A_i$ and we remember that it consists of an object of $\mathfrak{C}$ and of the system of morphisms $f_i: A_i \to \varinjlim A_i$ for which the Diagrams (5) are commutative. The inductive limit does not always exist and, when it exists, it is determined up to an isomorphism in $\mathfrak{C}_\leqslant^{A_i}$.

In the category of sets, there always exist inductive limits, and they can be obtained in the following manner. Let $\{A_i, f_j^i\}$ be an inductive family of sets and maps. We consider $\vee_i A_i$ and define on it a binary relation ρ such that $a_i \rho a_j$ ($a_i \in A_i$, $a_j \in A_j$) if and only if there is an index $k \in M$ with $k \geqslant i$, $k \geqslant j$ and $f_k^i(a_i) = f_k^j(a_j)$. Obviously, ρ is reflexive and symmetric. Hence, we get an equivalence relation on $\vee_i A_i$ by putting $a_i \sim a_j$ if and only if there is a finite sequence $a_{i_1}, \ldots, a_{i_n}$ such that $a_i \rho a_{i_1}, \ldots,$

$a_{i_n}\rho a_{i_{n+1}}, \ldots, a_{i_n}\rho a_j$. It is easy to see that the set of equivalence classes together with the canonical projections is just $\varinjlim A_i$.

If the previous A_i are in Top, $\varinjlim A_i$ with the topology coinduced by the canonical projections (i.e., the finest topology for which these projections are continuous) is the inductive limit in the category Top.

An important particular case is obtained when the set M of indices is *directed*, which means that for every pair $i, j \in M$ there exists a $k \in M$ such that $i \leqslant k$ and $j \leqslant k$. Now, in the above construction of $\varinjlim$ in Ens, the relation ρ is itself the necessary equivalence relation. If the set M is directed, $\{A_i, f_j^i\}_{i,j \in M}$ is called a *directed inductive family*.

In the category A-Mod (where A is a commutative unitary ring) we prove the existence of inductive limits for directed inductive families. In fact, this limit will be obtained just as for sets, the established equivalence relation being ρ and the module operations

$$[a_i] + [a_j] = [f_k^i(a_i)] + [f_k^j(a_j)] = [f_k^i(a_i) + f_k^j(a_j)], \quad (k \geqslant i, j)$$

$$\lambda[a_i] = [\lambda a_i].$$

Here, the brackets denote equivalence classes and the consistency of these definitions is easily shown.

Let $\{A_i, f_j^i\}$ and $\{B_i, g_j^i\}$ be two inductive families in a category $\mathfrak{C}$, having the same set M of indices. A *morphism* between these two families is by definition a family of morphisms in $\mathfrak{C}$, $h_i: A_i \to B_i$, defined for every $i \in M$ such that for $i \leqslant j$ the following diagram is commutative:

$$\begin{array}{ccc} A_i & \xrightarrow{h_i} & B_i \\ {\scriptstyle f_j^i} \downarrow & & \downarrow {\scriptstyle g_j^i} \\ A_j & \xrightarrow{h_j} & B_j \end{array} \qquad (6)$$

There is an obvious composition of such morphisms, and hence the inductive families in $\mathfrak{C}$ and their morphisms form, for a fixed set of indices, a new category.

Suppose that there exist in $\mathfrak{C}$ inductive limits and let $(\varinjlim A_i, f_i)$, $(\varinjlim B_i, g_i)$ be the limits of the two families considered above. The system of morphisms $g_i \circ h_i: A_i \to \varinjlim B_i$ is then an object of the category $\mathfrak{C}_{\leqslant}^{A_i}$ and, by Definition 9, there is a unique morphism $h: \varinjlim A_i \to \varinjlim B_i$ such that $hf_i = g_i h_i$, $i \in M$. This morphism will be denoted by $h = \varinjlim h_i$ and will be called the *inductive limit* of the morphisms h_i.

3. Morphisms, Objects and Operations

We now develop the dual notions.

First, if A_i and $f_j^i: A_i \to A_j$ are objects and morphisms in $\mathfrak{C}$, with indices in the partially ordered set M, f_j^i being defined for every pair $i \geqslant j$ and satisfying the conditions

$$f_i^i = 1_{A_i}, \quad f_k^j \circ f_j^i = f_k^i \quad i \geqslant j \geqslant k,$$

we shall say that $\{A_i, f_j^i\}_{i,j \in M}$ is a *projective family* of objects in $\mathfrak{C}$.

To every such family, we associate a category $\mathfrak{C}_{A_i}^{\leqslant}$ whose objects are systems of morphisms $X \xrightarrow{u_i} A_i$ such that for $i \geqslant j$ there is a commutative diagram

$$\begin{array}{c} & & A_i \\ & \nearrow^{u_i} & \\ X & & \downarrow f_j^i \\ & \searrow_{u_j} & \\ & & A_j \end{array} \qquad (7)$$

and whose morphisms are defined as those of $\mathfrak{C}_{A_i}$.

11 Definition A final object (if it exists) of the category $\mathfrak{C}_{A_i}^{\leqslant}$ is called the *projective limit* of the family A_i.

The projective limit is also called the *inverse limit*. It is denoted by $\varprojlim A_i$ and consists of an object of $\mathfrak{C}$ together with morphisms $f_i: \varprojlim A_i \to A_i$ which make Diagrams (7) commutative. The projective limit is determined up to an isomorphism in the category $\mathfrak{C}_{A_i}^{\leqslant}$.

In the category of sets we can always construct the projective limit. If we have a projective family $\{A_i, f_j^i\}$ we obtain $\varprojlim A_i$ as the subset of the Cartesian product $\times_i A_i$ of the elements of the form $(a_i \mid a_i \in A_i)$, for which $i \geqslant j$ implies $f_j^i(a_i) = a_j$ together with the restrictions of the projections of the Cartesian product. The necessary properties are easily verified.

The same is true for the category Top if the previous set projective limit $\varprojlim A_i$ is endowed with the subspace topology of the direct topological product, and for the category A-Mod if we take the structure of a submodule of the direct product of modules.

The case of a *directed projective family* can also be considered.

Finally, morphisms of projective families with the same set of indices can be defined by analogy with the inductive case, and one obtains the projective limit of such a morphism.

In the same manner one can define direct and inverse limits for general diagrams [33, 40].

4 Abelian Categories

The notions introduced in the previous section were inspired by set theory. Now, we shall see how one can obtain categories by generalizing the principal notions of the theory of Abelian groups.

Since we do not always work with Cantorian categories, we see that in order to obtain structures for objects in a category it is possible to give the sets of morphisms such a structure. In this connection, we consider

1 Definition An object A of a category $\mathfrak{C}$ is called a *pointed object* if for every set $\mathfrak{M}(X, A)$ there is given a base point in such a manner that the mappings u' ($u: X \to Y$) of Section 3 are base point preserving. Analogously, A will be called a $\mathfrak{C}$-*group* (*group in* $\mathfrak{C}$) if one gives every $\mathfrak{M}(X, A)$ a group structure such that the u' are homomorphisms; in particular we can consider *Abelian* $\mathfrak{C}$-*groups* in an obvious manner. The dual notions are called *copointed object* and $\mathfrak{C}$-*cogroup*, respectively.

For instance, in the categories Ens· and Top· all the objects are pointed and copointed and in Ab all the objects are Abelian groups and cogroups. In fact, in Ens· and Top· every object is a pointed set, and we can take for base-point in $\mathfrak{M}(A, B)$ the constant map which represents the points of A onto the base point of B. For Ab our assertion is well known in algebra and we leave it to the reader.

2 Definition A category for which all the objects are pointed and copointed with respect to the same base points is called a *pointed category* or a *category with base points*.

In a pointed category one has the important notion of the *kernel* of a morphism $u: A \to B$. Let p_{AB} be the base point of the set $\mathfrak{M}(A, B)$. We consider next the category $\mathfrak{C}_A$ of Section 3 and its full subcategory consisting of morphisms $X \xrightarrow{v} A$ such that $'u(v) = p_{XB}$. A final object of this subcategory, if there is one, is called the *kernel* of u. Hence, if the kernel of u exists, it is determined up to an isomorphism of the mentioned subcategory and consists of a pair $(K, i: K \to A)$ such that $'u(i) = p_{KB}$, and if $'u(v) = p_{XB}$ we have $v = if$ with a uniquely determined $f: X \to K$. The kernel of the morphism u will be denoted by ker u.

4. Abelian Categories

In Ens·, ker u is the inverse image in A of the base point of B together with its inclusion in A. In A-Mod (and in Grp too), the kernel defined here is the usual kernel in algebra if the base points considered are the 0 homomorphisms. Hence, in all these cases there is a representative class for kernels.

3 Proposition *If* ker $u = (K, i)$, i *is an injection*.

In fact, if $s_i: X \to K$ ($i = 1, 2$) and $is_1 = is_2 = v$, we have $uv = uis_1 = p_{KB}s_1 = p_{XB}$, so $v = if$ with $f: X \to K$ uniquely determined. Hence, $s_1 = s_2 = f$.   Q.E.D.

The dual notion is called the *cokernel*. It will be defined for $u: A \to B$ as an initial element of the full subcategory of $\mathfrak{C}^B$ (Section 3) whose objects are $B \xrightarrow{v} X$ with $u'(v) = p_{AX}$. We shall denote it by coker u and if coker $u = (C, j: B \to C)$ (up to an isomorphism) j is a surjection. In the category Ens·, coker u is the quotient set $B \mid u(A)$, i.e., the set of equivalence classes of the relation which identifies all the points of $u(A)$ with the base point, together with the canonical projection. The same holds in A-Mod. Hence, in these categories there is a representative class for cokernels.

We further extend the notions in Ab to pointed categories $\mathfrak{C}$.

4 Definition If $u: A \to B$ is a morphism in $\mathfrak{C}$, the *image* of u is im $u = $ ker coker u and the *coimage* is coim $u = $ coker ker u.

5 Lemma *If the morphism* $u: A \to B$ *has an image and a coimage, there is a canonical morphism* μ, *which makes the following diagram commutative*:

$$\begin{array}{ccc} A & \xrightarrow{u} & B \\ {\scriptstyle p}\downarrow & & \uparrow {\scriptstyle q} \\ \text{coim } u & \xrightarrow{\mu} & \text{im } u \end{array} \qquad (1)$$

p and q being the morphisms which appear by the definition of the coimage and image.

To prove this, we put ker $u = (K, i: K \to A)$, coker $u = (C, j: B \to C)$, coim $u = (M, p: A \to M)$ and im $u = (N, q: N \to B)$. From $ju = p_{AC}$ we deduce $u = q\lambda$ with a unique $\lambda: A \to $ im u. From $ui = q\lambda i = p_{KB}$ and the fact that q is an injection (since it is a kernel) we obtain a unique $\mu\colon M \to N$ such that $\lambda = \mu p$. It follows that $u = q\mu p$.   Q.E.D.

Diagram (1) is called the *canonical factorization* of the morphism u. We remark that $ui = p_{KB}$ implies $u = vp$ with $v\colon \operatorname{coim} u \to B$ uniquely determined. Hence Diagram (1) is completed to the following totally commutative diagram:

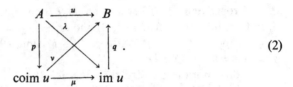
(2)

6   Definition   A pointed category $\mathfrak{C}$ is called *regular* if: (a) every morphism in $\mathfrak{C}$ has a kernel and a cokernel; (b) $\mu\colon \operatorname{coim} u \to \operatorname{im} u$ is a bijection, and (c) every bijection in $\mathfrak{C}$ is an isomorphism ($\mathfrak{C}$ is balanced).

Condition (b) above is meaningful because of (a) and Lemma 5, and Condition (c) implies that μ is an isomorphism. It follows easily that in Diagram (2), λ is a surjection and v an injection. Then, using the definitions of range u and corange u (Section 3) we obtain:

7   Proposition   *In a pointed regular category there are the isomorphisms*

$$s\colon \operatorname{range} u \to \operatorname{im} u, \qquad t\colon \operatorname{coim} u \to \operatorname{corange} u.$$

In fact, in the notation of Diagram (2), the relation $u = q\lambda$ shows that there is a unique morphism $s\colon \operatorname{range} u \to \operatorname{im} u$ such that $qs = a$, where $a\colon \operatorname{range} u \to B$ is the injection given by the definition of range u. It follows from Proposition 3.2 that s is an injection. Now, we must have a relation $u = au'$ ($u'\colon A \to \operatorname{range} u$), which gives $u = qsu' = q\lambda$ and, since q is an injection, $su' = \lambda$. But λ is also a surjection, so we infer from Proposition 3.2 that s is a surjection. Hence, s is a bijection and, the category being regular, it is an isomorphism. Next, there is a diagram containing these facts:

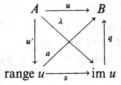

4. Abelian Categories

Finally, the isomorphism t defined in Proposition 7 is obtained by duality.

Now we give another important notion:

8 Definition Let $\mathfrak{C}$ be a pointed category and

$$A \xrightarrow{u} B \xrightarrow{v} C \tag{3}$$

a sequence of objects and morphisms of $\mathfrak{C}$. The sequence (3) is called *exact* if im u and ker v exist and im $u =$ ker v.

More generally, a sequence of objects and morphisms in $\mathfrak{C}$

$$\cdots \longrightarrow A_i \xrightarrow{u_i} A_{i+1} \xrightarrow{u_{i+1}} A_{i+2} \longrightarrow \cdots \tag{4}$$

is called *exact* if every part of it consisting of three terms is exact.

For instance, in Ens· a sequence (3) is exact if and only if the image of u is equal to the inverse image of the base point of C by v.

By defining morphisms of exact sequences to be totally commutative diagrams of the form

$$\begin{array}{ccccccc}
\cdots \longrightarrow & A_i & \xrightarrow{u_i} & A_{i+1} & \xrightarrow{u_{i+1}} & A_{i+2} & \longrightarrow \cdots \\
& \downarrow{f_i} & & \downarrow{f_{i+1}} & & \downarrow{f_{i+2}} & \\
\cdots \longrightarrow & B_i & \xrightarrow{v_i} & B_{i+1} & \xrightarrow{v_{i+1}} & B_{i+2} & \longrightarrow \cdots
\end{array}$$

we get the category of exact sequences in $\mathfrak{C}$.

9 Proposition *Let $\mathfrak{C}$ be a pointed regular category and* (3) *a sequence of $\mathfrak{C}$ where u is an injection. Then* (3) *is an exact sequence if and only if for any object X in $\mathfrak{C}$ the sequence*

$$\mathfrak{M}(X, A) \xrightarrow{'u} \mathfrak{M}(X, B) \xrightarrow{'v} \mathfrak{M}(X, C) \tag{5}$$

is exact in Ens·. Dually, if in (3) *v is a surjection, it is exact in $\mathfrak{C}$ if and only if*

$$\mathfrak{M}(C, X) \xrightarrow{v'} \mathfrak{M}(B, X) \xrightarrow{u'} \mathfrak{M}(A, X) \tag{6}$$

is exact in Ens·.

Because of the duality principle it suffices to prove only the first part. We shall use the notation of Diagram (2) and Lemma 5, so that $q: N \to B$

is im $u = \ker v$. Then, for $\alpha\colon X \to A$ we have

$$v u \alpha = v q \lambda \alpha = p_{NC} \lambda \alpha = p_{XC}.$$

If $\beta\colon X \to B$ and $v\beta = p_{XC}$, we have $\beta = q\alpha$ ($\alpha\colon X \to N$). But, μ being a bijection, λ is a surjection and, since u is an injection, λ is also an injection, since $u = q\lambda$. Hence, λ is an isomorphism and we have $q = u\lambda^{-1}$ and $\beta = u\lambda^{-1}\alpha$. It follows that the exactness of (3) implies the exactness of (5). Conversely, if (5) is exact for every X, we choose $X = N$, and from $q = u\lambda^{-1}$ we have $vq = p_{NC}$. Next, if $\beta\colon X \to B$ and $v\beta = p_{XC}$, it follows that $\beta = u\alpha = q\lambda\alpha$ ($\alpha\colon X \to A$), this factorization determining uniquely $\lambda\alpha$ because q is an injection. Hence, $q\colon N \to B$ is also $\ker v$. \quad Q.E.D.

10 Definition The sequence (3) is called *strictly exact* if for every X (5) is an exact sequence, and *strictly coexact* if (6) is an exact sequence.

Now, we consider a special type of category for which all the previous considerations hold.

11 Definition A category $\mathfrak{C}$ is called an *additive category* if the following conditions are satisfied[†]:

(a) Every object of $\mathfrak{C}$ is an abelian $\mathfrak{C}$-group and $\mathfrak{C}$-cogroup with respect to the same group structure on the morphism sets.
(b) $\mathfrak{C}$ has a zero object.
(c) $\mathfrak{C}$ has finite direct sums.

Clearly, an additive category is pointed, the base point in $\mathfrak{M}(A, B)$ being the identity element of the group structure on this set. For additive categories the notation $\mathrm{Hom}(A, B)$ instead of $\mathfrak{M}(A, B)$ is often used and also the terms monomorphism and epimorphism are used for injection and surjection, respectively. The zero object of an additive category will be denoted by 0 and p_{AB} will be replaced by 0_{AB} or simply by 0. It is easy to see that the zero object is characterized by the condition $0_{00} = 1_0$ and that, for every A, im $0_{0A} = 0$.

In an additive category $\mathfrak{C}$, the finite direct sums are also finite direct products. If A_h ($h = 1, \ldots, n$) are objects in $\mathfrak{C}$ and $\vee_h A_h$, with the morphisms $i_h\colon A_h \to \vee A_h$ is their direct sum, we define $p_k\colon \vee A_h \to A_k$ such that $p_k i_h = \delta_{kh}$, where

---

[†]Sometimes only Condition (a) is assumed [40].

4. Abelian Categories

$$\delta_{kh} = \begin{cases} 0 & \text{if } k \neq h \\ 1_{A_h} & \text{if } k = h. \end{cases}$$

It follows that the i_h are injections and the p_k surjections, whence for $X \xrightarrow{q_k} A_k$, $f = \sum i_k q_k : X \to \vee A_k$ is a unique morphism such that $p_k f = q_k$ and our assertion is proved. Also, we have $\sum i_k p_k = 1$ and it is easy to see that $B \xrightarrow{p_k} A_k$ is the direct product $\times A_k$ if and only if there are $i_h : A_h \to B$ such that $p_k i_h = \delta_{kh}$ and $\sum i_k p_k = 1_B$; if B is the product it is isomorphic with the direct sum and we have i_h. The converse follows just as in the proof given above for $\vee A_h$.

12 Definition A *short exact sequence* is an exact sequence of the form

$$0 \longrightarrow A' \xrightarrow{i} A \xrightarrow{j} A'' \longrightarrow 0. \tag{7}$$

In an obvious manner we talk of the category of short exact sequences of a given additive category $\mathfrak{C}$. If in this category there is an isomorphism

$$\begin{array}{ccccccccc} 0 & \longrightarrow & A' & \xrightarrow{i} & A & \xrightarrow{j} & A'' & \longrightarrow & 0 \\ & & \downarrow{1_{A'}} & & \downarrow{u} & & \downarrow{1_{A''}} & & \\ 0 & \longrightarrow & A' & \xrightarrow{i'} & A' \oplus A'' & \xrightarrow{j'} & A'' & \longrightarrow & 0, \end{array}$$

where i' is the *inclusion* of A' in the direct sum, we say that the exact sequence (7) *splits*, or that it is a *splitting*.

13 Definition An additive regular category is called an *Abelian category*.

It is easy to see that, in an Abelian category, the sequence $0 \to A \xrightarrow{u} B$ is exact if and only if u is a monomorphism and, dually, the sequence $A \xrightarrow{u} B \to 0$ is exact if and only if u is an epimorphism. The exact sequence (7) splits if and only if j has a right inverse or i has a left inverse.

The category Ab of Abelian groups is an Abelian category, the group structure of Hom(A, B) being given by the usual addition of homomorphisms, the zero object the group 0 and the direct sum the usual one in algebra. Kernels, cokernels, images, and coimages in this category are also the usual ones, and they exist. Every bijection is obviously an isomorphism. Finally, the homomorphism μ of Lemma 5 is a bijection by a classical isomorphism theorem in group theory. Observe that in Ab the exactness of the short sequence (7) means that A' is isomorphic with a subgroup of A and A'' with the quotient group A/A'.

All these statements also hold in the category A-Mod, for every commutative and unitary ring A. Hence, this is also an Abelian category.

Recall that a *graded A-module* is a family $\{M_n\}$ ($n \in J$, the ring of integers) of A-modules and that a homomorphism of graded A-modules is a family $f_n\colon M_n \to M'_n$ of A-module homomorphisms. Then it is obvious that the graded A-modules and their homomorphisms form an Abelian category.

Finally, we give one more example, namely, the differential A-modules. A *differential A-module* is an A-module M endowed with an endomorphism $d\colon M \to M$ such that $d^2 = 0$. A homomorphism f of the differential modules (M, d) and (M', d') is a homomorphism $f\colon M \to M'$ such that $fd = d'f$. It is easily seen that the differential A-modules and their homomorphisms form an Abelian category.

14   Proposition   *If $\mathfrak{C}$ is an Abelian category, A an object of $\mathfrak{C}$ and (A', i) a subobject of A, then (A', i) is the kernel of the cokernel (A'', j) of i.*

In fact, by applying Diagram (2) to $A' \xrightarrow{i} A$ and remarking that ker coker $i = \operatorname{im} i$, we get $i = g\lambda$, where λ is an epimorphism. But then λ is also a monomorphism, hence λ is an isomorphism.   Q.E.D.

Proposition 14 and its dual show that if A is an object of an Abelian category $\mathfrak{C}$, there is a one-to-one correspondence between the class of the subobjects of A and the class of quotient objects. In the above notation, with the subobject (A', i) there is associated the quotient object $(A'', j) = $ coker i, which is again denoted by A/A', and conversely, to (A'', j) there is associated $(A', i) = \ker j$. This correspondence is characterized by the exactness of the following short sequence:

$$0 \longrightarrow A' \xrightarrow{i} A \xrightarrow{j} A/A' \longrightarrow 0. \tag{8}$$

15   Proposition   *Consider the diagram*

$$\begin{array}{ccccccccc} 0 & \longrightarrow & A' & \xrightarrow{i} & A & \xrightarrow{j} & A/A' & \longrightarrow & 0 \\ & & \downarrow{\scriptstyle f'} & & \downarrow{\scriptstyle f} & & \downarrow{\scriptstyle f''} & & \\ 0 & \longrightarrow & B' & \xrightarrow{a} & B & \xrightarrow{b} & B/B' & \longrightarrow & 0, \end{array}$$

where all the arrows, except the dotted one, are defined. Suppose that the lines of the diagram are exact and that the first square is commutative. Then, there is a uniquely determined morphism f'' which makes the second square commutative.

In fact, we have $bfi = baf' = 0$; hence, because j is coker i, it follows that there is a unique $f'': A/A' \to B/B'$ such that $bf = f''j$.   Q.E.D.

5   Functors and Homology

In this section, we shall consider transformations between categories often met in mathematical operations.

1   Definition   Let $\mathfrak{P}_1$ and $\mathfrak{P}_2$ be pseudocategories. By a *covariant functor* F from $\mathfrak{P}_1$ to $\mathfrak{P}_2$, denoted by $F: \mathfrak{P}_1 \to \mathfrak{P}_2$, we understand a correspondence which associates with every object A of $\mathfrak{P}_1$ an object $F(A)$ in $\mathfrak{P}_2$ and with every morphism $u: A \to B$ of $\mathfrak{P}_1$ a morphism $F(u): F(A) \to F(B)$ in $\mathfrak{P}_2$ such that the following conditions are satisfied:

(a) $F(vu) = F(v)F(u), \quad u: A \to B, v: B \to C,$
(b) $F(1_A) = 1_{F(A)}.$

2   Definition   In the same notation, $F: \mathfrak{P}_1 \to \mathfrak{P}_2$ is a *contravariant functor*, if for $u: A \to B$, $F(u): F(B) \to F(A)$ and Condition (a) of Definition 1 is replaced by

(a') $F(vu) = F(u)F(v), \quad u: A \to B, v: B \to C.$

The two notions are obviously dual and a contravariant functor $F: \mathfrak{P}_1 \to \mathfrak{P}_2$ is at the same time a covariant functor from the dual pseudocategory $\mathfrak{P}_1^*$ to $\mathfrak{P}_2$ and from $\mathfrak{P}_1$ to $\mathfrak{P}_2^*$.

A covariant functor

$$F: \mathfrak{P}_1 \times \cdots \times \mathfrak{P}_h \times \mathfrak{P}_{h+1}^* \times \cdots \times \mathfrak{P}_{h+k}^* \to \mathfrak{P}$$

is also called a functor with $h + k$ arguments, which is covariant in the first h arguments and contravariant in the last k; of course, we could also have other places for the covariant and contravariant arguments. From a functor of several arguments, we can obtain different functors by fixing some of the arguments. For instance, if $F: \mathfrak{P}_1 \times \mathfrak{P}_2 \to \mathfrak{P}$ is a functor, we get functors $F_X: \mathfrak{P}_1 \to \mathfrak{P}$ by the relations $F_X(A) = F(A, X)$ and $F_X(u) = F(u, 1_X)$.

Let us consider some examples.

For every pseudocategory $\mathfrak{P}$ we have the contravariant functor Op: $\mathfrak{P} \to \mathfrak{P}^*$ which consists in changing the sense of the arrows.

For every cantorian pseudocategory $\mathfrak{P}$ we get a covariant functor $\mathfrak{P} \to \text{Ens}^p$ by attaching to every object and morphism of $\mathfrak{P}$ the underlying set and map. This is the so called *forgetful functor*.

If $\mathfrak{P}$ is a pseudocategory we get a covariant functor $\mathfrak{M}: \mathfrak{P}^* \underset{\times}{} \mathfrak{P} \to \text{Ens}^p$ which associates with every pair (A, B) of objects of $\mathfrak{P}$ the set $\mathfrak{M}(A, B)$ and with every pair of morphisms $(u: A' \to A, v: B \to B')$ the partial map

$$\mathfrak{M}(u, v): \mathfrak{M}(A, B) \dashrightarrow \mathfrak{M}(A', B')$$

defined by $\mathfrak{M}(u,v)(\alpha) = v\alpha u \, (\alpha: A \to B)$. If $\mathfrak{P}$ is a category, the values of the functor $\mathfrak{M}$ are in the category Ens. By fixing arguments, we get from $\mathfrak{M}$ two functors: $\mathfrak{M}^A$ which is a covariant functor from $\mathfrak{P}$ to Ens^p and $\mathfrak{M}_B$ which is a contravariant functor from $\mathfrak{P}$ to Ens^p. We remark that these functors allow us to characterize important types of objects of a category $\mathfrak{C}$. Thus, if $\mathfrak{M}^A$ is a surjection preserving A it is called a *projective object* in $\mathfrak{C}$ and if $\mathfrak{M}_A$ sends injections to surjections, A is an *injective object* in $\mathfrak{C}$. If for every object A in $\mathfrak{C}$ there is a surjection $P \to A$ with a projective object P, $\mathfrak{C}$ is said *to have enough projectives* and if there is an injection $A \to Q$ with an injective Q, $\mathfrak{C}$ is said *to have enough injectives*.

If we associate with every set the free Abelian group generated by it and with every map the corresponding induced homomorphism, we get a covariant functor from Ens to Ab.

Let P and Q be two A-modules (A is always a commutative unitary ring). It is known that their *tensor product* over A is the module $P \otimes Q$ defined by the quotient of the free Abelian group generated by $P \times Q$, through the subgroup generated by the elements of the form

$$(p, q_1 + q_2) - (p, q_1) - (p, q_2),$$

$$(p_1 + p_2, q) - (p_1, q) - (p_2, q),$$

$$(ap, q) - (p, aq), \qquad p_i \in P, q_i \in Q, a \in A.$$

We remark that this is an initial object in the category whose objects are

5. Functors and Homology

bilinear maps $P \times Q \to C$ and whose morphisms are commutative diagrams

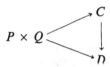

If P' and Q' are again two A-modules, and $f: P \to P'$, $g: Q \to Q'$ are homomorphisms, we get by the definition of the tensor product that there is an induced homomorphism $f \otimes g: P \otimes Q \to P' \otimes Q'$.

It is easily seen that if we associate with every pair (P, Q) the module $P \otimes Q$ and with every pair of homomorphisms (f, g) the associated homomorphism $f \otimes g$ we get a covariant functor from A-Mod $\times$ A-Mod to A-Mod. We assume that the properties of the tensor product are known.

An example of a covariant functor H_0 from the category Top to Ab is obtained if we associate with every topological space X the free Abelian group $H_0(X)$ generated by the set of the connected components of X and to each continuous map $f: X \to Y$ the homomorphism $H_0(f): H_0(X) \to H_0(Y)$ induced by the correspondence $C(x) \mapsto C(f(x))$ where C denotes the connected component of a point.

An inductive family of objects in a category $\mathfrak{C}$ having the indices in the partially ordered set M is the set of values of a covariant functor from the category M to $\mathfrak{C}$. A projective family is obtained in the same manner with a contravariant functor. The inductive limit is a covariant functor from the category of inductive families in $\mathfrak{C}$ (M-fixed) to $\mathfrak{C}$, and the projective limit is an analogous functor for the category of projective families.

Let $\mathfrak{P}$ be a topological pseudocategory, $\mathfrak{L}(\mathfrak{P})$ the associated local category and $\mathfrak{C}$ another category. A *derivative functor* or a $\mathfrak{C}$-*valued derivative* on $\mathfrak{P}$ is a covariant functor from $\mathfrak{L}(\mathfrak{P})$ to $\mathfrak{C}$. If we denote such a functor by D, the morphism $D([f]_x)$ (see the notation in Section 2) is called the *derivative* of the function f at the point x and we have the classical rule for the derivative of a composite function

$$D([f \circ g]_x) = D([f]_{g(x)}) \circ D([g]_x). \tag{1}$$

The classical derivative functor used in analysis and in differential geometry is obtained by the following schema [27]. Let E and F be locally convex topological vector spaces, U an open subset of E, and $f: U \to F$ a continuous function. f will be called a *differentiable function* at the point $x_0 \in U$ if there is a linear and continuous map $\lambda: E \to F$ such that the map

$$y \mapsto \varphi(y) = f(x_0 + y) - f(x_0) - \lambda(y),$$

which is defined in a neighborhood of the origin of E, has the property that for any neighborhood W of the origin of F there is a neighborhood V of the origin of E with

$$\varphi(tV) \subseteq \psi(t)W,$$

where t is a real variable and $\psi(t)$ a real function satisfying the condition

$$\lim_{t \to 0} \frac{\psi(t)}{t} = 0.$$

In that case, λ is called the *derivative* of f at x_0 and is denoted by $Df(x_0)$ or f'_{x_0}; it is uniquely determined and satisfies the rule (1) for the derivative of a composite function.

We now assume that the linear spaces considered are Banach spaces. Then, the function f is called *differentiable on U* or *of class C^1* if it is differentiable at all the points x_0 of U and if the derivative depends continuously on x_0. By iteration one defines higher order derivatives and functions of class C^p and C^∞, which have continuous derivatives up the order p or up to any order, respectively.

Hence, we get a derivative functor defined on the pseudocategory whose objects are Banach spaces and whose morphisms are functions of class C^1 given on open subsets of these spaces. The values of this functor are in the category of Banach spaces and of continuous linear maps. This functor associates with every space E the same space and with every germ of a differentiable function the derivative of that function. For the development of this theory see the work of J. Dieudonné [9]. The usual derivative of a real function of one real variable and the Jacobian matrix for functions of several variables obviously define derivatives in the above sense. In the fourth chapter we shall consider these derivatives in greater detail.

Now, let us denote by Abd the category of Abelian differential groups introduced in Section 4. We shall obtain an important functor from this category to Ab. Let A be an object of Abd and d the endomorphism such that $d^2 = 0$, called the *boundary operator* (or *differential*). Introduce the groups

5. Functors and Homology

$$Z(A) = \ker d, \quad B(A) = \operatorname{im} d, \quad H(A) = Z(A)/B(A), \qquad (2)$$

which are called the group of the *cycles*, the group of the *boundaries*, and the *homology group* of A, respectively. The last is defined because $d^2 = 0$, which implies $B(A) \subseteq Z(A)$.

Further, if $f: (A, d) \to (A', d')$ is a homomorphism commuting with the differentials, it induces homomorphisms $Z(A) \to Z(A')$ and $B(A) \to B(A')$, whence it induces a homomorphism

$$f_*: H(A) \to H(B). \qquad (3)$$

It follows that if we associate with A the homology group $H(A)$ and with f the homomorphism $H(f) = f_*$, we get a covariant functor $H: \text{Abd} \to \text{Ab}$ called the *homology functor*. This has an important exactness property.

3 Theorem *If*

$$0 \longrightarrow A' \xrightarrow{i} A \xrightarrow{j} A'' \longrightarrow 0 \qquad (4)$$

is an exact sequence in **Abd**, *there is a group homomorphism ∂ such that the sequence*

$$H(A') \xrightarrow{i_*} H(A) \xrightarrow{j_*} H(A'') \xrightarrow{\partial} H(A') \xrightarrow{i_*} H(A) \qquad (5)$$

is exact in **Ab**.

This is the so-called theorem *of the exact homology sequence*. The homomorphism ∂ is defined by

$$\partial\{z''\} = \{i^{-1} d j^{-1}(z'')\}, \quad z'' \in Z(A''), \qquad (6)$$

where $\{\ \}$ denote homology classes and we can choose any representative in $j^{-1}(z'')$. It is easily seen that ∂ does not depend on the cycle z'' chosen in $\{z''\}$ or on the representative used in $j^{-1}(z'')$. The homomorphism ∂ is called the *connecting* or *boundary homomorphism* of the sequence (4). Finally, it is a technical matter, which we leave to the reader, to verify the equality of kernels and images in the sequence (5).

The previous example has a generalization which is essential to our purpose.

4 Definition Let $\mathfrak{C}$ be an Abelian category. A *complex* K with values in $\mathfrak{C}$ is a system consisting of

(a) a sequence K^n ($n \in J$, the ring of integers) of objects in $\mathfrak{C}$, and
(b) the morphisms $d^n: K^n \to K^{n+1}$ ($n \in J$), such that $d^{n+1}d^n = 0$.

The complex K is called a *cochain complex* if $K^n = 0$, for $n < 0$ and a *chain complex* if $K^n = 0$, for $n > 0$.

If K and K' are two complexes in $\mathfrak{C}$, a morphism between them is defined as a sequence of morphisms $f^n: K^n \to K'^n$ such that for every $n \in J$ the diagram

$$\begin{array}{ccc} K^n & \xrightarrow{f^n} & K'^n \\ {\scriptstyle d^n}\downarrow & & \downarrow{\scriptstyle d'^n} \\ K^{n+1} & \xrightarrow{f^{n+1}} & K'^{n+1} \end{array} \qquad (7)$$

is commutative.

It follows that we can talk of the category of $\mathfrak{C}$-valued complexes and we denote this category by $\mathfrak{K}(\mathfrak{C})$. In particular, we have the category of the cochain complexes in $\mathfrak{C}$ and that of the chain complexes.

Given a complex K, we set

$$Z^n(K) = \ker d^n, \qquad B^n(K) = \operatorname{im} d^{n-1}, \qquad (8)$$

and call these the *object of cycles* and *object of boundaries* of K, respectively.

Using the diagram 4(2) we get a commutative diagram

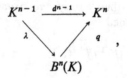

where λ is an epimorphism and q a monomorphism, and since $d^n d^{n-1} = 0$ we have

$$d^n q \lambda = 0 = 0\lambda,$$

whence $d^n q = 0$. This shows that, if $Z^n(K) \xrightarrow{h} K^n$ is the injection which defines $\ker d^n$, we have $q = h\alpha$ with a uniquely determined monomorphism $\alpha: B^n(K) \to Z^n(K)$.

5. Functors and Homology

It follows that there is a corresponding quotient object

$$H^n(K) = Z^n(K)/B^n(K) \tag{9}$$

and this is called the nth *homology object* of the complex K. If $H^n(K) = 0$, K is called *acyclic in dimension* n, and if K is acyclic for every n it is called *acyclic*. For the cochain complexes the previous terms are replaced by *cocycles, coboundaries*, and *cohomology*. For chain complexes the names introduced remain but the notation is changed by putting

$$K^{-n} = K_n, \quad Z^{-n} = Z_n, \quad B^{-n} = B_n,$$
$$H^{-n} = H_n, \, d^{-n} = d_n, \quad n > 0. \tag{10}$$

Next, let $f: K \to K'$ be a morphism of complexes, represented by the diagrams (7). In the above notation, we have $f^n h: Z^n(K) \to K'^n$ and $d'^n f^n h = f^{n+1} d^n h = 0$, from which the existence of a uniquely determined $f'^n: Z^n(K) \to Z^n(K')$ follows such that $f^n h = h' f'^n$ (h' being h for K'). Hence, for every n, Z^n is a covariant functor defined on $\mathfrak{K}(\mathfrak{C})$ with values in $\mathfrak{C}$.

Consider the diagram

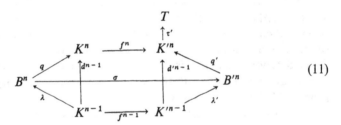

(11)

where q, q', λ, λ' are introduced by applying the diagram 4(2) and $\tau': K'^n \to T$ is coker d'^{n-1}. We have

$$\tau' f^n d^{n-1} = \tau' d'^{n-1} f^{n-1} = 0,$$

i.e., $\tau' f^n q \lambda = 0$ and, since λ is epimorphic, it follows that $\tau' f^n q = 0$. Hence, by the definition of the cokernel there exists a unique $\sigma: B^n \to B'^n$ such that $q'\sigma = f^n q$, which means that B^n is also a covariant functor from $\mathfrak{K}(\mathfrak{C})$ to $\mathfrak{C}$.

Finally, using Proposition 4.15 we see that H^n is a covariant functor from $\mathfrak{K}(\mathfrak{C})$ to $\mathfrak{C}$, since we have the *induced morphisms*

$$f^*: H^n(K) \to H^n(K').$$

Therefore we can talk of *homology* or *cohomology functors* of a complex.

Let $\mathfrak{C}$ be the category of Abelian groups. Then, $\mathfrak{K}(Ab)$ is an Abelian category and one obtains the following fundamental *theorem of the exact homology (cohomology) sequence*.

5 Theorem *There is a covariant functor defined on the category of short exact sequences in $\mathfrak{K}(Ab)$ with values in the category of exact sequences of Ab, which associates with the short exact sequence*

$$0 \longrightarrow K' \xrightarrow{i} K \xrightarrow{j} K'' \longrightarrow 0$$

of complexes in Ab *its exact homology (cohomology) sequence*

$$\cdots \xrightarrow{\partial} H^n(K') \xrightarrow{i_*} H^n(K) \xrightarrow{j_*} H^n(K'') \xrightarrow{\partial} H^{n+1}(K') \xrightarrow{i_*} \cdots$$

(where ∂ is a homomorphism to be defined) and with every morphism of short exact sequences the morphism of exact sequences obtained from the functorial property of H^n.

The definition of the *connecting homomorphism* ∂ is given in Theorem 3 and the verification of the exactness of the associated sequence and of its functorial character, which means the commutativity of the corresponding squares, is a matter of technical calculations left to the reader. (For a detailed proof, the reader is referred to [42], among other texts.)

This theorem holds, with the same proof, in the category A-Mod. For general Abelian categories see [3, 33].

We proceed now with the general considerations on functors.

6 Definition Let $F, G: \mathfrak{P}_1 \to \mathfrak{P}_2$ be two covariant functors between pseudocategories. A *functorial morphism* or *natural transformation* from F to G is a system $\mu: F \to G$ of morphisms $\mu(A): F(A) \to G(A)$ defined for every object A in $\mathfrak{P}_1$ which are such that for every $u: A \to B$ in $\mathfrak{P}_1$ the diagram

$$\begin{array}{ccc} F(A) & \xrightarrow{\mu(A)} & G(A) \\ {\scriptstyle F(u)}\downarrow & & \downarrow{\scriptstyle G(u)} \\ F(B) & \xrightarrow{\mu(B)} & G(B) \end{array} \qquad (12)$$

is commutative (i.e., if one of the composites $G(u)\mu(A)$ and $\mu(B)F(u)$ exists

5. Functors and Homology

the other also exists and they are equal). The same definition will apply to contravariant functors by changing the sense of the vertical arrows in (12). If all the morphisms $\mu(A)$ are isomorphisms, μ is called a *functorial isomorphism* or a *natural equivalence* of the two functors.

Obviously, the composition of functorial morphisms can be defined by the composition of the $\mu(A)$ and it follows that the covariant (contravariant) functors between two pseudocategories and their functorial morphisms are respectively the objects and morphisms of a hyperpseudocategory. The functors between categories form a hypercategory.

7 Proposition *If F is a contravariant functor from a category $\mathfrak{C}$ to Ens and X is an object in $\mathfrak{C}$, there is a bijection b: $\mathrm{Hom}(\mathfrak{M}_X, F) \to F(X)$, where $\mathrm{Hom}(\mathfrak{M}_X, F)$ is the class of functorial morphisms from $\mathfrak{M}_X$ to F.*

In fact, for $\mu \in \mathrm{Hom}(\mathfrak{M}_X, F)$ we put $b(\mu) = \mu(X)(1_X) \in F(X)$ and for $x \in F(X)$ and an object A in $\mathfrak{C}$, we take $\mu_x(A): \mathfrak{M}_X(A) \to F(A)$ defined by

$$\mu_x(A)(u) = F(u)(x), \qquad u: X \to A,$$

which gives a functorial morphism. We obtain the inverse map of b. The technical details of this proof consist in the verification of the commutativity of certain diagrams and we leave it to the reader.

Based on this proposition we consider the following definition.

8 Definition A contravariant functor $F: \mathfrak{C} \to \mathrm{Ens}$ is called *representable* if there is an object X in $\mathfrak{C}$ such that $\mathfrak{M}_X$ and F are isomorphic by some isomorphism γ. F is then *represented* by the pair $(X, b(\gamma))$.

The representable functors play a very important role in the theory of categories. They have the property of carrying inductive limits to projective limits, as well as other interesting properties. The corresponding definition and properties for covariant functors can also be stated [40].

Let $\mathfrak{C}_1$ and $\mathfrak{C}_2$ be two categories and $F: \mathfrak{C}_1 \to \mathfrak{C}_2$, $G: \mathfrak{C}_2 \to \mathfrak{C}_1$ two covariant functors. We associate with them two covariant functors from $\mathfrak{C}_1^* \times \mathfrak{C}_2$ to Ens by associating with a pair (A, B), with A an object in $\mathfrak{C}_1$ and B an object in $\mathfrak{C}_2$, first the set $\mathfrak{M}_2(F(A), B)$ and then the set $\mathfrak{M}_1(A, G(B))$, and with a pair $(u: A' \to A,\ v: B \to B')$ the mappings $\alpha \mapsto v\alpha F(u)$ ($\alpha \in \mathfrak{M}_2(F(A), B)$) and $\beta \mapsto G(v)\beta u$ ($\beta \in \mathfrak{M}_1(A, G(B))$), respectively. The two functors obtained will be denoted by $\mathfrak{M}_2(F)$ and $\mathfrak{M}_1(G)$.

9 Definition The functors F and G are called *adjoint functors* if $\mathfrak{M}_2(F)$ and $\mathfrak{M}_1(G)$ are naturally equivalent.

Consider the categories $\mathfrak{C}_1$ and $\mathfrak{C}_2$ and the functors F, G as above. We denote by $1_{\mathfrak{C}_1}$ and $1_{\mathfrak{C}_2}$ the functors which leave unchanged the objects and morphisms of the given categories.

10 Definition The categories $\mathfrak{C}_1$ and $\mathfrak{C}_2$ are called *equivalent* if there are functors $F: \mathfrak{C}_1 \to \mathfrak{C}_2$, $G: \mathfrak{C}_2 \to \mathfrak{C}_1$ and functorial isomorphisms $\varphi: 1_{\mathfrak{C}_1} \to GF$, $\psi: 1_{\mathfrak{C}_2} \to FG$.

Moreover, if the equivalence is by a one-to-one correspondence of the objects of the two categories, it is called an *isomorphism of categories*.

An important problem, which will not be treated here, is to study the behavior of the different notions and operations on categories under the action of functors and, especially, the conditions for two functors to commute.

For instance, it is easy to see that a functor associates with every isomorphism an isomorphism. If a functor preserves injections it is called a *monofunctor*, and if it preserves *surjections*, an *epifunctor*. If the correspondence $\mathfrak{M}(A, B) \to \mathfrak{M}(F(A), F(B))$ induced by the functor F is injective, F is called a *faithful* functor and if it is surjective, F is called a *full* functor. A faithful functor which associates with different objects different images is called an *imbedding*. There are important imbedding theorems for categories; e.g., any small Abelian category can be imbedded in Ab [33], etc.

There are, important special classes of functors in the case of additive and Abelian categories. Thus, a functor $F: \mathfrak{C}_1 \to \mathfrak{C}_2$ between two additive categories is called an *additive functor* if for every $u, v \in \text{Hom}_\mathfrak{C}(A, B)$ we have

$$F(u + v) = F(u) + F(v).$$

11 Proposition *The homology functors are additive functors.*

In fact, using the notation of the definition of these functors, we see that if (7) is a homomorphism of complexes it induces homomorphisms $f'^n: Z^n(K) \to Z^n(K')$ which are uniquely determined by the relation $f^n h = h' f'^n$. Hence, if $\varphi^n h = h' \varphi'^n$ for another morphism φ, we have $(f^n + \varphi^n)h = h'(f'^n + \varphi'^n)$, and we infer the additivity of the functors Z^n. By the same reasoning with the morphism σ in Diagram (11), we see that the B^n are also additive. The additivity of H^n follows by Proposition 4.15.

Next, in the case of Abelian categories the covariant functor F is called *left exact* if it associates with every exact sequence

$$0 \longrightarrow A' \xrightarrow{i} A \xrightarrow{j} A'' \tag{13}$$

5. Functors and Homology

the exact sequence

$$0 \longrightarrow F(A') \xrightarrow{F(i)} F(A) \xrightarrow{F(j)} F(A'').$$

F is a *right exact functor* if it associates with every exact sequence

$$A' \xrightarrow{i} A \xrightarrow{j} A'' \longrightarrow 0 \tag{14}$$

the exact sequence

$$F(A') \xrightarrow{F(i)} F(A) \xrightarrow{F(j)} F(A'') \longrightarrow 0.$$

If F is both left and right exact, it is called an *exact functor*. An exact functor associates with every exact sequence an exact sequence.

For contravariant additive functors, left exactness means (in this book) that the exactness of (14) implies that

$$0 \longrightarrow F(A'') \xrightarrow{F(j)} F(A) \xrightarrow{F(i)} F(A')$$

is exact and right exactness means that the exactness of (13) implies the exactness of

$$F(A'') \xrightarrow{F(j)} F(A) \xrightarrow{F(i)} F(A') \longrightarrow 0.$$

For instance, in A-Mod (and hence in Ab = J-Mod) the functors $\mathfrak{M}_X$ and $\mathfrak{M}^X$ are left exact functors and $X \otimes$ is right exact.

To conclude, let us indicate the possibility of extending the method used to introduce general notions in categories.

12 Definition By a *strengthened category* we mean a system consisting of a category $\mathfrak{C}$ and a family $\mathfrak{F} = \{F_i\}$ of functors on $\mathfrak{C}$ with values in Ens.

If $(\mathfrak{C}, \mathfrak{F})$ is a strengthened category we can define several notions. A *strengthened injection* is an injection u in $\mathfrak{C}$ such that, for every functor $F_i \in \mathfrak{F}$, $F_i(u)$ is injective in Ens; a *strengthened surjection* is a surjection u such that $F_i(u)$ are surjections; a *strengthened subobject*, etc., may be defined in an analogous manner. Generally, when strengthened categories are used the word strengthened will be omitted after the indication of the family $\mathfrak{F}$.

Hence in Sections 2–5 the possibility of using many general notions and operations was shown by merely indicating the category (strengthened category) to be considered, and we are not required to define them for every particular case.

6 Atlases

The problems discussed in Section 5 allow us to introduce the general theory of atlases, which is fundamental in differential geometry and in other fields of mathematics.

Let X be a set which is called *point space* and $Q(X)$ a family of nonempty subsets of X. $Q(X)$ can be made into a pseudocategory by taking as objects the subsets in $Q(X)$ and as morphisms the partial identity maps of these subsets. In other words, if $U_i \in Q(X)$ ($i = 1, 2$) and $U_1 \cap U_2 = \emptyset$, $\mathfrak{M}(U_1, U_2) = \emptyset$, and if $U_1 \cap U_2 \neq \emptyset$, $\mathfrak{M}(U_1, U_2)$ consists of only one element, the identity map of $U_1 \cap U_2$, considered as a partial map from U_1 to U_2. The verification of the conditions in Definition 2.1 is trivial and, generally, $Q(X)$ is not a category.

$Q(X)$ is a subpseudocategory of Ens^p and we shall denote by I the covariant imbedding functor of $Q(X)$ in Ens^p; it preserves all the objects and morphisms of $Q(X)$.

Let $\mathfrak{P}$ be an arbitrary subpseudocategory of Ens^p or, more precisely, a Cantorian pseudocategory. Then, we can identify the $\mathfrak{P}$-valued functors with Ens^p-valued functors, which means just the composition of those functors with the imbedding of $\mathfrak{P}$ in Ens^p.

1 Definition A $\mathfrak{P}$-*valued atlas* on X is a triple $\mathfrak{A} = (Q(X), \mathscr{A}, h)$ consisting of a pseudocategory $Q(X)$ as above, a covariant functor $\mathscr{A}: Q(X) \to \mathfrak{P}$ and a functorial isomorphism $h: I \to \mathscr{A}$, considered as Ens^p-valued functors.

Therefore, to obtain an atlas on X with values in $\mathfrak{P}$ we must consider a family of nonempty subsets Q(X) of X and associate with every element $U \in Q(X)$ an object $\mathscr{A}(U)$ of $\mathfrak{P}$ and a bijection $h_U: U \to \mathscr{A}(U)$ such that for any U, V in $Q(X)$ there is a commutative diagram

6. Atlases

$$\begin{array}{ccc} U & \xrightarrow{h_U} & \mathscr{A}(U) \\ {\scriptstyle i_V^U}\downarrow & & \downarrow{\scriptstyle a_V^U} \\ V & \xrightarrow{h^V} & \mathscr{A}(V) \end{array} \qquad (1)$$

Here, i_V^U denotes the partial inclusion from U to V and the $a_V^U = \mathscr{A}(i_V^U)$ are morphisms in $\mathfrak{P}$. There are also some functorial conditions to be stated later.

The functions a_V^U play a fundamental role in the structure of an atlas; they are called the *transition functions* of $\mathfrak{A}$. We make a few simple remarks concerning them. Thus, from Diagram (1) we get

$$a_V^U = h_V \, i_V^U h_U^{-1} = h_V h_U^{-1}, \qquad (2)$$

which shows that a_V^U is a bijection:

$$a_V^U \colon h_U(U \cap V) \to h_V(U \cap V). \qquad (3)$$

From the functorial property of $\mathscr{A}$ we get

$$a_U^U = 1_{\mathscr{A}(U)}, \qquad a_W^V a_V^U = a_W^U, \qquad U, V, W \in Q(X), \qquad (4)$$

where the last equality holds on $U \cap V \cap W \neq \varnothing$. Taking $U = W$ we see that on $U \cap V$ the bijections a_V^U and a_U^V are inverses of one another.

It is important to observe that the elements of $Q(X)$ are not necessarily distinct, so that it is possible for a subset of X to enter several times in $Q(X)$.

2   Definition   For an atlas $\mathfrak{A}$, the elements of $Q(X)$ are called *coordinate domains*, the bijections h_U are called *local charts*, and the value $h_U(x)$ for $x \in U$ is called the *coordinate* of the point x in the chart h_U.

The name "transition functions" comes from the property that they give the relation between the coordinates of a point in two different charts. In fact, if $x \in U \cap V$ and $y = h_U(x)$, $z = h_V(x)$, we have, using (2),

$$z = a_V^U(y). \qquad (5)$$

3   Definition   If X is a topological space an *atlas on X* is an atlas as in Definition 1 such that the coordinate domains are open subsets, $\mathfrak{P}$ is a topological pseudocategory, and h is an isomorphism of the Top^p-valued functors I and $\mathscr{A}$.

Therefore, in Diagram 1, the h_U are homeomorphisms and a_V^U are continuous maps, namely partial homeomorphisms.

4 Proposition *If $\mathfrak{A}$ is an atlas on a set X whose values are in a topological pseudocategory $\mathfrak{P}$, there is a uniquely determined topology on X such that $\mathfrak{A}$ is an atlas in the sense of Definition 3.*

In fact, for U in $Q(X)$ we define a topology by requiring h_U to be a homeomorphism. Clearly, the maps a_V^U of (3) are homeomorphisms; hence the topologies of U and V induce the same topology on $U \cap V$. Next, $h_U(U \cap V)$ is open in $\mathscr{A}(U)$ since it is the domain of a morphism of Top^p, so $U \cap V$ is open in U. It follows that the topology on X which is induced by the inclusion maps of U in X has the desired properties. Q.E.D.

5 Definition An atlas on a set X is called a *covering atlas* if the union of the subsets in $Q(X)$ is X.

Let $\mathfrak{A} = (Q(X), \mathscr{A}, h)$ and $\mathfrak{A}' = (Q'(X), \mathscr{A}', h')$ be two atlases on X with values in $\mathfrak{P}$. Their *sum* is defined as the system consisting of $Q(X) \vee Q'(X)$ with the structure of a pseudocategory obtained as for $Q(X)$, of the correspondence to Ens^p defined as $\mathscr{A} \vee \mathscr{A}'$ and of $\{h_U\} \vee \{h'_U\}$.

6 Definition The atlases $\mathfrak{A}$ and $\mathfrak{A}'$ are called *compatible* if their sum is again a $\mathfrak{P}$-valued atlas on X. The $\mathfrak{P}$-valued atlas $\mathfrak{A}$ on X is called *complete* if it is covering and contains all its compatible atlases.

7 Proposition *Any covering atlas of X can be imbedded in a complete atlas.*

This is a simple consequence of Zorn's lemma applied to a class of compatible atlases ordered by inclusion.

8 Definition A complete $\mathfrak{P}$-valued atlas on the space X is called a $\mathfrak{P}$-*structure* on X. The space X together with a $\mathfrak{P}$-structure on it is called a $\mathfrak{P}$-*manifold*.

For instance, let X and X' be two $\mathfrak{P}$-manifolds defined respectively by the atlases $(Q, \mathscr{A}, h)$ and $(Q', \mathscr{A}', h')$. Take on $X \times X'$ the subsets $\{U \times U' \mid U \in Q, \ U' \in Q'\}$ and define, if possible, $\mathscr{A}(U \times U') =$

$\mathscr{A}(U) \times \mathscr{A}'(U')$ and $h_{U \times U'} = h_U \times h'_{U'}$. "If possible" means here that the respective products exist in $\mathfrak{P}$. If this is the case, we obviously obtain a $\mathfrak{P}$-structure on $X \times X'$, and $X \times X'$ together with this structure is called the *product $\mathfrak{P}$-manifold* of the given manifolds.

The essential thing in a $\mathfrak{P}$-structure is the system of transition functions of its atlas. This is made clear by the following definition and theorem.

9 Definition By a *system of transition functions* in a Cantorian (topological) pseudocategory $\mathfrak{P}$ we mean a family of objects A_i ($i \in M$) of $\mathfrak{P}$ and a family of morphisms $a_j^i : A_i \to A_j$ defined for some $i, j \in M$, such that

$$a_i^i = 1_{A_i}, \quad a_k^j a_j^i = a_k^i, \tag{6}$$

when defined.

10 Theorem *Given a system of transition functions in a pseudocategory $\mathfrak{P}$, there is a $\mathfrak{P}$-manifold whose atlas has as transition functions the given system, and this manifold is determined up to an "isomorphism."*

Considering the given system, in the above notation, take $\tilde{X} = \vee_i A_i$, the sum being in Ens. Define on $\tilde{X}$ the equivalence relation $a_i \sim a_j$ ($a_i \in A_i$, $a_j \in A_j$) by $a_j = a_j^i(a_i)$. If X is the corresponding quotient set, take $Q(X)$ as the system of quotient sets of the A_i and the functor $\mathscr{A}$ defined by (A_i, a_j^i). Finally, consider the isomorphism h induced by the maps 1_{A_i}. In this manner, we obtain a $\mathfrak{P}$-structure on X, and X together with this structure is the required $\mathfrak{P}$-manifold. The meaning of "isomorphism" in Theorem 10 will become clear later.

If $\mathfrak{P}$ is a topological pseudocategory, we shall give X the topology of Proposition 4.

Now let X and Y be two $\mathfrak{P}$-manifolds with atlases $\mathfrak{A}$ and $\mathfrak{B}$ and let $f: X \dashrightarrow Y$ be a partial map. Then, for some coordinate domains U on X and V on Y we have the partial map

$$f_U^V = k_V f h_U^{-1} : \mathscr{A}(U) \dashrightarrow \mathscr{B}(V), \tag{7}$$

where h_U and k_V are the respective charts. The maps (7) are called the *coordinate expressions* of f. By a change of chart, f_V^U changes by the rule

$$f_{V'}^{U'} = b_{V'}^V f_V^U a_U^{U'}, \tag{8}$$

where a: and b: are the transition functions of $\mathfrak{A}$ and $\mathfrak{B}$, respectively. Equation (8) is a simple consequence of Formula (2).

11 Definition   The mapping $f: X \longrightarrow Y$ between two $\mathfrak{P}$-manifolds is called a *morphism* or a $\mathfrak{P}$-*mapping* if its coordinate expressions are morphisms in $\mathfrak{P}$.

By (8) this condition has an invariant meaning.

It follows that there is a pseudocategory whose objects are $\mathfrak{P}$-manifolds and whose morphisms are their $\mathfrak{P}$-mappings ($\mathfrak{P}$-fixed). If only mappings whose source and domain are equal are considered we have a corresponding category. In particular $\mathfrak{P}$-isomorphisms are isomorphisms of this category (see Theorem 10). We shall denote by $\mathfrak{P}$-Var[†] the previous pseudocategory and category. If $\mathfrak{P}$ is topological, $\mathfrak{P}$-Var is also topological, and we have the corresponding local category $\mathfrak{L}(\mathfrak{P}\text{-Var})$.

Let $\mathfrak{P}$ be a topological pseudocategory with a $\mathfrak{C}$-valued derivative functor D. Then, let X be a $\mathfrak{P}$-manifold and x a point of X. With this configuration we can associate the family $\{D(\mathscr{A}(U), h_U(x)) \mid x \in U\}$ of objects of $\mathfrak{C}$ with the family

$$D[a_V^U]_{h_U(x)}: D(\mathscr{A}(U), h_U(x)) \to D(\mathscr{A}(V), h_V(x))$$

of morphisms. Here, $(\mathscr{A}(U), h_U(x))$ is the corresponding object of $\mathfrak{L}(\mathfrak{P})$.

By the relations (4) and the functorial property of D this family can be considered to be an inductive family of objects in $\mathfrak{C}$ indexed by the set $\{U \mid x \in U\}$ and with the trivial partial ordering relation $U \leqslant V$ for every pair (U, V) (this relation is reflexive and transitive but not antisymmetrical.)

12 Definition   If the inductive limit in $\mathfrak{C}$ of the above family exists it will be called the *tangent object* of the $\mathfrak{P}$ manifold X at $x \in X$.

This object will be denoted by $T_x(X)$ or T_x if no confusion arises.

Now let Y be another $\mathfrak{P}$-manifold and $f: X \to Y$ a $\mathfrak{P}$-mapping such that $y = f(x)$ ($x \in X$). Let (7) be the coordinate expression of f in a chart of X whose domain contains x. We then get morphisms

$$D(\mathscr{A}(U), h_U(x)) \to T_y(Y)$$

by the composition of

[†] "Var" comes from the French *variété* = manifold.

6. Atlases

$$D[f_V^U]_x: D(\mathscr{A}(U), h_U(x)) \to D(\mathscr{B}(V), k_V(y))$$

with the morphism $D(\mathscr{B}(V), k_V(y)) \to T_y(Y)$ given by the definition of an inductive limit. These morphisms commute with the derivatives of the transition functions because of (8), and it follows that there is a uniquely determined induced morphism, called the *derivative morphism* of f:

$$T_x f = f^*: T_x(X) \to T_y(Y). \tag{9}$$

13 Theorem *If $\mathfrak{P}$ is a topological pseudocategory, endowed with a $\mathfrak{C}$-valued derivative functor, T_x is a covariant functor on $\mathfrak{L}(\mathfrak{P}\text{-Var})$ with values in $\mathfrak{C}$ and defines a derivative on $\mathfrak{P}$-Var.*

Only the rule for the derivation of composite functions remains to be established, but this is immediate.

14 Definition A $\mathfrak{P}$-mapping $f: X \to Y$ is called an *immersion* if, for every $x \in X$, f^* is an injection.

It follows that we can talk of the category of $\mathfrak{P}$-manifolds and immersions.

Because the category $\mathfrak{P}$-Var is Cantorian we can consider representative subobjects. These will be called $\mathfrak{P}$-*submanifolds*. We can also consider *immersed $\mathfrak{P}$-submanifolds* (for convenient $\mathfrak{P}$ of course) as subobjects in the category of $\mathfrak{P}$-manifolds and immersions. If, moreover, the topology of the immersed submanifold is the relative topology, the submanifold is called *imbedded*. Analogously, $f: X \to Y$ is a *submersion* if f^* is a surjection in $\mathfrak{C}$.

Let us consider a few examples of $\mathfrak{P}$-manifolds. Some of them will be considered in more detail later on.

If $\mathfrak{P}$ is the topological pseudocategory whose objects are open subsets in arithmetical spaces R^n (R is the real field and n are integers) and whose morphisms are partial continuous maps, a space X is a $\mathfrak{P}$-manifold if and only if there is an open covering of it by sets homeomorphic to open subsets of R^n. If, moreover, X is a Hausdorff space it is a topological manifold in the usual sense.

If $\mathfrak{P}$ has the same objects as above but its morphisms are differentiable maps (of class C^p, $1 \leq p \leq \infty$, or C^ω = real analytic), a Hausdorff space is a $\mathfrak{P}$-manifold if and only if it is a classical differentiable manifold. It is well known that in this case $\mathfrak{P}$ has a derivative functor with values in the category of finite dimensional real linear spaces, which associates with every open set in R^n the space R^n and with every differentiable map the linear map given by the Jacobian matrix. The tangent object of Definition 12 is then the classical tangent space at a point of a manifold. The derivatives of maps, immersions, and submanifolds are also the classical ones.

Analogously, if $\mathfrak{P}$ is the pseudocategory of the open sets in complex arithmetical spaces C^n (C being the complex field) and of the complex analytic maps between them, we get the category of complex analytic manifolds and maps. Using the Jacobian matrix we again obtain a tangent space.

If $\mathfrak{P}$ is the pseudocategory of the open subsets of Banach spaces and of their differentiable maps, the $\mathfrak{P}$-manifolds are manifolds modeled on Banach spaces and here we can also introduce tangent spaces and the other notions previously considered (see Chapter 4 of this book).

We shall also see that our schema contains the fiber bundles. Other very important mathematical objects such as polyhedra, Grothendieck's preschemes, etc., can also be introduced in the same schema.

Chapter 2

SHEAVES AND COHOMOLOGY

1 Presheaves on a Topological Space

This section is devoted to the study of some special kinds of functors associated with a topological space. Next, these functors are used in obtaining the cohomology functors which are essential in topology and geometry.

1 Definition Let X be a topological space and $\mathfrak{D}(X)$ the category of its open, nonempty subsets and inclusions. Then, a contravariant functor $P: \mathfrak{D}(X) \to \mathfrak{C}$, where $\mathfrak{C}$ is an arbitrary category, is called a *presheaf* on X with values in $\mathfrak{C}$.

It follows that we get a $\mathfrak{C}$-valued presheaf on X by assigning to each open nonempty subset $U \subseteq X$ an object $P(U)$ of $\mathfrak{C}$ and to each inclusion $U \subseteq V$ a *restriction morphism*

$$r_U^V: P(V) \to P(U), \tag{1}$$

such that

$$r_U^U = 1_{P(U)}, \qquad r_U^V r_V^W = r_U^W, \qquad U \subseteq V \subseteq W. \tag{2}$$

The object $P(U)$ is called *the object of the sections over U* and if $\mathfrak{C}$ is a Cantorian category the elements of $P(U)$ are called *sections over U*. The space X is called the *base space* or *basis* of the presheaf.

2 Definition If P and P' are two presheaves on X with values in $\mathfrak{C}$, a *homomorphism* $\varphi : P \to P'$ is a functorial morphism of the two functors. A functorial isomorphism between them is a *presheaf isomorphism*.

It follows that φ is given by a system of morphisms $\varphi_U : P(U) \to P'(U)$ such that if $U \subseteq V$ there is a commutative diagram

$$\begin{array}{ccc} P(U) & \xrightarrow{\varphi_U} & P'(U) \\ {\scriptstyle r^V_U} \uparrow & & \uparrow {\scriptstyle r'^V_U} \\ P(V) & \xrightarrow{\varphi_V} & P'(V) \end{array} \qquad (3)$$

3 Proposition *The $\mathfrak{C}$-valued presheaves on X as objects and their homomorphisms as morphisms make up a category.*

In fact, the only thing to be justified is that $\{\varphi \mid \varphi : P \longrightarrow P'\} = \mathfrak{M}(P, P')$ is a set, and this is a simple consequence of Axiom A7 of Section 1.1, taking into account that $\mathfrak{D}(X)$ is a small category.

We denote by $P(X, \mathfrak{C})$ the category given by Proposition 3. All the notions and operations on presheaves with a given basis are to be considered in this category. This refers to concepts such as monomorphisms, epimorphisms, direct sums and products, inductive and projective limits, subpresheaves and quotient sheaves, ranges, kernels, cokernels, exact sequences, etc., which, of course, are used only if they exist in the category $P(X, \mathfrak{C})$. Generally, the corresponding constructions can be made by using the operations in $\mathfrak{C}$. To give an example, we consider the following proposition.

4 Proposition *If for a morphism φ in $P(X, \mathfrak{C})$ all the φ_U are injections (surjections) then φ is an injection (surjection). If φ is an injection (surjection) and $\mathfrak{C}$ has an initial (final) object, then all the φ_U are injections (surjections). φ is an isomorphism if and only if all the φ_U are isomorphisms.*

It suffices to prove this for injections and then to use duality. Let $\varphi : P \to P'$ and $\alpha_i : Q \to P$ ($i = 1, 2$) be presheaf homomorphisms on X

1. Presheaves on a Topological Space

such that $\varphi\alpha_1 = \varphi\alpha_2$. Then for every U $\varphi_U\alpha_{1U} = \varphi_U\alpha_{2U}$ and, if the φ_U are injections, $\alpha_1 = \alpha_2$. Now, let φ be an injection and $a_i: A \to P(U)$ ($i = 1, 2$) be two morphisms in $\mathfrak{C}$. We define a presheaf Q on X by denoting by I the initial object of $\mathfrak{C}$ and putting

$$Q(V) = \begin{cases} A & \text{if } V \subseteq U \\ I & \text{if } V \not\subseteq U \end{cases}$$

$$\rho_{V_1}^{V_2} = \begin{cases} 1_A & \text{if } V_1 \subseteq V_2 \subseteq U \\ \text{the morphisms of source } I \text{ in the other cases.} \end{cases}$$

Also, we define two homomorphisms $\alpha_i: Q \to P$ ($i = 1, 2$) by

$$\alpha_{iU} = a_i, \quad \alpha_{iV} = r_V^U a_i, \quad V \subseteq U$$

$$\alpha_{iV} \in \mathfrak{M}(I, P(V)) \quad V \not\subseteq U,$$

for $i = 1, 2$. If now $\varphi_U a_1 = \varphi_U a_2$, we also have $\varphi\alpha_1 = \varphi\alpha_2$ whence $\alpha_1 = \alpha_2$ and $a_1 = a_2$. It follows that the φ_U are injections.   Q.E.D.

The last assertion of Proposition 4 follows immediately.

For instance, Proposition 4 holds if $\mathfrak{C}$ is an additive category, since we have a zero object. It also holds for $\mathfrak{C} = $ Ens with $\varnothing$ as initial object, and where the sets consisting of one element are final objects. Hence, if P is an object of $P(X, \text{Ens})$ its subobjects P' can be represented such that $P'(U) \subseteq P(U)$.

5   Proposition   *If $\mathfrak{C}$ is an additive category, $P(X, \mathfrak{C})$ is additive. If $\mathfrak{C}$ is Abelian, the same is true for $P(X, \mathfrak{C})$.*

In fact, if $\mathfrak{C}$ is additive and $\varphi_i: P \to P'$ ($i = 1, 2$) are morphisms in $P(X, \mathfrak{C})$, we define

$$\varphi_1 + \varphi_2 = \{\varphi_{1,U} + \varphi_{2,U}\},$$

which gives a group structure on $\mathfrak{M}(P, P')$, satisfying the condition (a) of Definition 1.4.11. Next, $P(X, \mathfrak{C})$ has as zero object the presheaf associating with every U the zero object of $\mathfrak{C}$. Finally, the direct sum of two presheaves is obtained by assigning to U the object $P_1(U) \oplus P_2(U)$ in $\mathfrak{C}$. All the conditions of Definition 1.4.11 are satisfied and our first assertion is proved.

Suppose that $\mathfrak{C}$ is also regular and therefore Abelian, and let $\varphi: P \to P'$

be a morphism in $P(X, \mathfrak{C})$. We obtain $\ker \varphi$ by assigning to every open subsets $U \subseteq X$ the object $\ker \varphi_U$ in $\mathfrak{C}$ and to a pair $V \subseteq U$ the morphism $\alpha_V^U \colon \ker \varphi_U \to \ker \varphi_V$, uniquely determined by the relations

$$\varphi_V r_V^U i_U = r_V'^U \varphi_U i_U = 0,$$

(where $i_U \colon \ker \varphi_U \to P(U)$) which imply the existence of unique α_V^U such that

$$r_V^U i_U = i_V \alpha_V^U.$$

The conditions (2) are satisfied since the morphisms α are uniquely determined.

In the same manner, assigning to U the object $\operatorname{coker} \varphi_U$ in $\mathfrak{C}$, we shall obtain $\operatorname{coker} \varphi$, and the canonical bijections

$$\mu_U \colon \operatorname{coim} \varphi_U \to \operatorname{im} \varphi_U$$

(see Diagram 1.4(2)) define the bijection $\mu \colon \operatorname{coim} \varphi \to \operatorname{im} \varphi$ in $P(X, \mathfrak{C})$.

Finally, the existence of the zero object in $\mathfrak{C}$, together with Proposition 4, show that if in $\mathfrak{C}$ every bijection is an isomorphism, the same is true in $P(X, \mathfrak{C})$. Hence, all the conditions of definition 1.4.6 are satisfied and Proposition 5 is proved.

It follows from the above considerations that if the category $\mathfrak{C}$ is Abelian it makes sense to talk of exact sequences of $\mathfrak{C}$-valued presheaves, and such a sequence

$$\cdots \xrightarrow{\varphi} P \xrightarrow{\psi} \cdots$$

is exact if and only if the sequence

$$\cdots \xrightarrow{\varphi_U} P(U) \xrightarrow{\psi_U} \cdots$$

is exact for every U.

This is the case, for instance, if $\mathfrak{C} = A\text{-Mod}$ (for A a commutative unitary ring). It should also be remarked that in this case a subpresheaf is obtained by taking for every U a submodule $P'(U) \subseteq P(U)$ such that $r_V^U(P'(U)) \subseteq P'(V)$, and that the corresponding quotient presheaf is defined by $P(U)/P'(U)$ with induced restrictions.

Consider a continuous mapping $f \colon X \to Y$ and let P be an object of

1. Presheaves on a Topological Space

$P(X, \mathfrak{C})$, where $\mathfrak{C}$ has a final object F. Assigning to every open subset $U \subseteq Y$ the object $P(f^{-1}(U))$ for $f^{-1}(U) \neq \emptyset$ and F otherwise, together with the corresponding restrictions, we get a presheaf $f(P)$ with base space Y, which is called the *image* of P by f. Clearly, if $\varphi: P \to P'$ is a morphism in $P(X, \mathfrak{C})$, $\psi = \{\varphi_{f^{-1}(U)}\}$ defines a morphism $f(P) \to f(P')$ in $P(Y, \mathfrak{C})$. Hence, we get the following proposition:

6   Proposition   *Any continuous map $f: X \to Y$ induces a covariant functor $f: P(X, \mathfrak{C}) \to P(Y, \mathfrak{C})$.*

Another simple result is

7   Proposition   *A covariant functor $F: \mathfrak{C} \to \mathfrak{C}'$ induces a covariant functor $F: P(X, \mathfrak{C}) \to P(X, \mathfrak{C}')$.*

In fact, given P in $P(X, \mathfrak{C})$, the system $\{F(P(U)), F(r_V^U)\}$ is an object $F(P)$ in $P(X, \mathfrak{C}')$ and, given $\varphi: P \to P'$, the system $F(\varphi_U)$ defines the corresponding morphism $F(\varphi)$.

We now consider some examples and constructions of presheaves.

If A is an object in $\mathfrak{C}$ and if we take $P(U) = A$ and $r_V^U = 1_A$, for every $V \subseteq U$ open subsets of the space X we obviously get a presheaf of $P(X, \mathfrak{C})$. It is called a *simple* or *constant presheaf*. Another example is obtained by taking $\mathfrak{C} = \text{Ens}$ and $P(U) = \{f \mid f: U \to M\}$, $r_V^U(f) = f|_V$, where M is an arbitrary set; this is a presheaf in $P(X, \text{Ens})$. If M is a topological space we can consider only continuous functions. Next, consider two $\mathfrak{P}$-manifolds X and Y and assign to every open subset $U \subseteq X$ the set $\mathfrak{F}(U)$ of the $\mathfrak{P}$-mappings with source X and target Y, whose domain is U, and to every pair $V \subseteq U$ the usual restriction of maps; we get obviously an Ens-valued presheaf with base space X, and this presheaf will be denoted by $\mathfrak{F}(X, Y)$.

Let $i: Y \subseteq X$ be the inclusion of a subspace Y of X in X and P be a presheaf with base space Y. By Proposition 6 we can consider the presheaf $i(P)$ with base space X which is obtained by associating with every open subset $U \subseteq X$ the object $P(U \cap Y)$ or the final object of $\mathfrak{C}$. This presheaf will usually be denoted by $\hat{P}$ and it is called the *trivial extension* of P to X. If a presheaf Q on X is equal to $\hat{P}$ for some P, Q is said to be *concentrated* on Y.

Let P be an object in $P(X, \mathfrak{C})$ and Y an open subspace of X. By associating with each open subset $U \subseteq Y$ the object $P(U)$ we get a presheaf $P|_Y$ with base space Y called the *restriction* of P to the open subspace Y. This restriction operation has a functorial character. Clearly, if we associate with every open $U \subseteq X$ the set $\mathfrak{M}(P|_U, P'|_U)$ for two objects P, P' in

$P(X, \mathbb{C})$, we get a presheaf on X which we denote by $\mathscr{P}(P, P')$.

An important presheaf category is that of the presheaf modules.

8 Definition Let $A = \{A(U), \rho_V^U\}$ be a presheaf of commutative unitary rings, with base space X and $P = \{P(U), r_V^U\}$ a presheaf of Abelian groups on X. P is called a *presheaf module* over A if every $P(U)$ has a structure of an $A(U)$-module such that if $V \subseteq U$ the diagram

$$
\begin{array}{ccc}
A(U) \times P(U) & \longrightarrow & P(U) \\
\rho_V^U \times r_V^U \downarrow & & \downarrow r_V^U, \\
A(V) \times P(V) & \longrightarrow & P(V)
\end{array}
\qquad (4)
$$

where the horizontal arrows represent the module operation, is commutative.

If P and P' are two presheaf modules over A, a homomorphism $\varphi: P \to P'$ is a homomorphism of the two group presheaves such that the corresponding φ_U are also $A(U)$-module homomorphisms. Hence, we can consider the category of presheaf modules over A with base space X, and this will be denoted by $P(X, A\text{-Mod})$. In the particular case when A is a constant presheaf we have just the presheaves of A-modules on X.

One can extend in an obvious manner the classical operations with modules to the category $P(X, A\text{-Mod})$, and one sees that this category is Abelian. The presheaf $\mathscr{P}(P, P')$ is also a presheaf module over A. One can define the tensor product of two A presheaf modules P and P' as the A-presheaf module which assigns to an open subset $U \subseteq X$ the module $P(U) \otimes_{A(U)} P'(U)$ and to a pair $U \subseteq V$ the tensor product of the restrictions of P and P'.

2 Sheaves of Sets

In this section the category Ens is assumed to contain the object $\varnothing$.

We introduce some terminology in $S(\text{Top})$, the arrow category of the category of the topological spaces. Let

2. Sheaves and Sets

$$p: E \to X \qquad (1)$$

be an object of $S(\text{Top})$; p is called a *projection* if it is surjective in Top. E is its *total space*, X is the *base space*, and the triple (1) is a *projected space*. If $x \in X$, $p^{-1}(x) = E_x$ is called the *fiber over* x, or the *local fiber*. If U is an open subset in X and $s: U \to E$ is a continuous map such that

$$p \cdot s = 1_U, \qquad (2)$$

s is called a *cross section* or a *section over* U. The cross section is *local* if $U \neq X$, and *global* otherwise.

The morphisms or homomorphisms of projected spaces will be the morphisms in $S(\text{Top})$, i.e., commutative diagrams

$$\begin{array}{ccc} E & \xrightarrow{\varphi} & E' \\ {\scriptstyle p}\downarrow & & \downarrow{\scriptstyle p'} \\ X & \xrightarrow{f} & X' \end{array} \qquad (3)$$

with continuous arrows.

If the base space X is fixed, we shall consider the subcategory $\mathfrak{E}(X)$ of $S(\text{Top})$ with projected spaces on X as objects and with commutative diagrams

$$\begin{array}{ccc} E & \xrightarrow{\varphi} & E' \\ {\scriptstyle p}\searrow & & \swarrow{\scriptstyle p'} \\ & X & \end{array} \qquad (4)$$

as morphisms. Then for $x \in X$ there is an induced map $\varphi_x: E_x \to E'_x$, and φ is determined by these maps.

If we associate with every object of $\mathfrak{E}(X)$ the total space, we get a covariant functor $\mathfrak{E}(X) \to \text{Ens}$, called the *total space functor*, which will be used to strengthen the category $\mathfrak{E}(X)$, and its subcategories, in the sense of Section 1.5. This strengthening will always be considered in the sequel. Also, subobjects will be considered as topological subspaces with the relative topology.

The direct products in $\mathfrak{E}(X)$ are called *fibered products* of the projected spaces. If $E \xrightarrow{p} X$ and $E' \xrightarrow{p'} X$ are objects in $\mathfrak{E}(X)$, their fibered product is $E'' \xrightarrow{p''} X$, where $E'' = \{(e, e') \mid e \in E, e' \in E', p(e) = p'(e')\}$, $p''(e, e') = p(e)$. A continuous mapping $f: X \to Y$ induces a covariant functor $f^*: \mathfrak{E}(Y) \to \mathfrak{E}(X)$ in the following manner. Let $F \xrightarrow{q} Y$ be an object of $\mathfrak{E}(Y)$. Consider $E = \{(x, \alpha) \mid x \in X, \alpha \in F, f(x) = q(\alpha)\}$ with the relative topology induced by $X \times F$ and $p: E \to X$ the continuous map $p(x, \alpha) = x$. We define $f^*(F) = E$ (considered as a projected space). We remark that there is a continuous mapping $\varphi: E \to F$ given by $\varphi(x, \alpha) = \alpha$ making the following diagram commutative

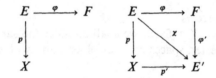 (5)

$f^*(F)$ is called the *inverse image* of the projected space F by f. It is easy to see that this is a final object of the category whose objects and morphisms are, respectively, the diagrams

$$\begin{array}{ccc} E & \xrightarrow{\varphi} & F \\ {\scriptstyle p}\downarrow & & \\ X & & \end{array} \qquad \begin{array}{ccc} E & \xrightarrow{\varphi} & F \\ {\scriptstyle p}\downarrow & \searrow^{\chi} & \downarrow^{\varphi'} \\ X & \xrightarrow{p'} & E' \end{array}$$

where the triangles of the second diagram are commutative. It follows that, if for some space E Diagram (5) commutes, and if φ and f are homeomorphisms, the space E is isomorphic in $\mathfrak{E}(X)$ to $f^*(F)$. Finally, if $\psi: F \to F'$ is a morphism in $\mathfrak{E}(Y)$ the corresponding morphism in $\mathfrak{E}(X)$ is given by the formula $f^*(\psi)(x, \alpha) = (x, \psi(\alpha))$ ($\alpha \in F$), the notation being as in the construction of $f^*(F)$.

The category $\mathfrak{E}(X)$ strengthened with the total space functor is to be considered always as the category where the operations with projected spaces on X are performed.

1 Definition The projected space $p: E \to X$ is called a *sheaf* (or, more precisely, a *sheaf of sets*) if the projection p is a local homeomorphism.

We recall that this condition means that every point of E has an open neighborhood mapped homeomorphically by p onto an open subset of X.

2. Sheaves and Sets

From this definition we get some obvious properties of sheaves. The projection p is an open map; the cross sections are open maps, and if two such sections are equal at a point they are equal in a neighborhood of that point; the topology induced in the local fibers is discrete; etc. The local fibers of a sheaf are called *stalks*.

A direct product $E = X \times A$, where A has the discrete topology and the projection is that on the first factor of the product, is an obvious example of a sheaf.

We denote by $\mathfrak{F}(X)$ the full subcategory of $\mathfrak{E}(X)$ whose objects are sheaves on X, and all the notions and operations on sheaves with base space X will be considered in $\mathfrak{F}(X)$ strengthened by the total space functor and, where the subobjects are given the usual relative topology. We can also use the full subcategory of $S(\text{Top})$ whose objects are sheaves. As examples of such notions we shall consider the following.

A morphism in $\mathfrak{F}(X)$, i.e., a *sheaf homomorphism*, $\varphi: E \to E'$ is a continuous map such that Diagram (4) is commutative. Since in this diagram p and p' are local homeomorphisms the same is true for φ. It is easily seen that φ is a strengthened injection (monomorphism) in $\mathfrak{F}(X)$ if and only if all the φ_x are injections in Ens, and that φ is a surjection (of course a strengthened one) if and only if all the φ_x are surjections in Ens. (As already agreed, the word "strengthened" will be omitted in the sequel.) Obviously, φ is an isomorphism in $\mathfrak{F}(X)$ if and only if it is a homeomorphism. A subsheaf of a sheaf E is an open subset of E which is projected onto X. We also consider quotient sheaves, sums, products, limits, etc. (For a version of these constructions the reader is referred to [15], among other works.)

We saw before that a continuous map $f: X \to Y$ induces a covariant functor $f^*: \mathfrak{E}(Y) \to \mathfrak{E}(X)$. Now we can see that the restriction of the functor f^* to $\mathfrak{F}(Y)$ has values in $\mathfrak{F}(X)$. In fact, if we use the notation from the construction of f^* and if $F \to Y$ is a sheaf, $E = f^*(F)$ and $(x, \alpha) \in E \subseteq X \times F$, there is an open neighborhood V of α in F homeomorphically projected by q, and it is easy to see that p projects homeomorphically the open neighborhood $(f^{-1}q(V) \times V) \cap E$ of (x, α) in E onto $f^{-1}q(V)$. Hence p is a local homeomorphism.   Q.E.D.

The sheaf $f^*(F)$ is called the *inverse image* of the sheaf F by f. If $f = i: X \subseteq Y$, where X is a subspace of Y, then $i^*(F) = q^{-1}(X)$ and this sheaf will be denoted by $F|_X$ and called the *restriction* of the sheaf F to the subspace X, or the sheaf *induced* by F on X.

We now study an important relation between sheaves and Ens-valued presheaves on a topological space X.

2   Theorem   *There are two covariant functors*

$$\text{Pr}: \mathfrak{F}(X) \to P(X, \text{Ens}) \quad \text{and} \quad \text{Fs}: P(X, \text{Ens}) \to \mathfrak{F}(X).$$

To get the first functor, we assign to every open subset $U \subseteq X$ the set $\Gamma(U)$ of all the cross sections of the sheaf $F \xrightarrow{p} X$ over U and to every pair $V \subseteq U$ the map $r_V^U: \Gamma(U) \to \Gamma(V)$ defined by the usual restriction of the cross sections. This construction defines the presheaf Pr F. As for Pr φ, where $\varphi: F \to F'$ is a morphism in $\mathfrak{F}(X)$, it is obtained by the correspondence $s \mapsto \varphi \circ s$ ($s \in \Gamma(U)$).

The definition of the second functor is more complicated. Let P be an object of $P(X, \text{Ens})$ (the notation is that of Section 1) and x a point of X. The system of sets and maps $\{P(U), r_V^U\}$, where U runs over the open subsets of X containing x is a directed inductive family where the indices U are ordered by $\supseteq$. Consider the inductive limit of this family, constructed as in section 1.3, and put

$$F_x = \varinjlim P(U). \tag{6}$$

The elements of F_x are classes of sections of $P(U)$ defined over open subsets containing x and which are equal over some neighborhood of that point. If $s \in P(U)$, we denote by $[s]_x$ the class of s at the point x and call it the *germ* of s at x.

Consider the set

$$F = \bigcup_{x \in X} F_x \tag{7}$$

and define on this set a topology as follows. For every open subset $U \subseteq X$ and every $s \in P(U)$ consider the set $\{[s]_x \mid x \in U\} = D_{s,U} \subseteq F$. Two sets $D_{s,U}$ and $D_{s',U'}$ have a nonempty intersection if and only if $U \cap U' \neq \emptyset$, and in that case an element of the intersection is of the form $[s]_x = [s']_x$, so that $[s]_x = [s']_x = [s'']_x$, where $s'' \in P(U'')$ and $U'' \subseteq U \cap U'$. The element is then contained in $D_{s'',U''} \subseteq D_{s,U} \cap D_{s',U'}$ and it follows that $\{D_{s,U}\}$ is the basis of a topology in F. This is just the topology of F required. In this topology, the map $p: F \to X$ defined by $p([s]_x) = x$ is a local homeomorphism, so we obtain a sheaf which is defined as $F = \text{Fs } P$. This is the so called *sheaf of germs* of the presheaf P.

To complete the proof of the theorem, we remark that if $\varphi: P \to P'$ is a morphism in $P(X, \text{Ens})$, then, by applying the previous procedure of passing to the inductive limit for the maps φ_U, we get a morphism Fs φ

2. Sheaves and Sets

between the corresponding sheaves of germs, and this correspondence has the functorial properties.

Theorem 2 suggests a possible equivalence of the categories $\mathfrak{F}(X)$ and $P(X, \text{Ens})$. One easily obtains a functorial isomorphism $\alpha: \text{Fs} \circ \text{Pr} \to 1_{\mathfrak{F}(X)}$ ($1_{\mathfrak{F}(X)}$ is the functor leaving the objects and morphisms of $\mathfrak{F}(X)$ invariant), which is given by the isomorphisms $\alpha_F: \text{Fs Pr } F \to F$ obtained by the correspondence $[s]_x \mapsto s(x)$. But, generally, there is no isomorphism between Pr Fs P and P, so the suggested equivalence does not exist. However, there is an equivalence between $\mathfrak{F}(X)$ and a subcategory of $P(X, \text{Ens})$, as we shall point out in the sequel.

3   Proposition   *There is a canonical presheaf homomorphism*

$$h: P \to \text{PrFs } P.$$

Moreover, h is an isomorphism if and only if (a) *for every family* $U_i (i \in M)$ *of open subsets of* X, *if* $s_\alpha \in P(U)$ ($\alpha = 1, 2$; $U = \bigcup_{i \in M} U_i$) *and if the restrictions of* s_1 *and* s_2 *to all the* U_i *are equal, then* $s_1 = s_2$, *and* (b) *if, for the mentioned family* U_i, *there are given* $s_i \in P(U_i)$ *such that for* $i, j \in M$, *the restrictions of* s_i *and* s_j *to* $U_i \cap U_j$ *are equal, then there exists* $s \in P(U)$ *(uniquely determined by* (a)*) such that for every* $i \in M$, *the restriction of* s *to* U_i *is* s_i.

The homomorphism h will be defined by the relation $h_U(s)(x) = [s]_x$, where $s \in P(U)$ and $x \in U$. If $h(s_1) = h(s_2)$, every point $x \in U$ has an open neighborhood $U_x \subseteq U$ such that the restrictions of s_1 and s_2 to U_x are equal. Hence, by condition (a), $s_1 = s_2$ on $U = \bigcup_{x \in U} U_x$. Finally, the property of h_U being a surjection follows analogously from (b). Hence, the sufficiency of the given conditions for h to be an isomorphism is proved. The necessity is obvious.

4   Definition   A presheaf of sets on X is called a *canonical presheaf* if it satisfies the conditions (a) and (b) of Proposition 3.

Denote by $P'(X, \text{Ens})$ the full subcategory of $P(X, \text{Ens})$ whose objects are the canonical presheaves. Clearly the functor Pr has its values in $P'(X, \text{Ens})$ and we can also consider the restriction of Fs to this subcategory (also denoted by Fs). The definition of the isomorphism α holds and, h is a functorial isomorphism of the functors $1_{P'(X, \text{Ens})}$ and $\text{Pr} \circ \text{Fs}$. Hence, we get the following theorem.

5   Theorem   *The categories* $\mathfrak{F}(X)$ *and* $P'(X, \text{Ens})$ *are equivalent.*

This is an important theorem, because it replaces problems regarding sheaves with problems of the corresponding canonical presheaves. Generally, the sheaves appearing are sheaves of germs of given presheaves (not necessarily canonical).

Before giving some examples, let us consider an example which shows that we have to be careful in applying the method indicated above. In fact, it is easy to see that an equivalence of categories leaves invariant the property of being an injection or a surjection. If we take a sheaf monomorphism, it has a corresponding monomorphism in the category $P'(X, \text{Ens})$, and one can see that this is also a monomorphism in $P(X, \text{Ens})$. If we take a sheaf epimorphism, the corresponding morphism is an epimorphism in $P'(X, \text{Ens})$ but, generally, it is not a presheaf epimorphism, i.e., it is not an epimorphism in $P(X, \text{Ens})$.

We consider a few other examples. If A is a set then the sheaf of germs of the simple presheaf defined on X by A is $X \times A$ where A has the discrete topology. If Y is another topological space, we get a presheaf of sets on X by assigning to every open subset U the set $P(U)$ of the continuous functions $f: U \to Y$ and the corresponding sheaf $F_c(X, Y)$ of germs of continuous Y-valued functions on X. If $Y = R$ is the field of real numbers we denote the previous sheaf by $C_r(X)$. In the same manner we can consider sheaves of germs of $\mathfrak{P}$-mappings between $\mathfrak{P}$-manifolds.

If $f: X \to Y$ is a continuous map and F is a sheaf on X, we construct $f(\text{Pr } F)$ and then Fs $f(\text{Pr } F)$ by applying Proposition 1.6. This is a sheaf on Y denoted by $f(F)$ and called the *direct image* of the sheaf F by the map f. The direct image has a covariant functorial character. In particular, we get the *trivial extension* of a sheaf defined on a closed subspace X of Y, denoted, by $\hat{F}$, whose stalks are those of F at the points of the subspace and consist of a single point over $Y - X$.

Finally, we remark that, for an open subspace, the restriction of a sheaf can also be obtained by considering the restriction of the corresponding canonical presheaf and that, for two given sheaves F and F' with base space X we can consider the sheaf Fs $\mathscr{P}(\text{Pr } F, \text{Pr } F')$.

3 Sheaves with Values in a Cantorian Category

We generalize the results of the previous section, defining sheaves with values in a Cantorian category $\mathfrak{C}$.

3. Sheaves with Values in a Cantorian Category

1   Definition   *A sheaf on X with values in the Cantorian category $\mathfrak{C}$ is a pair (F, σ) consisting of an object F in $\mathfrak{F}(X)$ and a structure σ of an object of $P(X, \mathfrak{C})$ defined on Pr F.*

The different notions introduced for set-valued sheaves can be transposed to $\mathfrak{C}$-valued sheaves. If no confusion arises a $\mathfrak{C}$-valued sheaf will be called F. Of course, not every sheaf in $\mathfrak{F}(X)$ has a structure σ of a $\mathfrak{C}$-valued sheaf, and we shall consider only the case where such a structure exists. In this connection, we have the following proposition.

2   Proposition   *Let F be an object of $\mathfrak{F}(X)$. Let $\mathfrak{C}$ be a Cantorian category with directed inductive limits equal to those in Ens and such that every set-bijection between an object of $\mathfrak{C}$ and a set can be interpreted uniquely as an isomorphism of $\mathfrak{C}$. If F has a structure of a $\mathfrak{C}$-valued sheaf, its stalks have structures of objects of $\mathfrak{C}$.*

In fact, from our hypotheses we get easily that the conclusion is true for the stalks of Fs ∘ Pr F, and hence, by the isomorphism α of Section 2, the conclusion is also true for F.

Generally, the conditions of Proposition 2 are not sufficient. Sufficient conditions for F to have a structure of a $\mathfrak{C}$-valued sheaf are obtained if one requires that the $\mathfrak{C}$-object structure of the stalks of F depend continuously on the point $x \in X$. The exact meaning of this continuity depends on the structure σ and we do not analyze it in the general case.

3   Definition   Let F and F' be two $\mathfrak{C}$-valued sheaves on X and $\varphi: F \to F'$ be a morphism in $\mathfrak{F}(X)$; φ is called a $\mathfrak{C}$-*valued sheaf homomorphism* if Pr φ is a morphism in $P(X, \mathfrak{C})$.

For such homomorphisms one easily gets a result similar to Proposition 2.

We see that the $\mathfrak{C}$-valued sheaves as objects and their homomorphisms as morphisms form a category which we denote by $\mathfrak{F}(X, \mathfrak{C})$, and all the notions and operations regarding $\mathfrak{C}$-valued sheaves will be considered in this category strengthened with the total space functor. If $\mathfrak{C} = $ Ens, we have $\mathfrak{F}(X, \text{Ens}) = \mathfrak{F}(X)$.

Clearly, Definition 2.4 of a canonical presheaf is meaningful for presheaves with values in an arbitrary Cantorian category $\mathfrak{C}$, so we can again introduce the full subcategory $P'(X, \mathfrak{C})$ of $P(X, \mathfrak{C})$ consisting of canonical presheaves. Just as for Theorem 2.5 we get here

4   Theorem   *If the category $\mathfrak{C}$ satisfies the hypotheses of proposition 2, the categories $\mathfrak{F}(X, \mathfrak{C})$ and $P'(X, \mathfrak{C})$ are equivalent.*

The proof is given by introducing the functors Fs and Pr and the isomorphisms α and h of Section 2. The details are left to the reader.

With the given definitions and results one could make a general study of the categories $\mathfrak{F}(X, \mathfrak{C})$, but we shall only consider the cases required in the sequel. First, let $\mathfrak{C} = A\text{-Mod}$, for a commutative unitary ring A. Then, we have the following proposition.

5   Proposition   *An object of the category $\mathfrak{F}(X, A\text{-Mod})$ is characterized by the following properties*: (a) *it is an object of $\mathfrak{F}(X)$*, (b) *its stalks are A-modules, and* (c) *the module operations are continuous.*

In fact, if F is an object in $\mathfrak{F}(X, A\text{-Mod})$ it satisfies the conditions (a) and (b) because of Definition 1 and Proposition 2. Then, by recalling the definition of the topology in Fs Pr F (Section 2,) it is immediate that the module operations are continuous on F (when defined) which is just the condition (c). Conversely, it is obvious that the conditions (a), (b), and (c) above define a unique structure of a presheaf of A-modules on Pr F.

Corollary: Every sheaf of A-modules has a global cross section, the zero cross section.

Let $\varphi: F \to F'$ be a morphism in $\mathfrak{F}(X, A\text{-Mod})$. Then Pr φ is a homomorphism of presheaves of modules and, taking the inductive limit composed with α of Section 2, it follows that the $\varphi_x: F_x \to F_x$ are A-module homomorphisms. Conversely, if F and F' are sheaves of A-modules and $\varphi: F \to F'$ is a morphism in $\mathfrak{F}(X)$ such that the φ_x are module homomorphisms, φ is a morphism in $\mathfrak{F}(X, A\text{-Mod})$. This remark together with Proposition 5 give a new definition of the category $\mathfrak{F}(X, A\text{-Mod})$ which is more practical. From this definition we immediately get the following propostion.

6   Proposition   *If the functor Fs is considered on the category $P(X, A\text{-Mod})$, its values are in the category $\mathfrak{F}(X, A\text{-Mod})$.*

With the same new definitions we see that $\varphi: F \to F'$ (in $\mathfrak{F}(X)$) is a monomorphism in $\mathfrak{F}(X, A\text{-Mod})$ if and only if all the φ_x are monomorphisms in A-Mod, and it is an epimorphism if and only if the φ_x are epimorphisms. We also see that the inverse images of A-Mod valued sheaves are again A-Mod-valued. The same is true for the direct images, since this is obviously true for every $\mathfrak{F}(X, \mathfrak{C})$. Hence, we can talk of the trivial extension of a sheaf F of modules, defined on a closed subspace $Y \subseteq X$, which we shall denote by $\hat{F}$. Its stalks are the stalks of F over Y and the zero module over $X - Y$.

3. Sheaves with Values in a Cantorian Category

7 Proposition $\mathfrak{F}(X, A\text{-Mod})$ *is an Abelian category.*

In fact, $\mathfrak{F}(X, A\text{-Mod})$ is additive since we can add its morphisms by adding the homomorphisms of the stalks; it has a zero object, namely, the sheaf 0 whose stalks are zero modules; and since the direct sum of two sheaves can be obtained from the direct sums of the stalks, or by taking the direct sum of the corresponding canonical presheaves. Then, if $\varphi: F \to F'$ is a morphism in $\mathfrak{F}(X, A\text{-Mod})$ we get ker φ by taking $\bigcup_{x \in X} \ker \varphi_x$ with the relative topology and coker $\varphi = $ Fs coker Pr $\varphi = \bigcup_{x \in X} \operatorname{coker} \varphi_x$ (and also im $\varphi = \bigcup_{x \in X} \operatorname{im} \varphi_x$, coim $\varphi = \bigcup_{x \in X} \operatorname{coim} \varphi_x$). Because of some known properties of modules all the conditions of Definition 1.4.6 are satisfied. Hence, $\mathfrak{F}(X, A\text{-Mod})$ is a regular, i.e., an Abelian category.   Q.E.D.

It follows that we can talk of exact sequences of sheaves of A-modules, and such sequences are characterized by the property that, for each point of the base space, the corresponding sequence of stalks is exact. But generally the corresponding sequence of canonical presheaves is not exact. However, the converse is true. More generally, *if we have an exact sequence of presheaves of modules, the corresponding sequence of sheaves of germs is exact.* This is a consequence of the following lemma, the proof of which is left to the reader.

8 Lemma *The inductive limit of an exact sequence of directed inductive families of modules is an exact sequence of modules.*

An important example of an exact sequence can be obtained for a pair (X, Y) of topological spaces, when Y is a closed subspace of X, and for a sheaf F of A-modules on X. In order to find this sequence, consider the subset $F' = p^{-1}(X - Y) \cup s_0(X) \subseteq F$, where p is the projection and s_0 the zero section of F. F' is open and defines a subsheaf of F with the inclusion $i: F' \to F$. We have $F'|_{X-Y} = F|_{X-Y}$ and $F'|_Y = 0$. Next, consider on X the sheaf $F'' = \widehat{F|_Y}$ which induces $F|_Y$ on Y and 0 on $X - Y$. We have then an epimorphism $j: F \to F''$ which is the identity over Y and 0 over $X - Y$. Now it is easily seen that the following sequence of sheaves is exact:

$$0 \to F' \xrightarrow{i} F \xrightarrow{j} F'' \to 0.$$

This is called the *exact sequence of sheaves of the triple* (X, Y, F).

According to Proposition 1.4.14, we can also consider the quotient sheaf $F'' = F/F'$, where F' is a subsheaf of F. Then we have the exact sequence

$$0 \to F' \xrightarrow{i} F \xrightarrow{p} F/F' \to 0. \qquad (1)$$

A simple analysis shows that F/F' is the sheaf of germs of the quotient presheaf Pr F/Pr F′, which is defined by the modules $\Gamma(U)/\Gamma'(U)$ and the induced restrictions, but this is not, generally, the canonical presheaf of F/F'.

The previous results hold of course for sheaves of Abelian groups, where $A = J$, the ring of integers, and for sheaves of linear spaces when A is a commutative field K. Similar results can be obtained for sheaves of graded modules (*graded sheaves*), sheaves of differential modules (*differential sheaves*), etc.

A similar analysis can be made for the case $\mathfrak{C} = A$-Alg, the category of algebras over A and in particular for the case $\mathfrak{C} = $ Ann, the category of rings. Thus, for the categories $\mathfrak{F}(X, A$-Alg) and $\mathfrak{F}(X, $ Ann) we have a proposition similar to Proposition 5, which says that an object F of the category is characterized by the facts that it is a sheaf, it has as stalks algebras or rings, and the algebraic operations are continuous. The morphisms of $\mathfrak{F}(X, A$-Alg) ($\mathfrak{F}(X, $ Ann)) will be sheaf morphisms $\varphi: F \to F'$ such that the induced maps φ_x are algebra(ring) homomorphisms and they will be monomorphisms or epimorphisms if all the φ_x are monomorphisms or epimorphisms, respectively. Next, a proposition like Proposition 6 also holds, etc.

Let A be a sheaf of unitary commutative rings with base space X.

9   Definition   A *sheaf module over A* (or A-*sheaf module*) is a pair (F, σ) consisting of an object F of $\mathfrak{F}(X)$ and a structure σ of an A-presheaf module on Pr F.

This means that for every open subset $U \subseteq X$, $F(U)$ (the set of sections over U) has a structure of an $A(U)$-module, where $A(U)$ denotes the sections of A over U, and the restrictions are homomorphisms which make the diagrams of the form 1(4) commutative.

The homomorphisms of A-sheaf modules are obtained as in Definition 3, and it follows that we can consider the category $\mathfrak{F}(X, A$-Mod) of A-sheaf modules. It is easy to extend for this category the results previously given for the sheaves of modules. Thus, passing to the inductive limit and using the isomorphism of Section 2 one sees that the stalks of F (the notation is that considered above) are modules over the stalks of A at the same points, and that Proposition 5 holds. We have also the corresponding

3. Sheaves with Values in a Cantorian Category

propositions 6, 7, etc., which are proved as in the case when A is a ring, but taking into account Definition 9. With the method indicated at the end of Section 1 for the construction of the tensor product of two presheaf modules and passing to the inductive limit we get the tensor product of sheaf modules; its stalks are the tensor products of the stalks of the given sheaves.

Let us indicate some important notions connected with sheaf modules.

10 Definition An A-sheaf module F is called *finitely generated* if it has a finite number of global cross sections whose values at each point generate the corresponding stalk. F is *locally finitely generated* (or of *finite type*) if every point x of the base space X has an open neighborhood U such that $F|_U$ is finitely generated.

Now let F be an A-sheaf module and $s_1, \ldots, s_h$ be global cross sections of it. We say that there is a *relation* between them at $x \in X$ if there are elements $r_1, \ldots, r_h \in A_x$, not all zero, and such that

$$\sum_{i=1}^{h} r_i s_i(x) = 0.$$

The relations between $s_1, \ldots, s_h$ can be organized as an A-sheaf module which can be identified with the kernel of the homomorphism

$$\varphi: \underbrace{A \times \cdots \times A}_{h \text{ times}} \to F$$

defined by

$$\varphi(r_1, \ldots, r_h) = \sum_{i=1}^{h} r_i s_i(x)$$

(the product is in $\mathfrak{F}(X)$).

11 Definition An A-sheaf module F is called *free* if it is finitely generated and the sheaf of the relations of its generators is trivial (zero). It is *locally free* if for each point $x \in X$ there is an open neighborhood U such that $F|_U$ is free.

12 Definition The A-sheaf module F is called *coherent* [41] if it is locally finitely generated and if for every finite system of cross sections of

F over an open subset $U \subseteq X$, the sheaf of the relations is locally finitely generated.

We conclude with some simple examples. Let A be an object of a Cantorian category $\mathfrak{C}$. The (*simple* or *constant*) sheaf $X \times A$ is a sheaf with values in $\mathfrak{C}$. The sheaf of germs of continuous functions on X with values in a topological group Y (see the examples in Section 2) is a sheaf of groups on X. The sheaf $C_r(X)$ of germs of real valued continuous functions on X (and $C_c(X)$ of germs of complex-valued continuous functions) is a sheaf of unitary commutative rings and over it we can consider sheaf modules. Very important examples are obtained from differential geometry, from the theory of complex spaces and from algebraic geometry (where one studies some *ringed spaces*, i.e., topological spaces together with a sheaf of rings on them). The theory of coherent sheaves is also very important in complex functions theory. Some of these sheaves will be considered in succeeding chapters.

4 Cohomology with Coefficients in Presheaves

In this section, we obtain the cohomology functors associated with topological spaces endowed with presheaves with values in Abelian categories. They are very important for many theories and can be constructed in different ways. We shall consider the Čech construction which has the clearest geometrical meaning and is quite satisfactory for paracompact Hausdorff spaces, the only case which we need in the sequel. The more general cases are covered by the theories of Grothendieck and Godement [18, 15].

Let X be a topological space, without restrictive conditions, and $P(X, \mathfrak{C})$ be the category of presheaves on X with values in an Abelian category $\mathfrak{C}$, which is assumed to have arbitrary direct products and directed inductive limits. Then, by Proposition 1.5, we know that $P(X, \mathfrak{C})$ is also Abelian.

Let $\mathfrak{U} = \{U_i\}_{i \in I}$ be an open covering of X with an arbitrary set I of indices. We shall put

$$U_{S_q} = U_{s_0} \cap \cdots \cap U_{s_q}, \quad S_q = (s_0, \ldots, s_q), \quad q = 0, 1, 2, \ldots \quad (1)$$

4. Cohomology with Coefficients in Presheaves

and

$$\Sigma_q = \{S_q \mid U_{S_q} \neq \emptyset\} = \Sigma_q(\mathfrak{U}). \tag{2}$$

Then, $\{\Sigma_q\}$ is a simplicial complex [42] called the *nerve* of the covering $\mathfrak{U}$. Since the category $\mathfrak{C}$ has direct products, we can consider

$$C^q(\mathfrak{U}, P) = \prod_{S_q \in \Sigma_q} P(U_{S_q}), \qquad q = 0, 1, 2, \ldots \tag{3}$$

for a given presheaf P on X with values in $\mathfrak{C}$ (the notation is that of Section 1). The object $C^q(\mathfrak{U}, P)$ of $\mathfrak{C}$ is called *the object of q-dimensional cochains of the covering $\mathfrak{U}$ with coefficients, or values, in $\mathfrak{C}$*. By the definition of the direct product, we have the projection morphisms

$$p_{S_q}: C^q(\mathfrak{U}, P) \to P(U_{S_q}). \tag{4}$$

Following Section 1.5, in order to define cohomology functors we need a cochain complex, hence, we shall look for a *coboundary morphism*

$$\delta: C^q(\mathfrak{U}, P) \to C^{q+1}(\mathfrak{U}, P) \tag{5}$$

such that $\delta\delta = \delta^2 = 0$ (when no confusion arises, the index q of δ is omitted). We obtain δ as follows.

Let $S_{q+1} = (s_0, \ldots, s_{q+1}) \in \Sigma_{q+1}$. Put

$$S_{q+1}^i = (s_0, \ldots, \hat{s}_i, \ldots, s_{q+1}) = (s_0, \ldots, s_{i-1}, s_{i+1}, \ldots, s_{q+1})$$

and consider the morphisms

$$\sum_{i=0}^{q+1} (-1)^i r_{U_{S_{q+1}}}^{U_{S_{q+1}^i}} p_{S_{q+1}^i}: C^q(\mathfrak{U}, P) \to P(U_{S_{q+1}}), \tag{6}$$

where the r: are the restriction morphisms of the presheaf P. From the definition of the direct product, we define a unique morphism (5) which we take as the sought after coboundary morphism. Note that the morphism of Formula (6) exists since $\mathfrak{C}$ is Abelian and, hence, additive. Remark also that the definition of δ implies

$$p_{S_{q+1}} \delta = \sum_{i=0}^{q+1} (-1)^i r_{U_{S_{q+1}}}^{U_{S_{q+1}^i}} p_{S_{q+1}^i}. \tag{7}$$

Now, we consider the following calculation:

$$p_{S_{q+2}}\delta^2 = \sum_{i=0}^{q+2}(-1)^i r_{US_{q+2}}^{US_{q+2}^i} p_{S_{q+2}^i}\delta$$

$$= \sum_{i=0}^{q+2}(-1)^i r_{US_{q+2}}^{US_{q+2}^i}\left[\sum_{h=0}^{i-1}(-1)^h r_{US_{q+2}^i}^{US_{q+2}^{ih}} p_{S_{q+2}^{ih}}\right.$$

$$\left.+\sum_{h=i}^{q+1}(-1)^h r_{US_{q+2}^i}^{US_{q+2}^{i,h+1}} p_{S_{q+2}^{i,h+1}}\right]$$

$$= \sum_{h<i=0}^{q+2}(-1)^{i+h} r_{US_{q+2}}^{US_{q+2}^{ih}} p_{S_{q+2}^{ih}} + \sum_{h<i=0}^{q+2}(-1)^{i+h+1} r_{US_{q+2}}^{US_{q+2}^{ih}} p_{S_{q+2}^{ih}}$$

$$= 0.$$

which, in view of the definition of the direct product, implies

$$\delta^2 = 0 \tag{8}$$

(0 being the only morphism α such that $p_{S_{q+2}}\alpha = 0$).

In this manner, we get the system $C(\mathfrak{U}, P) = (C^q(\mathfrak{U}, P), \delta)$ which is a $\mathfrak{C}$-valued cochain complex. We remark that the definition of δ is, in some sense, the dual of the definition usually given for the boundary of a simplex in Euclidean space [42], and this can be considered as the geometrical meaning of the coboundary.

Now, let $h: P \to P'$ be a morphism in $P(X, \mathfrak{C})$. It induces a uniquely determined morphism h^q such that the diagram

$$\begin{array}{ccc} C^q(\mathfrak{U}, P) & \xrightarrow{h^q} & C^q(\mathfrak{U}, P') \\ {\scriptstyle p_{S_q}}\downarrow & & \downarrow{\scriptstyle p'_{S_q}} \\ P(U_{S_q}) & \xrightarrow{h_{U_{S_q}}} & P'(U_{S_q}) \end{array} \tag{9}$$

is commutative and we get

$$p'_{S_{q+1}}\delta' h^q = \sum_{i=0}^{q+1}(-1)^i r'^{US_{q+1}^i}_{US_{q+1}} p'_{S_{q+1}^i} h^q$$

$$= \sum_{i=0}^{q+1}(-1)^i r'^{US_{q+1}^i}_{US_{q+1}} h_{US_{q+1}^i} p_{S_{q+1}^i}$$

4. Cohomology with Coefficients in Presheaves

$$= \sum_{i=0}^{q+1} (-1)^i h_{U_{S_{q+1}}} r^{U^i_{S_{q+1}}}_{U_{S_{q+1}}} p_{S^i_{q+1}}$$

$$= h_{U_{S_{q+1}}} p_{S_{q+1}} \delta$$

$$= p'_{S_{q+1}} r'_{S_{q+1}} h^{q+1} \delta.$$

From the definition of the direct product,

$$\delta' h^q = h^{q+1} \delta, \tag{10}$$

which means that we assigned to the morphism h a morphism of the corresponding cochain complexes of P and P'. Hence, we have proved the following proposition.

1 Proposition *For every open covering $\mathfrak{U}$ of the topological space X, there is a covariant functor $C(\mathfrak{U}, \cdot)$ from the category $P(X, \mathfrak{C})$ to the category $\mathfrak{K}(\mathfrak{C})$ of the $\mathfrak{C}$-valued cochain complexes.*

By composition of the functor in Proposition 1 with the cohomology functors of Section 1.5 we get covariant functors from $P(X, \mathfrak{C})$ to $\mathfrak{C}$ which depend on the covering $\mathfrak{U}$ considered above and which will be denoted by $H^q(\mathfrak{U}, P)$. In other words, for a given $\mathfrak{U}$ and $q = 0, 1, 2, \ldots$, every P has corresponding cohomology objects $H^q(\mathfrak{U}, P)$ in $\mathfrak{C}$, *called the cohomology objects of the covering $\mathfrak{U}$ with coefficient (or values) in P*. They are the quotient objects of the objects of cocycles $Z^q(\mathfrak{U}, P)$ with respect to the objects $B^q(\mathfrak{U}, P)$ of the coboundaries. Also, for every morphism $h: P \to P'$ there are the corresponding morphisms

$$h^{*q}: H^q(\mathfrak{U}, P) \to H^q(\mathfrak{U}, P') \tag{11}$$

of the category $\mathfrak{C}$ and these correspondences have a functorial character.

Our next purpose is to obtain from $H^q(\mathfrak{U}, P)$ other functors depending only on the space X and on the presheaf P. In order to get them, we shall consider the system of objects of $\mathfrak{C}\{H^q(\mathfrak{U}, P)\}$, with fixed q, as a system indexed by $\{\mathfrak{U}\}$, we shall make it into a directed inductive family and take the corresponding limit.

First, recall that the open covering $\mathfrak{V}$ of the topological space X is called *finer* then the covering $\mathfrak{U}$, denoted $\mathfrak{V} \prec \mathfrak{U}$, if every set of $\mathfrak{V}$ is contained in at least one set of $\mathfrak{U}$. The relation $\succ$ is a partial ordering of the set $\{\mathfrak{U}\}$ of open coverings of X and this set is directed by the relation.

Consider $\mathfrak{U}$ and $\mathfrak{V}$ as two open coverings of the space X having the sets of indices I and J, respectively, such that $\mathfrak{U} \succ \mathfrak{V}$. Then, there is a mapping $\tau: J \to I$ such that for every $j \in J$, $V_j \subseteq U_{\tau j}$. We denote by τJ_q the set of indices which correspond by τ to the indices of $J_q \in \Sigma_q(\mathfrak{V})$. By choosing such a mapping τ, we get a uniquely determined morphism τ^* which makes the following diagram commutative:

$$\begin{array}{ccc} C^q(\mathfrak{U}, P) & \xrightarrow{\tau^*} & C^q(\mathfrak{V}, P) \\ {\scriptstyle p_{\tau J_q}}\downarrow & & \downarrow{\scriptstyle p_{J_q}} \\ P(U_{\tau J_q}) & \xrightarrow[r^{U_{\tau J_q}}_{V_{J_q}}]{} & P(V_{J_q}) \end{array} \qquad (12)$$

Now by calculations as for h^q it follows that τ^* and δ commute. Hence τ^* is a morphism in $\mathfrak{K}(\mathbb{C})$ and induces the morphisms

$$t^{\mathfrak{U}}_{\mathfrak{V}}: H^q(\mathfrak{U}, P) \to H^q(\mathfrak{V}, P), \qquad q = 0, 1, 2, \ldots. \qquad (13)$$

We shall prove that the morphisms (13) do not depend on the choice of τ. To get this, let $\tau': J \to I$ be another mapping such that $V_j \subseteq U_{\tau' j}$. If $J_{q-1} \in \Sigma_{q-1}(\mathfrak{V})$ then, for $h = 0, \ldots, q-1$, we can consider the elements of $\Sigma_q(\mathfrak{U})$ defined by

$$I_q^{(h)} = (\tau j_0, \ldots, \tau j_h, \tau' j_h, \ldots, \tau' j_{q-1}).$$

Consider the morphisms

$$k^q: C^q(\mathfrak{U}, P) \to C^{q-1}(\mathfrak{V}, P), \qquad q \geqslant 1, \qquad (14)$$

uniquely determined by the definition of the direct product and the condition

$$p_{J_{q-1}} k^q = \sum_{h=0}^{q-1} (-1)^h r^{U_{I_q^{(h)}}}_{V_{J_{q-1}}} p_{U_{I_q^{(h)}}}. \qquad (15)$$

For k^0 we shall take the morphism 0.

The following calculation, with obvious notation, is now valid:

4. Cohomology with Coefficients in Presheaves

$$p_{J_q}\delta k^q = \sum_{i=0}^{q}(-1)^i r_{V_{J_q}}^{V_{J_q}^i} p_{J_q^i} k^q$$

$$= \sum_{h<i=0}^{q}(-1)^{i+h} r_{V_{J_q}}^{U_{I_q}{}^{i,(h)}} p_{I_q{}^{i,(h)}} + \sum_{h>i=0}^{q}(-1)^{i+h-1} r_{V_{J_q}}^{U_{I_q}{}^{i,(h)}} p_{I_q{}^{i,(h)}},$$

and

$$p_{J_q} k^{q+1}\delta = \sum_{h=0}^{q}(-1)^h r_{V_{J_q}}^{U_{I_{q+1}}^{(h)}} p_{I_{q+1}^{(h)}} \delta$$

$$= \sum_{i\leqslant h=0}^{q}(-1)^{i+h} r_{V_{J_q}}^{U_{I_{q+1}}^{(h),\tau J_i}} p_{I_{q+1}^{(h),\tau J_i}}$$

$$+ \sum_{i\geqslant h=0}^{q}(-1)^{i+h+1} r_{V_{J_q}}^{U_{I_{q+1}}^{(h),\tau' J_i}} p_{I_{q+1}^{(h),\tau' J_i}},$$

from which we get

$$p_{J_q}(\delta k^q + k^{q+1}\delta) = p_{J_q}(\tau'^* - \tau^*).$$

In fact, by developing the above one sees that all the terms cancel except for those with $i = h$. Then, from these terms, after further reductions, there remain only those with $i = h = 0$ in the first sum and $i = h = q$ in the second.

As in the similar situations already met, we get

$$\delta k^q + k^{q+1}\delta = \tau'^* - \tau^*, \quad q \geqslant 0. \tag{16}$$

We now establish the following lemma.

2 Lemma *Let $f, g: K \to K'$ be two morphisms in $\Re(\mathfrak{C})$ and suppose that there are morphisms $k^q: K^q \to K'^{q-1}$ ($q \in J$) in $\mathfrak{C}$ such that*

$$g^q - f^q = d'^{q-1} k^q + k^{q+1} d^q. \tag{17}$$

Then, f and g induce the same morphisms of the homology objects of the two complexes.

Here d and d' are the boundary morphisms of the two complexes and the notation of the following proof is that of Section 1.5. Obviously, the morphisms

$$t^q = d'^{q-1}k^q + k^{q+1}d^q \colon K^q \to K'^q$$

define a morphism of complexes and this induces uniquely determined morphisms of the homology objects such that the diagram

$$
\begin{array}{ccccccccc}
0 & \longrightarrow & B^q & \xrightarrow{\alpha} & Z^q & \xrightarrow{p} & H^q & \longrightarrow & 0 \\
& & {\scriptstyle \sigma}\Big\downarrow & & {\scriptstyle t'^q}\Big\downarrow & & {\scriptstyle t^{*q}}\Big\downarrow & & \\
0 & \longrightarrow & B'^q & \xrightarrow{\alpha'} & Z'^q & \xrightarrow{p'} & H'^q & \longrightarrow & 0
\end{array}
\qquad (18)
$$

is commutative.

Let h and h' be the morphisms defining the kernels of d^q and d'^q, respectively, and $d^{q-1} = q\lambda$, $d'^{q-1} = q'\lambda'$ the decompositions given for d and d' in Section 1.5. It is known from Section 1.5 that $q = h\alpha$, $q' = h'\alpha'$ and that $t^q h = h' t'^q$. It follows, by the definition of t^q that

$$h' t'^q = d'^{q-1} k^q h' = h'\alpha'\lambda' k^q h$$

and, since h' is a monomorphism,

$$t'^q = \alpha'\lambda' k^q h.$$

This result shows that if in Diagram (18) we replace t^{*q} by 0 the diagram will be commutative. Hence, $t^{*q} = 0$ and Formula (17) together with the known additivity property of the homology functors completes the proof of the lemma.

3 Definition The system of morphisms $\{k^q\} = k$ of Lemma 2 is called a *homotopy of complexes*, and if f and g satisfy the conditions (17) they are called *homotopic morphisms* of $\mathfrak{K}(\mathfrak{C})$.

Hence, we have proved that homotopic morphisms of $\mathfrak{K}(\mathfrak{C})$ induce the same morphisms of the homology functors. This is a very frequently used fact in homological algebra and in algebraic topology.

Formula (16) shows that τ^* and τ'^* are homotopic morphisms of the complexes $C(\mathfrak{U}, P)$ and $C(\mathfrak{V}, P)$, and it follows that the morphisms (13) are uniquely determined and do not depend on the choice of τ.

Finally let $\mathfrak{W}$ be another covering indexed by the set K and such that $\mathfrak{W} < \mathfrak{V} < \mathfrak{U}$. Choose $\tau' \colon K \to J$ and $\tau'' \colon K \to I$ to be maps like $\tau \colon J \to I$ such that $\tau'' = \tau\tau'$. Then $\tau''^* = \tau'^*\tau^*$ and we get

4. Cohomology with Coefficients in Presheaves

$$t_{\mathfrak{W}}^{\mathfrak{V}} t_{\mathfrak{V}}^{\mathfrak{U}} = t_{\mathfrak{W}}^{\mathfrak{U}}.$$

If for $\mathfrak{V} = \mathfrak{U}$ we take $\tau = \text{id}$, we find $t_{\mathfrak{U}}^{\mathfrak{U}} = 1_{H^q(\mathfrak{U}, P)}$.

In the category $\mathfrak{C}$, we have the directed inductive family $\{H^q(\mathfrak{U}, P), t_{\mathfrak{V}}^{\mathfrak{U}}\}$ indexed by the open coverings of the space X and defined for every $q \geqslant 0$. Since we supposed that $\mathfrak{C}$ has directed inductive limits, we can give the following definition.

4 Definition The object

$$\varinjlim H^q(\mathfrak{U}, P) = H^q(X, P), \qquad q \geqslant 0, \tag{19}$$

is called the *q-dimensional cohomology object of the space X with coefficients (or values) in the presheaf P*.

Let $h: P \to P'$ be a presheaf homomorphism. By combining Diagrams (9) and (12), we see that the morphisms h and τ^* of $\mathfrak{K}(\mathfrak{C})$ commute. It follows that the morphisms h^{*q} of (11) define a morphism of the inductive families $\{H^q(\mathfrak{U}, P)\}$ and $\{H^q(\mathfrak{U}, P')\}$ and, hence, there are in $\mathfrak{C}$ the limit morphisms

$$h^{*q}: H^q(X, P) \to H^q(X, P'). \tag{20}$$

Thus we have obtained the following theorem.

5 Theorem *There is a sequence of covariant functors $P \to H^q(X, P)$ ($q \geqslant 0$) from the category $P(X, \mathfrak{C})$ to $\mathfrak{C}$.*

These are just the functors which we intended to introduce.

Now, let $f: X \to Y$ be a continuous mapping of the topological spaces X and Y and let P be a presheaf on X with values in the same category $\mathfrak{C}$ as before. Denote by $f(P)$ the direct image of P by f. Next, let $\mathfrak{V}$ be an open covering of Y and $\mathfrak{U} = f^{-1}(\mathfrak{V})$ the corresponding covering of X. By the definition of $f(P)$, it follows that there are some determined morphisms of source $C^q(\mathfrak{V}, f(P))$ and with the objects as target whose product is $C^q(\mathfrak{U}, P)$. Thus, we obtain a unique morphism of complexes denoted by

$$f'^q: C^q(\mathfrak{V}, f(P)) \to C^q(\mathfrak{U}, P)$$

and induced morphisms

$$f^{*q}: H^q(\mathfrak{V}, f(P)) \to H^q(\mathfrak{U}, P). \tag{21}$$

By the composition of these morphisms with the inductive limit and the definition of the inductive limit, we obtain the induced morphisms

$$f^{*q}: H^q(Y, f(P)) \to H^q(X, P). \tag{22}$$

Hence, we can say that the cohomology functors have "contravariant behavior" with respect to the space X.

In particular, if $i: X \subseteq Y$, we have the induced morphisms

$$i^{*q}: H^q(Y, \hat{P}) \to H^q(X, P). \tag{23}$$

If, moreover, X is closed, to every open covering $\mathfrak{U} = \{D_i \cap X\}$ of X, with the D_i open subsets of X, we can associate the open covering $\{D_i, Y - X\}$ of Y and the previous procedure can be inverted, giving an inverse of (23). Hence, we get the following proposition.

6 Proposition *If X is a closed subspace of Y, P a presheaf in $P(X,\mathfrak{C})$, and $\hat{P}$ its trivial extension to Y, the induced morphisms (23) are isomorphisms.*

5 The Case $\mathfrak{C} = A$-Mod

In this section, we consider only the case of the presheaves of A-modules, for a commutative unitary ring A, which is the case to be used in our book. We shall look for properties of the cohomology functors and also for a method to calculate them. It will be shown that these functors are essentially determined by the functor H^0.

Note that for the category A-Mod and more generally for a Cantorian category $\mathfrak{C}$, the q-dimensional cochain object is the set of functions defined on Σ_q and which assign to every element $S_q \in \Sigma_q$ an element of $P(U_{S_q})$. Then p_{S_q} for a given cochain is just the value of the respective cochain for S_q. This interpretation will be used in formula 4(7) and in other analogous formulas.

We begin with the important theorem of the exact cohomology sequence.

1 Theorem *There is a covariant functor from the category of short exact sequences of $P(X, A\text{-Mod})$ to the category of exact sequences of*

5. The Case $\mathfrak{C} = A$-Mod

A-modules, which associates with every exact sequence of presheaves

$$0 \to P' \xrightarrow{i} P \xrightarrow{p} P'' \to 0 \tag{1}$$

its exact cohomology sequence

$$0 \to \cdots \xrightarrow{\delta^*} H^q(X, P') \xrightarrow{i^*} H^q(X, P) \xrightarrow{p^*} H^q(Y, P'')$$
$$\xrightarrow{\delta^*} H^{q+1}(X, P') \xrightarrow{i^*} \cdots, \tag{2}$$

where i^* and p^* are induced by i and p, and δ^* is a specially determined homomorphism.

In fact, since the functor of Proposition 4.1 is exact, if $\mathfrak{U}$ is an open covering of X, we get a covariant functor which assigns to the exact sequence (1) the exact sequence

$$0 \to C(\mathfrak{U}, P') \xrightarrow{i} C(\mathfrak{U}, P) \xrightarrow{p} C(\mathfrak{U}, P'') \to 0. \tag{3}$$

By Theorem 1.5.5, we can associate in a functorial manner with the exact sequence (3) an exact sequence which is just the sequence (2) with $\mathfrak{U}$ instead of X and with δ^* given by ∂ of 1.5.5.

The proof of Theorem 1 is finished by applying Lemma 3.8.

The previous theorem is very important. We shall use it to calculate the cohomology functors.

2 Definition A *resolution* of the presheaf P is an exact sequence of $P(X, A\text{-Mod})$, of the form

$$\rho \equiv 0 \to P \xrightarrow{i} P_0 \xrightarrow{h_0} P_1 \xrightarrow{h_1} \cdots, \tag{4}$$

where the presheaves P_α ($\alpha \geq 0$) satisfy the conditions[†]

$$H^n(X, P_\alpha) = 0, \quad n \geq 1. \tag{5}$$

The presheaves which satisfy the conditions (5) will be called *presheaves with trivial cohomology*.

Suppose that (4) is a resolution of P. Then, we can construct the cochain complex $\mathfrak{R}(\rho)$, defined by the A-modules $H^0(X, P_\alpha)$ ($\alpha \geq 0$) and by the induced homomorphisms h_α^*; the relation $h_\alpha^* h_{\alpha-1}^* = 0$ follows from the

---

[†] Sometimes (4) without any other conditions than its exactness is also called a resolution.

exactness of the sequence (4). We now have the following important theorem.

3 Theorem (The abstract de Rham's theorem for presheaves.) *The cohomology modules of the space X with coefficients in P are given by*

$$H^q(X, P) \approx H^q(\mathfrak{R}(\rho)), \qquad q \geqslant 0. \tag{6}$$

To prove this, consider the sequences

$$0 \to \ker h_\alpha \subseteq P_\alpha \xrightarrow{h_\alpha} \ker h_{\alpha+1} \to 0, \qquad \alpha \geqslant 0, \tag{7}$$

which are exact, since (4) is an exact sequence. Writing down the corresponding exact cohomology sequences (2) and using the conditions (5) we immediately get

$$H^{i+1}(X, \ker h_\alpha) \approx H^i(X, \ker h_{\alpha+1}), \qquad i \geqslant 1, \qquad \alpha \geqslant 0. \tag{8}$$

Repeated applications of (8) give

$$H^{i+1}(X, \ker h_0) \approx H^1(X, \ker h_i), \qquad i \geqslant 0,$$

and, noting that, by the exactness of (4), we have $\ker h_0 \approx P$,

$$H^{i+1}(X, P) \approx H^1(X, \ker h_i). \tag{9}$$

Consider the sequence (7) for $\alpha = i$. The first six terms of its exact cohomology sequence are

$$0 \to H^0(X, \ker h_i) \to H^0(X, P_i) \xrightarrow{h_i^*} H^0(X, \ker h_{i+1})$$
$$\to H^1(X, \ker h_i) \to 0, \tag{10}$$

whence one derives easily

$$H^1(X, \ker h_i) \approx H^0(X, \ker h_{i+1})/\operatorname{im} h_i^* \approx \ker h_{i+1}^*/\operatorname{im} h_i^*. \tag{11}$$

Here, the last isomorphism can also be seen from the sequence (10) written for $i + 1$ instead of i.

Formulas (9) and (11) yield the desired result for $q > 0$, and for $q = 0$

5. The Case $\mathfrak{C} = A$-Mod

the result follows directly, hence the theorem is completely proved.

In order to give a formal expression for the isomorphism (6), we denote by ∂_α^i the connecting homomorphisms of the exact sequence (7). Now, (8) is $(\partial_\alpha^i)^{-1}$ and (9) is $(\partial_{i-1}^1)^{-1} \circ \cdots \circ (\partial_0^i)^{-1} \circ i^*$, where i^* is induced by i of (4). ∂_i^0 is not an isomorphism but it induces the isomorphism $(\bar{\partial}_i^0)^{-1}$ which is the first isomorphism of (11). Hence, if the last isomorphism of (11) is considered as an identification we get for the isomorphism (6) the expression

$$\rho^q = (\bar{\partial}_{q-1}^0)^{-1} \circ (\partial_{q-2}^1)^{-1} \circ \cdots \circ (\partial_0^{q-1})^{-1} \circ i^*. \tag{6'}$$

Instead of ρ^q we can also consider the isomorphisms

$$r^q = (-1)^q \rho^q, \tag{6''}$$

which is sometimes more convenient.

The determination of the functors H^q ($q > 0$) by H^0 is justified by Theorem 3 above, but we shall not consider here the problem of the existence of a resolution for an arbitrary presheaf.

One can obtain important results for the case $\mathfrak{C} = A$-Mod from the case $\mathfrak{C} = \mathfrak{K}(A\text{-Mod})$, the category of cochain complexes of A-Mod where the complexes at the same time are considered as graded modules by taking the respective direct sum. We only wish to give here an interpretation of the modules $H^q(\mathfrak{U}, P^*)$ for a $\mathfrak{K}(A\text{-Mod})$-valued presheaf P^* by using the theory of *spectral sequences*. To get this, we give a brief exposition of spectral sequences of modules.

Recall that the homology modules are terms of a system of modules associated by an algebraic procedure to a complex of modules. In just the same manner, a spectral sequence is a more complicated system of modules, associated with a so-called filtered complex.

4 Definition A *filtered complex* of modules is a complex (K^n, d) together with a decomposition

$$K = \bigoplus_{n \in J} K^n = \bigcup_{p \in J} K_p, \qquad K_p \supseteq K_{p+1}, \tag{12}$$

where J is the ring of integers, such that

$$K_p = \bigoplus_{q \in J} K_p \cap K^q, \qquad d(K_p) \subseteq K_p. \tag{13}$$

The decomposition (12) is called a *filtration*[†] of the complex K.

Now, an element of K has two indices: one upper index, which is uniquely determined and is called the *degree*; and one lower index, which is not uniquely determined and is called the *filtered degree*. The first condition (13) shows that the homogeneous parts of an element (i.e., its components of fixed degree) with a given filtered degree have the same filtered degree. The second condition (13) shows that the K_p are subcomplexes of K.

Introduce the modules

$$Z_r^{pq} = \{x \mid x \in K_p \cap K^{p+q} \text{ and } dx \in K_{p+r}\}$$
$$B_r^{pq} = \{x \mid x \in K_p \cap K^{p+q} \text{ and } x = dy \text{ for some } y \in K_{p-r}\} \quad (14)$$
$$= K_p \cap dK_{p-r} \cap K^{p+q}, \qquad p, q, r \in J,$$
$$K_\infty = 0, \quad K_{-\infty} = K, \quad Z_\infty^{pq} = \{x \mid x \in K_p \cap K^{p+q} \text{ and } dx = 0\}, \quad (15)$$
$$B_\infty^{pq} = \{x \mid x \in K_p \cap K^{p+q} \text{ and } x = dy \text{ for some } y \in K\}.$$

Remark that B_{r-1}^{pq} and $Z_{r-1}^{p+1,q-1}$ are submodules of Z_r^{pq}, hence their union (sum) is again such a submodule. This union will be denoted by $+$ (not the same as $\oplus$, the direct sum). It follows that there are modules

$$E_r^{pq} = Z_r^{pq}/(B_{r-1}^{pq} + Z_{r-1}^{p+1,q-1}), \qquad r \in J \cup \{\infty\}. \quad (16)$$

5 Definition The *spectral sequence* of the filtered complex K consists of the modules (16).

Now, if $x \in Z_r^{pq}$ we obtain $dx \in Z_r^{p+r,q-r+1}$ and it follows that there are induced homomorphisms

$$d_r: E_r^{pq} \to E_r^{p+r,q-r+1}, \qquad p, q, r \in J, \quad (17)$$

such that $d_r^2 = 0$. (E_r, d_r) is a complex such that

$$E_r^n = \bigoplus_{p+q=n} E_r^{pq},$$

where n is the *degree* of an element in E_r^{pq}.

The following result is an essential property of the spectral sequence.

[†] More precisely it is a decreasing filtration and one can also consider in an obvious manner increasing filtrations.

5. The Case $\mathfrak{C} = A$-Mod

6 Theorem $H^n(E_r) \approx E^n_{r+1}$, $n \in J$.

Put

$$H^{pq}(E_r) = Z^{pq}(E_r)/B^{pq}(E_r),$$

where $Z^{..}$ is the module of cycles and $B^{..}$ the module of boundaries. We find that $\alpha = [x]$, $x \in Z^{pq}_r$ is in $Z^{pq}(E_r)$ if and only if $dx \in B^{p+r,q-r+1}_{r-1} | Z^{p+r+1,q-r}_{r-1}$ whence, $dx = d\lambda + \mu$, the terms belonging to the respective submodules. It follows that $x - \lambda = \tau \in Z^{pq}_{r+1}$, since λ is in $Z^{p+1,q-1}_{r-1}$. Finally,

$$Z^{pq}(E_r) = (Z^{pq}_{r+1} + Z^{p+1,q-1}_{r-1})/(B^{pq}_{r-1} + Z^{p+1,q-1}_{r-1}). \tag{18}$$

Next, by direct considerations we get

$$B^{pq}(E_r) = (B^{pq}_r + Z^{p+1,q-1}_{r-1})/(B^{pq}_{r-1} + Z^{p+1,q-1}_{r-1}). \tag{19}$$

Using classical isomorphism theorems (see for instance the introduction of [42]) it follows that

$$H^{pq}(E_r) = (Z^{pq}_{r+1} + Z^{p+1,q-1}_{r-1})/(B^{pq}_r + Z^{p+1,q-1}_{r-1})$$
$$= Z^{pq}_{r+1}/Z^{pq}_{r+1} \cap (B^{pq}_r + Z^{p+1,q-1}_{r-1}) = E^{pq}_{r+1}.$$

Now, Theorem 6 follows from the fact that $H^n(E_r)$ is given by $\bigoplus_{p+q=n} H^{pq}(E_r)$.

As an application, we shall calculate the terms E_0, E_1, and E_2 of the spectral sequence. First, we have by definition

$$E^{pq}_0 = K_p \cap K^{p+q}/K_{p+1} \cap K^{p+q} = (\mathfrak{R}_p/\mathfrak{R}_{p+1})^{p+q}, \tag{20}$$

where $\mathfrak{R}_p$ is the complex given by $K_p \cap K^{p+q}$ and the operator d.

Now, by the previous theorem we have

$$E^{pq}_1 = H^{pq}(E_0) = H^{p+q}(\mathfrak{R}_p/\mathfrak{R}_{p+1}). \tag{21}$$

Consider the exact sequence of complexes

$$0 \to \mathfrak{R}_{p+1}/\mathfrak{R}_{p+2} \to \mathfrak{R}_p/\mathfrak{R}_{p+2} \to \mathfrak{R}_p/\mathfrak{R}_{p+1} \to 0. \tag{22}$$

The corresponding exact cohomology sequence defines the homomorphisms

$$\partial: H^{p+q}(\mathfrak{R}_p/\mathfrak{R}_{p+1}) \to H^{p+q+1}(\mathfrak{R}_{p+1}/\mathfrak{R}_{p+2})$$

and, from the definitions of ∂ and of $d_1: E_1^{pq} \to E_1^{p+1,q}$, one gets $d_1 = \partial$. The modules E_2^{pq} are obtained by Theorem 6.

The role of the term E_∞ of the spectral sequence is shown by the following considerations. Denote

$$H^q(K)_p = H^q(K) \cap H(K)_p,$$

where $H(K)_p$ is the image of $H(K_p)$ in $H(K)$. It follows that

$$E_\infty^{pq} \approx (H^{p+q}(K))_p/(H^{p+q}(K))_{p+1}. \tag{23}$$

Moreover, if the filtered complex K is *regular*, which means that it has the property $K_p \cap K^q = 0$ for $p > n(q)$, one sees that E_∞^{pq} is an inductive limit of the modules E_r^{pq}. In fact, from the regularity condition

$$Z_r^{p+r,q-r+1} = 0$$

for r sufficiently large ($r > n(p + q + 1) - p$). Hence, for these r, $d_r = 0$ and, by Theorem 6, there is an epimorphism $E_r^{pq} \to E_{r+1}^{pq}$. By the composition of these epimorphisms we obtain, for s and r sufficiently large, epimorphisms

$$\theta_s^r: E_r^{pq} \to E_s^{pq}, \quad r < s,$$

such that (E_r^{pq}, θ_s^r) (with fixed p and q) is a directed inductive family of modules. The regularity condition also implies that for r sufficiently large we have

$$Z_r^{pq} = Z_\infty^{pq}, \quad Z_{r-1}^{p+1,q-1} = Z_\infty^{p+1,q-1}.$$

Finally,

$$B_\infty^{pq} = \bigcup_r B_r^{pq}.$$

This leads to the epimorphisms

$$\theta_\infty^r: E_r^{pq} \to E_\infty^{pq},$$

5. The Case $\mathbb{C} = \text{A-Mod}$

which are such that $\theta_\infty^r = \theta_\infty^s \circ \theta_s^r$ ($r < s$) and E_∞^{pq} has the universality property of the inductive limit, which proves the above assertion.

These results mean that there is a kind of approximation of the homology of K by the terms of the spectral sequence, which explains the importance of the latter.

7 Proposition *Let K be a regular filtered complex and suppose that there is an index r and an index q_0 such that*

$$E_r^{n-q,q} = 0 \quad \text{for} \quad q \neq q_0. \tag{24}$$

Then

$$H^n(K) \approx E_r^{n-q_0,q_0}. \tag{25}$$

By hypothesis (24), Theorem 6, and the fact that E_∞ is the limit of E_r, we immediately get

$$E_\infty^{n-q,q} = \begin{cases} 0 & \text{if} \quad q \neq q_0, \\ E_r^{n-q_0,q_0} & \text{if} \quad q = q_0. \end{cases}$$

Formula (23) shows that the filtration of $H^n(K)$ has only two distinguished terms

$$H^n(K)_{n-q_0} \supseteq H^n(K)_{n-q_0+1},$$

all the other terms being equal to one of them. But, since K is regular,

$$\bigcap_p H^n(K)_p = 0,$$

which implies

$$H^n(K)_{n-q_0+1} = 0$$

and

$$H^n(K) = H^n(K)_{n-q_0}.$$

The result follows using again the formula (23).

Finally, we consider an important example of a spectral sequence. Let (K, d) be a cochain complex such that

$$K^n = \bigoplus_{p+q=n} K^{pq}, \qquad p, q \geq 0. \tag{26}$$

Then, we say that $K = \bigoplus_n K^n$ is a *bigraded module*.

Applying d to an element of K^{pq} we get a decomposition

$$d = \sum_{h+k=1} d_{hk}, \tag{27}$$

where d_{hk} is an *operator of the type* (h, k), which means $d_{hk}: K^{pq} \to K^{p+h, k+q}$.

If

$$d_{hk} = 0 \quad \text{for} \quad h < 0 \tag{28}$$

the complex K is called a *double semipositive complex*. If

$$d_{hk} = 0 \quad \text{for} \quad k < 0$$

also, K is a *double positive complex* or, simply, a *double complex*. It follows that a double complex is a system of modules K^{pq} together with two operators

$$d': K^{pq} \to K^{p+1, q}, \qquad d'': K^{pq} \to K^{p, q+1}$$

such that

$$d'^2 = d''^2 = d'd'' + d''d' = 0 \tag{29}$$

(since $d = d' + d''$ and $d^2 = 0$). Now, negative indices p, q could also be admitted. But we are not interested in them here.

For a double semipositive complex K, we get a regular filtration by putting

$$\begin{aligned} K_p &= \bigoplus_{i \geq p} \bigoplus_q K^{iq}, \quad p \geq 0, \\ K_p &= K, \quad p < 0. \end{aligned} \tag{30}$$

For this filtration, there is an associated spectral sequence E_r^{pq} and, using

5. The Case $\mathfrak{C} = A$-Mod

Formulas (20) and (21), we have

$$E_0^{pq} = K^{pq}, \quad d_0 = d_{01},$$
$$E_1^{pq} = {}''H^q(K^{p*}), \quad K^{p*} = (K^{pq}, d_{01}), \tag{31}$$

where $''H$ is the cohomology of the complex K with respect to the second index as the degree and to the operator d_{01} as the coboundary.

To obtain the operator d_1, we must consider the exact sequence (22) which in this case is

$$0 \to K^{p+1,*} \to K^{p*} \oplus K^{p+1,*} \to K^{p*} \to 0,$$

with the respective coboundaries d_{01}, $(d_{01} + d_{10}) \oplus d_{01}$, and d_{01}. The corresponding ∂ is then induced by d_{10} and, by Theorem 6, it follows that

$$E_2^{pq} = {}'H^p({}''H^q(K)), \tag{32}$$

where $'H$ is the cohomology with respect to the coboundary induced by d_{10}.

In the case of a double complex, the notations above are replaced by $'K_p$ and $'E_r^{pq}$ and one says that this is the *first spectral sequence* of the complex. A *second spectral sequence* is obtained with the modules $''K_p$ and $''E_r^{pq}$ given in the same manner but changing the role and the place of the indices p and q. One gets

$$''E_1^{pq} = {}'H^q(K^{*p}), \quad ''E_2^{pq} = {}''H^p({}'H^q(K)). \tag{33}$$

Returning to our main object, we remark that if we have a presheaf P^* of complexes over a topological space X and an open covering $\mathfrak{U}$ of X, we can take the modules $C^p(\mathfrak{U}, P^q)$. They define a double cochain complex (with $K^{pq} = C^p(\mathfrak{U}, P^q)$) such that d' is the coboundary operator of the complex $C^p(\mathfrak{U}, P^q)$ with fixed q, and d'' is induced by the coboundary of P^* with the opposite sign. This complex will be denoted by $C^*(\mathfrak{U}, P^*)$ and it has the two corresponding spectral sequences.

By formulas (33),

$$''E_1^{pq} = H^q(\mathfrak{U}, P^p), \quad ''E_2^{pq} = H^p(H^q(\mathfrak{U}, P^*)), \tag{34}$$

and this is just the interpretation which we had in mind.

Important results are also obtained for the other spectral sequence. To indicate them, we remark first that objects similar to the presheaves are obtained if one assigns a module to every simplex S_q of the nerve of a covering $\mathfrak{U}$ and if one defines restrictions from the module of a subsimplex to that of the simplex. Such objects are called *systems of coefficients* and they define, just as the presheaves do, cochain complexes and cohomology modules.

P^* defines the coefficient systems $S_q \mapsto H^i(P^*(U_{S_q}))$, with obviously induced restrictions, which will be denoted by $\mathfrak{S}^i(P^*)$.

Using Formulas (31) and (32) we get

$$'E_1^{pq} = C^p(\mathfrak{S}^q(P^*)), \qquad 'E_2^{pq} = H^p(\mathfrak{S}^q(P^*)), \tag{35}$$

where C^p is the module of the p-dimensional cochains and H^p the p-dimensional cohomology module associated to the system $\mathfrak{S}^q(P^*)$ of coefficients.

A more complete treatment of these matters is given in [15]. For arbitrary Abelian categories see, for instance, [3].

6 Cohomology with Coefficients in a Sheaf

Applying the theory presented in Sections 4 and 5 to the canonical presheaves we get the cohomology functors with coefficients in sheaves, which are the most important for applications.

Let $\mathfrak{C}$ be a Cantorian Abelian category having direct products and directed inductive limits and let F be an object of $\mathfrak{F}(X, \mathfrak{C})$ (X is a topological space).

1 Definition *The cohomology objects of X with coefficients (or values) in F are defined as the objects $H^q(X, \text{Pr } F)$.*

These objects are denoted by $H^q(X, F)$.

Consequently, in order to obtain $H^q(X, F)$, we must consider the open coverings $\mathfrak{U}$ of X, the complexes $C(\mathfrak{U}, F) = C(\mathfrak{U}, \text{Pr}, F)$ which consist of the modules $C^q(\mathfrak{U}, \text{Pr } F)$ ($q = 0, 1, 2, \ldots$) of functions assigning to every simplex S_q of the nerve of $\mathfrak{U}$ a cross section in F over U_{S_q} (a cochain is interpreted as such a function) and of the usual coboundary operators. We

6. Cohomology with Coefficients in a Sheaf

next consider the cohomology objects of these complexes and take the inductive limit as in Section 4. Then, for every q, $H^q(X, \cdot)$ is a covariant functor from $\mathfrak{F}(X, \mathfrak{C})$ to $\mathfrak{C}$.

Suppose that $\mathfrak{C}$ also satisfies the following condition: if $\alpha = \beta\gamma$ where α is a morphism in $\mathfrak{C}$, β a monomorphism in $\mathfrak{C}$, and γ a morphism in Ens, then γ is a morphism in $\mathfrak{C}$. It is easily seen that if $\mathfrak{C} = A\text{-Mod}$ this condition holds. Now, we have the following important result.

2 Proposition $H^0(X, F) \approx \Gamma(X, F)$, *the object of the global cross sections of F.*

In fact, we shall prove that $H^0(\mathfrak{U}, F) \approx \Gamma(X, F)$ for every open covering $\mathfrak{U}$. We have $H^0(\mathfrak{U}, F) = Z^0(\mathfrak{U}, F) = \ker \delta$. There is a monomorphism $i: \Gamma(X, F) \to C^0(\mathfrak{U}, F)$ determined by the condition $p_s i = r^X_{U_s}$ ($s \in I$, which is the set of indices of $\mathfrak{U}$) and we have $p_{S_1} \delta i = 0$ for every $S_1 \in \Sigma_1$, hence, $\delta i = 0$. Now, if $\sigma: M \to C^0(\mathfrak{U}, F)$ is a morphism such that $\delta\sigma = 0$, we consider the cross sections $p_s \sigma(a)$ ($a \in M$) over U_s and obtain

$$r^{U_{s_1}}_{U_{S_1}} p_{s_1} \sigma(a) = r^{U_{s_0}}_{U_{S_1}} p_{S_0} \sigma(a).$$

Hence, there is a corresponding element $\lambda(a) \in \Gamma(X, F)$, which is uniquely determined. We obtained in this manner a uniquely determined morphism $\lambda: M \to \Gamma(X, F)$, such that $\sigma = i\lambda$. It follows that $(\Gamma(X, F), i) = \ker \delta$.
Q.E.D.

We also observe that the induced morphisms 4(22) can be obtained for sheaf cohomology and that proposition 4.6 also holds.

In the remaining part of this section, we shall consider the case $\mathfrak{C} = A\text{-Mod}$ for a commutative unitary ring A.

First, we study the exact cohomology sequence. The transposition of Theorem 5.1 is not straightforward, since for an exact sequence of sheaves the sequence of the canonical presheaves is not generally exact. We shall see that the complete transposition of that theorem is possible for the case of a base space X which is Hausdorff and paracompact.

3 Lemma *Let X be a paracompact Hausdorff space and P a presheaf of modules on X. Then, if Fs $P = 0$, $H^q(X, P) = 0$ for $q \geq 0$.*

To prove this, let $\alpha \in H^q(X, P)$ and let $a \in H^q(\mathfrak{U}, P)$ be a representative of α for the covering $\mathfrak{U} = \{U_i\}_{i \in I}$. Next, a is represented by an element $f \in C^q(\mathfrak{U}, P)$. We shall see that there is a covering $\mathfrak{V} = \{V_j\}_{j \in J}$, finer than $\mathfrak{U}$, and a mapping $\tau: J \to I$ with $V_j \subseteq U_{\tau j}$ such that $\tau^* f = 0$. Obviously, this proves the announced result.

Since X is paracompact, we can suppose that $\mathfrak{U}$ is locally finite. Then, there is an open covering $\mathfrak{W} = \{W_i\}_{i \in I}$ such that $\overline{W}_i \subseteq U_i$ ($i \in I$).

To obtain $\mathfrak{V}$, choose $J = X$ and impose step by step the following conditions.

(a) If $x \in U_i$, $V_x \subseteq U_i$, and if $x \in W_i$, $V_x \subseteq W_i$ (where $x \in V_x$). This condition can be satisfied, since U_i and W_i are open.

(b) If $V_x \cap W_i \neq \emptyset$ then $V_x \subseteq U_i$. For this we shall consider first that V_x satisfies (a) and it has nonempty intersection with the $W_1, \ldots, W_h$. Then, if $x \notin U_a$ ($1 \leqslant a \leqslant h$) we have also $x \notin \overline{W}_a$ and we can take a smaller V_x such that $V_x \cap W_a = \emptyset$, so the set W_a will be eliminated. After a finite number of steps, we shall have only the case $x \in W_a$ ($1 \leqslant a \leqslant h$), which together with condition (a) implies (b).

(c) If $x \in U_{S_q}$, then $r^{U_{S_q}}_{V_x} f = 0$. This can be obtained by taking a smaller V_x because of the hypothesis Fs $P = 0$.

With the chosen V_x, we have a covering $\mathfrak{V} \prec \mathfrak{U}$ and we can consider a map τ such that $V_x \subseteq W_{\tau x}$. Now,

$$\tau^* f(x_0, \ldots, x_q) = r^{U_{S_q}}_{V_{X_q}} f(\tau x_0, \ldots, \tau x_q), \qquad X_q = (x_0, \ldots, x_q).$$

But, from $V_{x_0} \cap \cdots \cap V_{x_q} \neq \emptyset$ it follows $V_{x_0} \cap V_{x_k} \neq \emptyset$ for $k = 0, \ldots, q$; hence, $V_{x_0} \cap W_{\tau x_q} \neq \emptyset$ which, because of condition (b) above, implies $V_{x_0} \subseteq U_{\tau x_0}$ and also $V_{x_0} \subseteq U_{\tau x_q}$. We then derive from condition (c) that $\tau^* f(x_0, \ldots, x_q) = 0$ and the lemma is proved.

It is important to remark that for $q = 0$ we do not use the condition (b), which means that we can work without the covering $\mathfrak{W}$ and without the condition that $\mathfrak{U}$ is locally finite. Hence, for $q = 0$, Lemma 3 holds for any topological space.

4   Lemma   *If the space X is Hausdorff and paracompact, the induced homomorphisms*

$$h^{*q}: H^q(X, P) \to H^q(X, \text{Fs } P)$$

of the canonical homomorphism (Section 2)

$$h: P \to \text{Pr Fs } P,$$

are isomorphisms.

Consider the presheaves ker h and (Pr Fs P)/im h. It is easily seen that they have zero sheaves of germs. Consider the exact sequences of pre-

6. Cohomology with Coefficients in a Sheaf

sheaves

$$0 \to \ker h \to P \to \operatorname{im} h \to 0,$$

$$0 \to \operatorname{im} h \to \operatorname{Pr} \operatorname{Fs} P \to (\operatorname{Pr} \operatorname{Fs} P)/\operatorname{im} h \to 0.$$

Writing down the exact cohomology sequences and applying Lemma 3 we obtain the required result. It follows also if h is a monomorphism, that Lemma 4 for $q = 0$ holds for an arbitrary space X and in this case h^{*1} is injective.

Let

$$0 \to F' \xrightarrow{i} F \xrightarrow{j} F'' \to 0 \tag{1}$$

be an exact sequence in $\mathfrak{F}(X, A\text{-Mod})$. It has the associated exact sequence of presheaves

$$0 \to \operatorname{Pr} F' \to \operatorname{Pr} F \to \operatorname{Pr} F/\operatorname{Pr} F' \to 0$$

and the corresponding cohomology exact sequence. Since

$$\operatorname{Fs}(\operatorname{Pr} F/\operatorname{Pr} F') \approx F'',$$

we can apply Lemma 4, if X is Hausdorff paracompact, and we get the exact sequence

$$0 \to \cdots \xrightarrow{\delta^*} H^q(X, F') \xrightarrow{j^*} H^q(X, F) \xrightarrow{j^*} H^q(X, F'')$$
$$\xrightarrow{\delta^*} H^{q+1}(X, F') \xrightarrow{i^*} \cdots, \tag{2}$$

with a determined homomorphism δ^*. This is called the *exact cohomology sequence* of (1).

The remarks made before for an arbitrary space X allow us in the general case to derive from the exact sequence (1) the exact sequence

$$0 \to H^0(X, F') \xrightarrow{i^*} H^0(X, F) \xrightarrow{j^*} H^0(X, F'') \xrightarrow{\delta^*} H^1(X, F')$$
$$\xrightarrow{i^*} H^1(X, F) \xrightarrow{j^*} H^1(X, F''). \tag{3}$$

One easily sees from the previous considerations the functorial dependence of the sequences (2) and (3) on the sequence (1). Hence, we get the following theorem.

5 Theorem *There is a covariant functor from the category of short exact sequences of $\mathfrak{F}(X, A\text{-Mod})$ to the category of exact sequences of A-Mod which assigns to the exact sequence of sheaves the exact sequence (3). If X is Hausdorff and paracompact this functor assigns to the sequence (1) the exact sequence (2).*

As an example, consider for (1) the exact sequence of sheaves of a triple (X, Y, F) where X is Hausdorff and paracompact, Y is a closed subspace of X and F is a sheaf on X. We shall have a corresponding exact cohomology sequence, which, by Proposition 4.6, for sheaves, takes the form

$$\cdots \xrightarrow{\delta^*} H^q(X, F') \xrightarrow{i^*} H^q(X, F) \xrightarrow{j^*} H^q(Y, F|_Y) \xrightarrow{\delta^*} H^{q+1}(X, F') \xrightarrow{i^*} \cdots$$

We define the resolutions of a sheaf and give a corresponding de Rham theorem. From now on the base space will be assumed Hausdorff and paracompact.

6 Definition A *resolution* of a sheaf F is an exact sequence of sheaves, on the same space X, of the form

$$\rho \equiv 0 \to F \xrightarrow{i} F_0 \xrightarrow{h_0} F_1 \xrightarrow{h_1} \cdots, \tag{4}$$

such that

$$H^n(X, F_\alpha) = 0, \quad \alpha \geqslant 0, n \geqslant 1. \tag{5}$$

Hence, the sheaves F_α have trivial cohomology.

If in (4) we go over to the corresponding sequence of canonical presheaves, we do not generally obtain a resolution of $\mathrm{Pr}\, F$, since the resulting sequence is not exact. The complex $\mathfrak{K}(\rho)$ of Section 4 can however be constructed, and a theorem similar to Theorem 4.3 holds, since in the proof of that theorem we used only the exact cohomology sequence, which is also valid here as shown by Theorem 5 above. Moreover, the modules $H^0(X, F_\alpha)$ of $\mathfrak{K}(\rho)$ will be replaced, because of Proposition 2, by $\Gamma(X, F_\alpha)$, the modules of the global cross sections. Hence, we have the following theorem.

7 Theorem (Abstract de Rham's theorem for sheaves.) *If X is a paracompact Hausdorff space and ρ is a resolution of the sheaf F on X, we have*

6. Cohomology with Coefficients in a Sheaf

$$H^q(X, F) \approx H^q(\Re(\rho)), \quad q \geq 0, \tag{6}$$

where $\Re(\rho)$ is the complex $(\Gamma(X, F_\alpha), h_\alpha^*)$.

Because of this theorem, we say that the cohomology of paracompact Hausdorff spaces with coefficients in a sheaf of modules is determined by the functor Γ, which consists in taking the global cross sections.

It is important to remark that the isomorphisms (6) have a natural character. In other words, if we have a diagram

$$(\rho) \quad 0 \longrightarrow F \xrightarrow{i} F_0 \xrightarrow{h_0} F_1 \xrightarrow{h_1} \cdots$$
$$\downarrow \alpha \quad \downarrow \alpha_0 \quad \downarrow \alpha_1$$
$$(\rho') \quad 0 \longrightarrow F' \xrightarrow{i'} F'_0 \xrightarrow{h_0'} F'_1 \xrightarrow{h_1'} \cdots$$

where the two lines are, respectively, resolutions of F and F' on the space X and every square is commutative, then the α_i induce a homomorphism $a: \Re(\rho) \to \Re(\rho')$ such that the following diagram is commutative:

$$\begin{array}{c} H^q(X, F) \approx H^q(\Re(\rho)) \\ \downarrow \alpha^* \quad \downarrow a^* \\ H^q(X, F') \approx H^q(\Re(\rho')) \end{array} \quad q \geq 0.$$

This is seen by using the functorial character of the exact cohomology sequence along the lines of the proof of Theorem 7.

Finally, the expression 5(6″) of the isomorphism given by de Rham's theorem is also valid for Theorem 7.

The efficiency of Theorem 7 in calculating cohomology functors is very high, since we can give readily applicable conditions for verifying property (5).

8 Proposition *If F is a $C_r(X)$-sheaf module on a paracompact Hausdorff space X, then it has trivial cohomology.*

Recall that $C_r(X)$ is the sheaf of germs of real continuous functions. We have to show that the groups $H^q(X, F)$ ($q \geq 1$) (F being considered as a sheaf of Abelian groups) vanish. This will be proved with the help of a homotopy of complexes which will establish that for an open, locally finite covering $\mathfrak{U} = \{U_i\}_{i \in I}$ of X, $H^q(\mathfrak{U}, F) = 0$.

Let φ_i ($i \in I$) be a partition of unity subordinated to the covering $\mathfrak{U}$ (which exists since X is paracompact). Let $S_{q-1} = (i_0, \ldots, i_{q-1})$ be a simplex of the nerve of $\mathfrak{U}$. For $f \in C^q(\mathfrak{U}, F)$, define the cross section of F over $U_{S_{q-1}}$,

$$t(i, i_0, \ldots, i_{q-1}) = \begin{cases} \varphi_i f(i, S_{q-1}) & \text{on} \quad U_i \cap U_{S_{q-1}} \\ 0 & \text{on} \quad U_{S_{q-1}} - U_i \cap U_{S_{q-1}} \end{cases} \quad i \in I, \quad (7)$$

where $\varphi_i f(i, S_{q-1})$ is the operation given by the structure of $C_r(X)$-sheaf module of F. Finally define the homomorphisms

$$k^q: C^q(\mathfrak{U}, F) \to C^{q-1}(\mathfrak{U}, F), \quad q \geq 1$$

by

$$(k^q f)(i_0, \ldots, i_{q-1}) = \sum_{i \in I} t(i, i_0, \ldots, i_{q-1}),$$

the sum having a finite number of terms at every point of X.

A direct calculation and the use of the relation $\sum_{i \in I} \varphi_i(x) = 1$ ($x \in X$) gives

$$\delta k^q + k^{q+1} \delta = \text{id}, \quad q \geq 1. \quad (8)$$

Hence, the identity homomorphism of the complex $C(\mathfrak{U}, F)$ is homotopic to the zero homomorphism, from which $H^q(X, F) = 0$ for $q \geq 1$. For $q = 0$, this is no longer true, since (8) does not hold.

It is important to remark that if X is a differentiable manifold and F is a sheaf module over the sheaf of germs of differentiable functions, then, because of the existence of differentiable partitions of unity, Proposition 8 and its proof remain valid, so that $H^q(X, F) = 0$ for $q \geq 1$.

Proposition 8 can be generalized as follows.

9 Definition Let F be a sheaf of modules on the paracompact Hausdorff space X. F is called a *fine sheaf* if, for every open, locally finite covering of X, $\mathfrak{U} = \{U_i\}_{i \in I}$, there are the endomorphisms $h_i: F \to F$ ($i \in I$) such that (a) supp $h_i \subseteq U_i$, (b) $\sum_{i \in I} h_i = \text{id}$.

We recall that supp $h_i = \{x \mid x \in X, h_i(F_x) \neq 0\}$. The condition (b) is meaningful since, $\mathfrak{U}$ being locally finite, the sum is finite for every $x \in X$.

If F is a fine sheaf on X, the proof of Proposition 8 remains valid if in

(7) we replace multiplication by φ_i with applications of h_i. Hence, we have the following proposition.

10   Proposition   *If F is a fine sheaf on X, which is Hausdorff and paracompact, then $H^q(X, F) = 0$ for $q \geqslant 1$.*

Of course, differentiable manifolds are included in this proposition.

Now, if (4) is an exact sequence of sheaves and if all the sheaves F_α are fine, (5) is satisfied and we have a resolution of F. This is called a *fine resolution*.

Another matter to be discussed is the behavior of the functors H^q by the operation of taking inverse images of a sheaf.

Let $\varphi: X \to Y$ be a continuous map, G a sheaf of modules on Y, and $F = \varphi^{-1}(G)$ its inverse image by φ. Then, let $f \in C^q(\mathfrak{U}, G)$ be a cochain of a covering $\mathfrak{U}$ of Y. We can associate with it the element $\varphi(f) \in C^q(\varphi^{-1}(\mathfrak{U}), F)$ given by

$$\varphi(f)(i_0, \ldots, i_q)(x) = (x, f(i_0, \ldots, i_q)(\varphi(x))).$$

Now, considering the necessary inductive limit, we get some induced homomorphisms

$$\varphi^*: H^q(Y, G) \to H^q(X, f^{-1}(G)), \tag{9}$$

which shows, once more, the contravariant behavior of these functors with respect to the base space.

If G is the constant sheaf $Y \times M$, F is also isomorphic to the constant sheaf $X \times M$. The modules $H^q(X, X \times M)$ are denoted simply by $H^q(X, M)$ and are called the *cohomology modules of the space X with coefficients in M*. The relation (9) offers induced homomorphisms

$$\varphi^*: H^q(Y, M) \to H^q(X, M) \tag{10}$$

and we see that $H^q(\cdot, M)$ are contravariant functors on Top. These are the classical Čech cohomology functors considered in algebraic topology.

If $M = J$ is the additive group of integers and if $H^q(X, J)$ are finitely generated, their ranks and torsion numbers are called respectively the *Betti and torsion numbers* of the space X. The Betti numbers $b^q = $ rank $H^q(X, J)$ are also equal to dim $H^q(X, R)$, where R is the real field, and with the complex dimension of $H^q(X, C)$, where C is the complex field. The sum $\chi(X) = \sum_q (-1)^q b^q$, when finite, is called the *Euler–Poincaré char-*

acteristic of X. More generally, if F is a sheaf of linear spaces on X and if

$$\chi(X, F) = \sum_q (-1)^q \dim H^q(X, F) \tag{11}$$

exists, it is called the *Euler–Poincaré characteristic of the sheaf F*. An immediate consequence of the cohomology exact sequence theorem follows.

11 Proposition *Consider on X (Hausdorff and paracompact) the exact sequence of sheaves of linear spaces*

$$0 \to F' \to F \to F'' \to 0.$$

If two of them have Euler-Poincaré characteristics, the third also has, and

$$\chi(X, F) = \chi(X, F') + \chi(X, F'').$$

The proof is left to the reader.

The objects $H^q(X, F)$ were obtained as the inductive limit of the $H^q(\mathfrak{U}, F)$. It is of course interesting to know whether there are coverings $\mathfrak{U}$ of X such that $H^q(\mathfrak{U}, F) \approx H^q(X, F)$. We give an example for which this is the case.

Let (4) be a fine resolution of F such that $F_\alpha|_U$ are fine for every open subset U of X. Let P^* be the object of $P(X, \mathfrak{K}(A\text{-Mod}))$ defined by the canonical presheaves of the sheaves F_α of the resolution and $\mathfrak{U}$ an open covering of X. Supposing the U_{S_q} paracompact, we have, by de Rham's theorem $H^i(P^*(U_{S_q})) \approx H^i(U_{S_q}, F)$, which determines the system of coefficients $\mathfrak{S}^i(P^*)$ of Section 5.

Assume

$$H^i(U_{S_q}, F) = 0, \quad i \geqslant 1. \tag{12}$$

Then the previous remark and the formulas 5(35) give for the first spectral sequence

$$\begin{aligned}'E_2^{pq} &= 0 \quad \text{for} \quad q \neq 0, \\ 'E_2^{p0} &= H^p(\mathfrak{U}, F),\end{aligned} \tag{13}$$

whence, by proposition 5.7,

$$H^p(K) \approx H^p(\mathfrak{U}, F), \tag{14}$$

6. Cohomology with Coefficients in a Sheaf

where K is the double complex associated with $\mathfrak{U}$ and P^*.

But, for the same double complex we have, by 5(34), and since the F_α have trivial cohomology,

$$''E_2^{pq} = 0, \quad q \geqslant 1,$$
$$''E_2^{p0} = H^p(X, F). \tag{15}$$

Hence, by an application of Proposition 5.7,

$$H^p(K) \approx H^p(X, F). \tag{16}$$

From (14) and (16) we have the following theorem (given here under more restrictive conditions then usually).

12 Theorem (Leray.) *Let F be a sheaf on the Hausdorff paracompact space X which has a resolution of the type considered and let $\mathfrak{U}$ be an open covering of X such that the U_{S_q} are paracompact and satisfy (12). Then*

$$H^p(\mathfrak{U}, F) \approx H^p(X, F).$$

For instance, if X is also perfectly normal the U_{S_q} are paracompact and Theorem 12 can be applied if (12) holds. One can prove Leray's theorem without any other conditions on F or on $\mathfrak{U}$ than the conditions (12) [15].

We indicate, without proof, another important application of the spectral sequence. Let $\varphi: X \to Y$ be a continuous map and F a sheaf of A-modules on X. For every $q \geqslant 0$, we assign to the open subsets $U \subseteq Y$ the modules $H^q(\varphi^{-1}(U), F)$ together with the obvious restrictions for $V \subseteq U$. In this manner, we obtain presheaves on Y and let Φ^q be the corresponding sheaves of germs. Then one can prove the following theorem [15].

13 Theorem *There is a spectral sequence associated with the previously described configuration such that $E_2^{pq} = H^p(Y, \Phi^q)$ and E_∞ is the bigraded module corresponding to the filtered module $H^*(X, F)$.*

All the previous results hold, of course, for the cohomology modules of a sheaf of A-algebras. But now there is also a new important operation.

Let X be a topological space and F a sheaf of A-algebras on X. Consider the direct sum

$$C^*(\mathfrak{U}, F) = \bigoplus_q C^q(\mathfrak{U}, F) \qquad (17)$$

and let $f \in C^p(\mathfrak{U}, F)$ and $g \in C^q(\mathfrak{U}, F)$. Define a product $fg \in C^{p+q}(\mathfrak{U}, F)$ by the cross sections

$$(fg)(i_0, \ldots, i_{p+q}) = r^{U(i_0,\ldots,i_p)}_{U(i_0,\ldots,i_{p+q})} f(i_0, \ldots, i_p)$$
$$\cdot r^{U(i_p,\ldots,i_{p+q})}_{U(i_0,\ldots,i_{p+q})} g(i_p, \ldots, i_{p+q}), \qquad (18)$$

the product of the second member being the A-algebra operation.

Since this product is sometimes denoted by $f \cup g$, it is called the *cup product*. It is easy to see that this product is associative if the stalks of F are associative. In this manner the module (17) has a structure of a *graded A-algebra* [42].

By straightforward calculations one obtains the formula

$$\delta(fg) = (\delta f)g + (-1)^p f(\delta g), \qquad (19)$$

which shows that

$$Z^*(\mathfrak{U}, F) = \bigoplus_q Z^q(\mathfrak{U}, F)$$

(the cocycles) is a subalgebra of $C^*(\mathfrak{U}, F)$ and that

$$B^*(\mathfrak{U}, F) = \bigoplus_q B^q(\mathfrak{U}, F)$$

(the coboundaries) is an ideal (both right and left) of $Z^*(\mathfrak{U}, F)$. Hence

$$H^*(\mathfrak{U}, F) = \bigoplus_q H^q(\mathfrak{U}, F)$$

has an induced structure of a graded A-algebra and, passing to the inductive limit, we get such a structure on

$$H^*(X, F) = \bigoplus_q H^q(X, F).$$

This is called the *cohomology algebra (or ring) of the space X with coefficients in F*. The corresponding product operation is also called the *cup product*. If the stalks of F are commutative algebras, one can establish

6. Cohomology with Coefficients in a Sheaf

that

$$\xi \cup \eta = (-1)^{pq}\eta \cup \xi, \quad p = \deg \xi, q = \deg \eta,$$

hence, the cup product is *anticommutative*. Also, one can see that the cup product commutes with the induced homomorphisms of continuous maps.

Observing the isomorphisms given by de Rham's abstract theorem, we see that the cup product operation can be calculated by using a convenient resolution.

In fact, suppose that (4) is such a resolution of the sheaf F of A-algebras that $\mathfrak{F} = \oplus_\alpha F_\alpha$ has a structure of a sheaf of graded A-algebras, i.e., for $\lambda \in F_\alpha$ and $\mu \in F_\beta$ (at the same point of X) the product $\lambda \times \mu$ in $\mathfrak{F}$ is in $F_{\alpha+\beta}$. Suppose also that

$$h_{\alpha+\beta}(\lambda \times \mu) = (h_\alpha \lambda) \times \mu + (-1)^\alpha \lambda \times (h_\beta \mu).$$

For the cochains with coefficients in $\mathfrak{F}$, we have the usual cup product given by Formula (18).

Let λ_α^p be a p-dimensional cocycle of a sufficiently fine covering of X with coefficients in F_α (hence also in $\mathfrak{F}$) and μ_β^q be a q-dimensional cocycle of F_β, such that $h_{\alpha-1}^{-1}\lambda_\alpha^p \neq \emptyset$ and $h_{\beta-1}^{-1}\mu_\beta^q \neq \emptyset$, where the operators h are applied componentwise. It is easy to see that we can take as a representative for $h_{\alpha+\beta-1}^{-1}(\lambda_\alpha^p \cup \mu_\beta^q)$ either $(h_{\alpha-1}^{-1}\lambda_\alpha^p) \cup \mu_\beta^q$ or $(-1)^\alpha \lambda_\alpha^p \cup (h_{\beta-1}^{-1}\mu_\beta^q)$, whence, by a straightforward calculation we get, using the notation of Formulas 5(6') and (19),

$$\partial_{\alpha+\beta-1}^{p+q}(\lambda_\alpha^p \cup \mu_\beta^q) = (\partial_{\alpha-1}^p \lambda_\alpha^p) \cup \mu_\beta^q = (-1)^{\alpha+\beta}\lambda_\alpha^p \cup (\partial_{\beta-1}^q \mu_\beta^q).$$

Now, by repeated applications of this formula one obtains finally

$$(r^{\alpha+\beta})^{-1}(\xi \times \eta) = (r^\alpha)^{-1}(\xi) \cup (r^\beta)^{-1}(\eta),$$

where $\xi \in H^\alpha(\mathfrak{R}(\rho))$, $\eta \in H^\beta(\mathfrak{R}(\rho))$, the r^α are the isomorphisms 5(6''), and the product $\times$ is induced by the product of $\mathfrak{F}$, which is meaningful because of the behavior of h with respect to the operation $\times$.

Hence, *by the isomorphisms of de Rham's theorem (when valid) the cup product is represented by the product of the corresponding cross sections of the sheaves F_α of the resolution.*

For a complete treatment of the cup product see, for instance, [15].

The last subject to be discussed here is the definition of the cohomology with non-Abelian coefficients. In this case one cannot obtain a complete theory as in the Abelian case, but one also has some useful concepts.

Let X be a topological space and G a sheaf of multiplicative, generally non-Abelian groups on X. Let $\mathfrak{U}$ be an open covering of X. Introduce the group

$$C^1(\mathfrak{U}, G) = \prod_{S_1 \in \Sigma_1} G(U_{S_1}), \qquad (20)$$

whose elements will be called *1-dimensional cochains of* $\mathfrak{U}$ *with coefficients in* G. If $U_i \cap U_j \neq \varnothing$, we denote by f_{ij} the corresponding cross section over $U_i \cap U_j$.

The cochain $\{f_{ij}\}$ is called a *cocycle* (*1-cocycle*) of the covering $\mathfrak{U}$ if, for every triple (i, j, k) such that $U_i \cap U_j \cap U_k \neq \varnothing$,

$$(r^{U(i,j)}_{U(i,j,k)} f_{ij})(r^{U(j,k)}_{U(i,j,k)} f_{jk})(r^{U(k,i)}_{U(i,j,k)} f_{ik}^{-1}) = 1. \qquad (21)$$

For the sake of simplicity we write (21) as

$$f_{ij} f_{jk} f_{ik}^{-1} = 1 \quad \text{on} \quad U_i \cap U_j \cap U_k. \qquad (22)$$

This is equivalent to

$$f_{ij} f_{jk} = f_{ik}, \qquad (23)$$

which immediately gives

$$f_{ii} = 1, \quad f_{ji} = f_{ij}^{-1}. \qquad (24)$$

Denote by $Z^1(\mathfrak{U}, G)$ the set of the cocycles of $C^1(\mathfrak{U}, G)$. This set has a distinguished element given by $f_{ij} = 1$, but, generally, it is not a group.

Let $\{f_{ij}^\alpha\}$ ($\alpha = 1, 2$) be two elements of $Z^1(\mathfrak{U}, G)$. We say that they are *cohomologous* if there is a system of cross sections $h_i \in \Gamma(U_i, G)$ such that

$$f_{ij}^2 = h_i^{-1} f_{ij}^1 h_j \qquad (25)$$

(the restrictions to $U_{(i,j)}$ again being omitted). Obviously, this is an equivalence relation on the set $Z^1(\mathfrak{U}, G)$.

14 Definition The *1-dimensional cohomology set* of the covering $\mathfrak{U}$, with coefficients in G is the set $H^1(\mathfrak{U}, G)$ of the classes of cohomologous cocycles introduced above.

Generally, $H^1(\mathfrak{U}, G)$ is not a group but it has a distinguished element defined by the cocycle 1.

Now let $\mathfrak{V}$ be a covering such that $\mathfrak{V} \prec \mathfrak{U}$ and let J be its set of indices; the set of indices of $\mathfrak{U}$ is denoted as usual by I. Let $\tau: J \to I$ be a mapping such that $V_j \subseteq U_{\tau j}$. We get the map $\tau^*: C^1(\mathfrak{U}, G) \to C^1(\mathfrak{V}, G)$ by the relation $(\tau^* f)_{ij} = f_{\tau i, \tau j}$ restricted to $V_{(i,j)}$, which sends cocycles into cocycles and preserves cohomology. Hence, it induces a mapping

$$t_{\mathfrak{V}}^{\mathfrak{U}}: H^1(\mathfrak{U}, G) \to H^1(\mathfrak{V}, G)$$

which preserves the distinguished element.

If $\tau': J \to I$ is another such mapping, we see that $\tau'^* f$ and $\tau^* f$ are cohomologous by taking $h_j = f_{\tau j, \tau' j}$ in (25) and, hence, $t_{\mathfrak{V}}^{\mathfrak{U}}$ does not depend on the choice of τ. It follows easily that $\{H^1(\mathfrak{U}, G), t_{\mathfrak{V}}^{\mathfrak{U}}\}$ is a directed inductive family of pointed sets. By taking the inductive limit, we get a pointed set $H^1(X, G)$ which is defined as the *1-dimensional cohomology set of the space X, with coefficients in G*.

The previous construction obviously has a covariant functorial character with respect to G and a contravariant behavior with respect to the base space X. Hence, we have the following theorem.

15 Theorem *There is a covariant functor $H^1(X, \cdot)$ from the category of sheaves of groups on X to the category* Ens· *of pointed sets.*

We can also define $H^0(X, G) = \Gamma(X, G)$ (the set of global cross sections) and this is, of course, a group, but it is complicated to define higher-dimensional cohomology sets.

If G is a sheaf of Abelian groups, then by exchanging the multiplicative notation with the additive one, it becomes obvious that we have the functor H^1 of Section 4 and $H^1(X, G)$ is an Abelian group.

The cohomology with coefficients in sheaves of arbitrary groups has a geometric equivalent which is very important, namely, the fiber bundles with structure group or sheaf. They are studied in the next chapter.

Chapter 3

FIBER AND VECTOR BUNDLES

1 Fiber Bundles

For the most part, the sheaves which appear in differential geometry are sheaves of germs of cross sections of some special kind of projected spaces called vector bundles, which are particular fiber bundles. This chapter is devoted to the study of fiber and vector bundles.

1 Definition The projected space $p: E \to X$ is called *trivial* if there is a topological space F and a homeomorphism $h: E \to X \times F$ such that $\operatorname{pr}_X h = p$. The homeomorphism h is called a *trivialization* of E and F is its *fiber*. The same projected space is called *locally trivial* if every point $x \in X$ has an open neighborhood U and a homeomorphism

$$h_U: p^{-1}(U) \to U \times F$$

such that $\operatorname{pr}_U h_U = p|_{p^{-1}(U)}$. F is again called the *fiber* and $\{h_U\}$ is a *local trivialization* of E.

We remark that if the space E is locally trivial and $\{h_U\}$ is a corresponding trivialization, then $h_U|_{p^{-1}(u)} = h_{U,u}$, $u \in U$, is a homeomorphism between $p^{-1}(u)$ and $\{u\} \times F$, with inverse $h_U^{-1}|_{\{u\} \times F}$, and it follows that all the local fibers of E are homeomorphic to the fiber F.

One can say that a projected space is locally trivial if it behaves locally like a direct product or if, locally, p is like the projection of a direct product onto the first factor. The system $\{h_U, F\}$ defines a *locally trivial structure* on E and such a structure, if it exists, is, generally, not uniquely determined. In the sequel, we shall see that such a structure is an atlas on E with values in some convenient pseudocategory.

Let U and V be two neighborhoods on X, belonging to a local trivialization of E and such that $U \cap V \neq \emptyset$. Then, clearly, for $u \in U \cap V$, we have

$$h_V h_U^{-1}(u, f) = (u, f'), \qquad f, f' \in F \tag{1}$$

and the correspondence $f \mapsto f'$ is a continuous function $g_V^U(u): F \to F$ such that $g_V^U(u) = h_{V,u} h_{U,u}^{-1}$, and this function depends continuously on u, i.e., $(u, f) \mapsto f'$ is continuous in both arguments. It follows from the definition of g_U^V that

$$g_V^U g_U^W = g_V^W, \qquad g_U^U = 1_F. \tag{2}$$

Thus g_U^V is the inverse mapping of g_V^U and it follows that the $g_V^U(u)$ are homeomorphisms of F onto F.

These considerations suggest the definition of a pseudocategory $\mathfrak{P}(X, F)$ by taking as objects the products $U \times F$, with open subsets U of X, and as morphisms $\alpha: U \times F \to V \times F$ the partial continuous maps α such that $\mathrm{pr}_V \circ \alpha = \mathrm{id} \circ \mathrm{pr}_U$, if $U \cap V \neq \emptyset$ (there will be no morphism if $U \cap V = \emptyset$).

Then, if we have a local trivialization of E we can consider it to be an atlas on E with values in $\mathfrak{P}(X, F)$, as defined in 1.6.3, having the pseudocategory $Q(E)$ of coordinate domains given by the system $\{p^{-1}(U)\}$ where U runs through the neighborhoods of the trivialization. The functor $\mathscr{A}$ of 1.6.3 will be the one which assigns to $p^{-1}(U)$ the object $U \times F$ in $\mathfrak{P}(X, F)$ and to the morphism $p^{-1}(U) \to p^{-1}(V)$ of $Q(E)$ the morphism $a_V^U = h_V h_U^{-1}$ of $\mathfrak{P}(X, F)$ defined by (1). Finally, the local charts will be the homeomorphisms h_U of the trivialization. The conditions of Definition 1.6.3 are satisfied, and we also note that the transition functions a_V^U are defined for every $f \in F$. These transition functions are determined by the homeomorphisms $g_V^U(u)$ introduced previously, and by an abuse of language we say that the $g_V^U(u)$ are the *transition functions of the atlas*. Also, if

$$h_U(e) = (u, f) (e \in E)$$

1. Fiber Bundles

we shall sometimes say that f is the *coordinate* or *fiber coordinate* of e in the chart h_U (in fact, (u, f) is this coordinate) and the coordinate expressions which do not depend on the point u will be considered only for f in just the same manner as we considered g_V^U instead of a_V^U.

Conversely, let $(Q(E), \mathscr{A}, h)$ be a $\mathfrak{P}(X, F)$-valued atlas on E and suppose that it is covering and *compatible with the projection p*, which means that it satisfies the following conditions: (a) if $T \in Q(E)$ and $\mathscr{A}(T) = U \times F$, then $T = p^{-1}(U)$; (b) $p \circ h_T^{-1} = \mathrm{pr}_U$. Then, taking $h_U = h_T$, we obviously get a local trivialization of E. In the sequel, all the $\mathfrak{P}(X, F)$-valued atlases considered are compatible with the projection. Now we can give the following definition.

2 Definition Any $\mathfrak{P}(X, F)$-manifold E (with compatible atlas) is called a *fiber bundle*.

It is understood here that E is a projected space with base space X. For the fiber bundles, we shall use the terminology used for general projected spaces and that of the theory of $\mathfrak{P}$-manifolds.

Hence, a fiber bundle is a projected space $p: E \to X$ together with a $\mathfrak{P}(X, F)$-structure on E, i.e., a complete atlas on E, compatible with the projection and with values in $\mathfrak{P}(X, F)$. This structure is determined by a local trivialization of E. The simplest example of a fiber bundle is that of a trivial space, where there is an atlas consisting of a single chart.

Now, we apply the considerations of Section 1.6 to the case of fiber bundles.

In the study of fiber bundles we use coordinates. If $e \in E$ and $p(e) = u \in U$, the coordinates of e in the chart h_U are

$$h_U(e) = (p(e), f_U) = (u, f_U).$$

The transition from the chart h_U to the chart h_V is given by

$$(u, f_V) = (u, g_V^U(u) f_U). \tag{3}$$

Since the construction of $\mathfrak{P}(X, F)$ does not use the space E, we can apply Theorem 1.6.10. In other words, we can consider systems of transition functions in $\mathfrak{P}(X, F)$, which consist of objects $\{U_i \times F\}_{i \in I}$, determined by an open covering $\{U_i\}_{i \in I}$ of X (this is obviously a necessary supplementary condition) and of morphisms $a_j^i: U_i \times F \to U_j \times F$, which can be represented by homeomorphisms $g_j^i(u): F \to F$ obtained just as the

g_V^U were before. Such a system defines, up to an isomorphism, a fiber bundle with transition functions g_j^i which is obtained by the canonical construction of Theorem 1.6.10.

We can also apply Proposition 1.6.4. To obtain this, we first consider projected sets and fiber bundles of sets, by applying the previous definitions to sets and maps instead of spaces and continuous maps. If for such a bundle X and F are topological spaces, and if the transition functions are continuous, Proposition 1.6.4 shows that there is on F a uniquely defined topology which makes it into a topological fiber bundle.

By Definition 1.6.11 and the definition of the morphisms of $\mathfrak{P}(X, F)$, it follows immediately that a partial map $\varphi \colon E \relbar\joinrel\rightarrow E'$ of two fiber bundles with the same base space and fiber is a $\mathfrak{P}(X, F)$-mapping if and only if it commutes with the projections of the two bundles. With such mappings as morphisms, we can talk of the pseudocategory of fiber bundles on X with fiber F. Moreover, we see that the morphisms in the category $\mathfrak{E}(X)$ of projected spaces, if considered for bundles, are $\mathfrak{P}(X, F)$-mappings and, hence, that the *category of fiber bundles on X with fiber F* is the full subcategory of $\mathfrak{E}(X)$ whose objects are bundles. In the sequel, we shall consider the notions and operations regarding fiber bundles with given base space X and fiber F to be performed in this category, strengthened by the total space functor. This category will be denoted by $S(X, F)$.

Let $\varphi \colon E \to E'$ be a morphism in $S(X, F)$. Then its local coordinate expression is of the form

$$\varphi h_U^{-1}(u, f) = h_{U'}^{-1}(u, f'),$$

where the prime denotes the elements of the second bundle. But it should be observed that we can always use for the two bundles the same covering of X. In fact, if we have a fiber bundle whose atlas is defined for an open covering $\mathfrak{U}$ of the base space, and if $\mathfrak{V}$ is another open covering finer then $\mathfrak{U}$, we get another atlas of the same bundle structure defined for $\mathfrak{V}$ by taking $h_V = h_U|_{p^{-1}(V)}$ for $V \subseteq U$. Then, our remark follows from the existence of an open covering finer than two given coverings, which is an obvious fact. The coordinate expression of φ then becomes

$$\varphi h_U^{-1}(u, f) = h_U'^{-1}(u, f'), \qquad u \in U, \tag{4}$$

and it is determined by the continuous map $\varphi_U(u) \colon F \to F$ defined by $\varphi_U(u)(f) = f'$. According to a previous convention, this last relation is also called the coordinate expression of φ. In the intersection $U \cap V \neq \varnothing$,

1. Fiber Bundles

we have, by (3),

$$g'^V_U \varphi_V = \varphi_U g^V_U. \tag{5}$$

Hence, the morphism φ is determined by the maps $\{\varphi_U\}$ which satisfy the compatibility conditions (5).

Obviously, φ is an isomorphism in $S(X, F)$ if and only if $\varphi_U(u)$ are homeomorphisms of F and, because of the strengthening of this category by the total space functor, φ is a monomorphism or an epimorphism if and only if $\varphi_U(u)$ is a monomorphism or an epimorphism, respectively.

Sometimes it is also necessary to consider bundles with different fibers. For instance, if we consider the fibered product of an object of $S(X, F)$ and an object of $S(X, F')$, both considered as projected spaces, we get an object of $S(X, F \times F')$. This suggests the consideration of the category $S(X)$ defined as the full subcategory of $\mathfrak{E}(X)$ whose objects are fiber bundles, with different fibers. This will always be used when the base space is fixed but the fibers are not. Clearly, $S(X, F)$ is a full subtactegory of $S(X)$. Of course, $S(X)$ will also be strengthened by the total space functor. Then, the fibered product is the direct product in $S(X)$. The subobjects and quotient objects in the category $S(X)$ will be defined as *subbundles* and *quotient fiber bundles*. If bundles with different base spaces are needed they can be considered as objects of a full subcategory of $S(\text{Top})$.

3 Proposition *The inverse image of a fiber bundle by a continuous map is a fiber bundle.*

Consider the bundle $p: E \to X$, the continuous map $\alpha: Y \to X$, and the inverse image $\alpha^*(E) \subseteq Y \times E$ with the projection q onto Y. If U is a neighborhood of the local trivialization of E, there is a homeomorphism

$$q^{-1}(\alpha^{-1}(U)) \approx \alpha^{-1}(U) \times F$$

given by $(y, e) \mapsto (y, f)$ $(\alpha(y) = p(e))$, where $h^{-1}(\alpha(y), f) = e$. Now we immediately get a local trivialization of $\alpha^*(E)$, which proves Proposition 3.

Note that the transition functions of $\alpha^*(E)$ are given by the relation

$$g^{*\alpha^{-1}(V)}_{\alpha^{-1}(U)}(y) = g^V_U(\alpha(y)). \tag{6}$$

The fiber bundle $\alpha^*(E)$ is called the *α-induced bundle of E*. The correspondence assigning to every bundle the α-induced bundle extends obviously to a covariant functor $\alpha': S(X, F) \to S(Y, F)$ so that we get the following proposition.

4   Proposition   *Every continuous mapping $\alpha: Y \to X$ induces a covariant functor $\alpha': S(X, F) \to S(Y, F)$.*

There is also another interpretation of the previous results, which is obtained by considering $S(., F)$ as a functor Top $\to$ Ens; by Proposition 3, this is indeed a contravariant functor.

In particular, if $Y \subseteq X$ and α is the inclusion, $\alpha^*(E)$ is isomorphic to the bundle $p^{-1}(Y) \xrightarrow{p|} Y$ ($|$ is the restriction to $p^{-1}(Y)$) which is called the *restriction* of E to the subspace Y and is denoted by $E|_Y$.

We have already seen an example of a fiber bundle, namely, the trivial bundle E, isomorphic to $X \times F$. In this connection, we can ask how to recognize that a fiber bundle given by a general atlas is trivial. Then, there must exist $\varphi: E \to X \times F$ defined by some system $\{\varphi_U\}$ of homeomorphisms of F satisfying (5) and such that $g_U^V = $ id. (On $X \times F$, we get an atlas with $\{U\}$ as coordinate domains by the restriction of the atlas with a single chart and this restriction has identity transition functions.) The following proposition results.

5   Proposition   *Let E be a fiber bundle given by some general atlas. Then, E is trivial if and only if there exist homeomorphisms $\varphi_U(u): F \to F$ such that in $U \cap V$*

$$g_U^V = \varphi_U^{-1} \varphi_V \qquad (7)$$

where $\varphi_U(u)f$ are continuous in both variables u and f.

Another simple example is given by a surjective differentiable map $\varphi: R^n \to D \subseteq R^m$ $(m < n)$ of class C^k with $k \geqslant 1$, D being an open subset of R^m (R is the real field). If we suppose that the rank of the Jacobian matrix of φ is everywhere m, the classical implicit function theorem determines obviously a local trivialization of φ with fiber R^{n-m}.

An important class of fiber bundles will now be defined.

6   Definition   A fiber bundle $p: E \to X$ whose fiber is a discrete topological space is called a *covering* of X. E is then the *covering space* and p the *covering projection*.

This is equivalent to the classical definition [42] which says that every point of X has an open neighborhood U such that $p^{-1}(U)$ is a disjoint

1. Fiber Bundles

union of open subsets each projected homeomorphically by p onto U. The coverings of X form a full subcategory of $S(X)$ which we denote by $A(X)$, and a total source of this category, if it exists, is called a *universal covering* of X. One can show that if X is a connected and locally path-connected space, then a simply connected covering space is a universal covering. If, moreover, X is semilocally simply connected, i.e., every point has an open neighborhood such that its closed paths are deformable in X to a point, the universal covering exists and is uniquely determined up to an isomorphism. For instance, R^n is the universal covering space of the n-dimensional torus $T^n = R^n/\Gamma_n$, where Γ_n is the group of transformations of R^n generated by the translations defined by n fixed linearly independent vectors. For the general theory of covering spaces we refer the reader to Spanier's book [42].

An important problem is that of the existence of global cross sections of a fiber bundle or, more generally, the problem of extending a cross section which is given over a closed subspace of the base space. This problem can be treated with general topological methods and one gets results like: *if $p\colon E \to X$ is a fiber bundle where X is a normal and final compact space and F is complete in Tietze's sense, a cross section of E over a closed subspace of X can be extended over X; in particular, such a bundle has global cross sections* [44]. Another approach is by the so-called obstruction theory which uses techniques of algebraic topology and whose principal idea is to consider the base space as a CW-complex and to construct the cross section step by step over the cells of different dimensions of that complex [42].

Another important problem is the problem of *lifting paths* from X to E. Let $p\colon E \to X$ be a projected space and X^I the path space of X with the compact open topology (a path on X is a continuous map $\varphi\colon I = [0, 1] \to X$). Consider the product $E \times X^I$ and let $T = \{(e, \varphi) \mid e \in E, \varphi \in X^I, \varphi(0) = p(e)\} \subseteq E \times X^I$ with the relative topology.

7 Definition A *connection* or a *lifting function* on E is a continuous function $\omega\colon T \to E^I$ such that

$$\omega(e, \varphi)(0) = e, \qquad p(\omega(e, \varphi)(t)) = \varphi(t).$$

$\omega(e, \varphi)$ is called the *lift* of φ by ω.

Remark that the name *lift* is always used for the case when elements associated with the base space are sent by some functions into corresponding elements of the total space in such a manner that by projection we again get the given elements of the base space.

It is obvious that for a trivial space we always have a connection. If a fiber bundle which is only locally trivial is given, one can under some conditions also obtain a connection by patching together local connections. For instance, *a fiber bundle with a paracompact base space has a connection* [42]. The existence of a connection on a projected space is equivalent to the fact that the projection is a *Hurewicz fibration*, i.e., it has the *homotopy lifting property* with respect to any space Y, which means that if $f: Y \to E$ and $F: Y \times I \to X$ are continuous maps such that $F(y, 0) = pf(y)$, there is a map $G: Y \times I \to E$ such that $G(y, 0) = f(y)$ and $p \circ G = F$ [42]. Hence, a fiber bundle with paracompact base space has the homotopy lifting property with respect to any space Y. Generally, a fiber bundle has the homotopy lifting property with respect to paracompact spaces Y and, of course, to a polyhedron. Hence it is a *Serre fibration* (i.e., a projected space with the homotopy lifting property with respect to polyhedra).

Definition 7 (of a connection) was recently changed by C. Teleman [46] in the following manner. Consider a projected space whose base space X has a fixed base point x and is weakly locally contractible (i.e., if $y \in X$, there is a neighborhood U of y which is contractible in X to x). A *good family of paths* in X is a subset $P \subset X^I$ such that: (a) if $f_i \in P$ ($i = 1, 2$) and $f_1(1) = f_2(0)$, then

$$t \mapsto \begin{cases} f_1(2t) & \text{for } t \in [0, \tfrac{1}{2}], \\ f_2(2t - 1) & \text{for } t \in [\tfrac{1}{2}, 1] \end{cases}$$

is in P; (b) if $f \in P$ and $\Phi: I \to I$ is continuous and piecewise linear, then $f \circ \Phi$ is in P; (c) if $y \in X$ there is an open set $U \subseteq X$ containing y and a map $h: U \times I \to X$ such that $h(z, 0) = x$, $h(z, 1) = z$ ($z \in U$). Now, one introduces the *P-connections* as functions ω as in Definition 7, with T replaced by $T \cap (E \times P)$ and which, besides the conditions of that definition, satisfy also the condition

$$\omega(e, \varphi) \circ \Phi = \omega[\omega(e, \varphi)(\Phi(0)), \varphi \circ \Phi]$$

for any function Φ as above.

Using this notion, one obtains a characterization of the fiber bundles in the class of projected spaces when the base space satisfies some restrictive conditions. One can actually show [46] that $p: E \to X$ is a fiber bundle if and only if it has a *P*-connection. For a *P*-connection one can introduce, as in differential geometry, the notion of holonomy.

2   Fiber Bundles with Structure Group

In Section 1, we defined fiber bundles as $\mathfrak{P}(X, F)$-manifolds and their transition functions were represented by the homeomorphisms $g_V^U(u)$ of F. Because, generally, the transition functions determine a $\mathfrak{P}$-manifold, we get different kinds of bundles by imposing conditions on these functions. If we ask that the transition functions belong to some group, we get the fiber bundles with structure group, to be studied here.

Let G be a topological group, i.e., a topological space which has a group structure such that the group operations are continuous. Suppose that we have an *effective action* of G as a left transformation group on F, which means that there is given a continuous map $G \times F \to F$ such that $(g, f) \mapsto gf$,

$$g(g'f) = (gg')f, \qquad 1f = f, \qquad g, g' \in G, f \in F,$$

where 1 is the unity of G, and if $gf = f$ for all $f \in F$ then $g = 1$. It follows immediately that $f \mapsto gf$ is, for any $g \in G$, a homeomorphism of F.

1   Definition   A fiber bundle on X with fiber F has the *structure group* G if G has a left effective action on F, as defined above, and if there is a covering $\mathfrak{P}(X, F)$-atlas, compatible with the projection, such that $g_V^U(u)$ are the values of a continuous map

$$g_V^U: U \cap V \to G.$$

Of course, if $G \subseteq G'$ and G' acts effectively on F, G' is also a structure group for the bundle. Hence, it will always be necessary to fix the structure group.

For fiber bundles with structure group, we can consider again the matters discussed in the previous paragraph. Thus, we can construct a bundle if its transition functions are given; we can turn a set bundle whose atlas is "topological" into a topological bundle with group G; we can go over to a finer trivializing covering and, hence, we can always consider the same trivialization neighborhood for two bundles; etc.

If E and E' are objects of the category $S(X, F)$, which have the structure group G, a morphism $\varphi: E \to E'$ in $S(X, F)$ is called *compatible with the structure group* if the corresponding maps $\varphi_U(u)$ are the values of a continuous map $U \to G$. It follows that $\varphi_U(u)$ are homeomorphisms of F. Hence, every morphism which is compatible with the structure group is an isomorphism.

Since compatible morphisms can be composed in the obvious manner and the identity morphism is compatible, it follows that the bundles with structure group G and their compatible morphisms define a subcategory of $S(X, F)$, which will be denoted by $S(X, F, G)$ and is a groupoid. For introducing notions and operations regarding fiber bundles with fixed X, F, and G we shall refer to the category $S(X, F, G)$.

Formula 1(6) shows that, if $\alpha: Y \to X$ and if $p: E \to X$ is an object of $S(X, F, G)$, the induced bundle $\alpha^*(E)$ is an object of $S(Y, F, G)$ and we derive the following proposition.

2   Proposition $S(\cdot, F, G)$ *is a contravariant functor from Top to Ens.*

A trivial bundle is a fiber bundle with trivial structure group, hence, one for which every group acting on F is a structure group.

We remark that if G is a topological group acting on F from the left, but not in an effective manner, the elements $g \in G$ such that $gf = f$ for every $f \in F$ form a closed normal subgroup $N \subseteq G$, and we obviously have an effective action of G/N on F. It is sometimes convenient to say that the fiber bundles with structure group G/N are fiber bundles with structure group G.

In the sequel, the structure group is assumed to act effectively on the fiber.

Let $p: E \to X$ be an object of $S(X, F, G)$ and $g_j^i: U_i \cap U_j \longrightarrow G$ the transition functions of a corresponding atlas, with trivialization neighborhoods U_i, $i \in I$. The relations 1(2) are now

$$g_k^j g_j^i = g_k^i, \qquad g_i^i = 1, \tag{1}$$

in $U_i \cap U_j \cap U_k$ and U_i, respectively. Hence, these functions define a 1-dimensional cocycle of the covering $\mathfrak{U} = \{U_i\}$, with coefficients in the sheaf of germs of continuous G-valued functions on X. This is a sheaf of generally non-Abelian groups and will be denoted by G_c.

Obviously, if we consider a finer covering $\mathfrak{V} = \{V_j\}_{j \in J}$ and choose $\tau: J \to I$ such that $V_j \subseteq U_{\tau j}$, the cocycle of the induced atlas is $\tau^* g_j^i$ as defined in Section 2.6. If we consider two isomorphic bundles with the same trivialization neighborhoods, Formula 1(5) shows that the corresponding cocycles are cohomologous.

Hence, if we consider two atlases of a bundle E, there is a sufficiently fine open covering of the base space such that the corresponding cocycles are cohomologous. It follows that the cohomology classes of the cocycles of transition functions of the two atlases are equal and we can assign to every object of $S(X, F, G)$ a uniquely determined class in $H^1(X, G_c)$. Moreover, two isomorphic fiber bundles have the same corresponding cohomology class.

Conversely, given an element $\xi \in H^1(X, G_c)$ we shall represent it by a cocycle of some covering $\mathfrak{U}$ and construct the bundle for which that cocycle is the cocycle of transition functions. This fiber bundle is determined up to an isomorphism and has ξ as its corresponding cohomology class. The class of trivial bundles corresponds to the distinguished element of $H^1(X, G_c)$ as defined in Section 2.6.

Denoting by $[S(X, F, G)]$ the set of isomorphism classes of fiber bundles of $S(X, F, G)$ and considering that this is the value on X of a contravariant functor Top $\to$ Ens$\cdot$ (the base point is the class of trivial bundles), which is possible because of Proposition 2, our previous considerations give the following theorem.

3 Theorem *There is a functorial isomorphism of the functors* $[S(X, F, G)]$ *and* $H^1(X, G_c)$ *between the categories* Top *and* Ens$\cdot$.

Because of this theorem, when classes of isomorphic bundles are used, we shall talk of an element of $H^1(X, G_c)$ as an *abstract bundle*. Such an abstract bundle defines a class of $[S(X, F, G)]$ for every representation of G as an effective group of left transformations on F.

Now, we shall consider an important class of fiber bundles with structure group which in some sense generates all the others.

4 Definition An object of $S(X, F, G)$ where $F = G$ and the action of G on itself is by *left translations* (i.e., gf is the product of these elements in G) is called a *principal fiber bundle with structure group* G.

The category of the principal bundles with group G will be denoted by $Sp(X, G)$. Of course, it is a groupoid and all the previous results hold. In particular, there is a natural equivalence of the functors $[S(\cdot, G)]$ and $H^1(\cdot, G_c)$.

There is an important characterization of the principal bundles in the category $S(X, G)$ of fiber bundles with fiber G. Recall, first, that a group G

acts as a right transformation group on a space T if there is given a map $T \times G \to T$ such that $(t, g) \mapsto tg$ and $(tg)g' = t(gg')$, $t1 = t$ $(t \in T)$. This action is called *free* if $tg = t$ for some $t \in T$ implies $g = 1$ (of course, this definition is valid for a left action too).

Now, let $p: E \to X$ be an object of $Sp(X, G)$ whose structure is given by an atlas for which we use the same notation as before. We can define a free right action of the group G on E by the formula

$$[h_U^{-1}(u, f)]g = h_U^{-1}(u, fg), \qquad f, g \in G. \tag{2}$$

This definition is correct since it follows immediately from the associativity of the product of G that the given correspondence does not depend on the local charts.

This action of G on E is called *right translation* and is denoted by $D_g: E \to E$. Clearly, D_g preserves the local fibers of E, which are the *transitivity domains* of G in E and on these domains the action of G is simple transitive.

Conversely, if $p: E \to X$ is an object of $S(X, G)$ and if there is given an action of G as a right transformation group on E, called the right translation, which is *compatible with a local trivialization* of E, i.e., every point $x \in X$ has an open neighborhood U, belonging to the local trivialization, such that Formula (2) holds, then E is a principal bundle with group G. In fact, denote by $g_V^U(u): G \to G$ the transition functions of E as an object of $S(X, G)$ and define

$$\tilde{g}_V^U: U \cap V \to G$$

by

$$\tilde{g}_V^U(u) = g_V^U(u)(1).$$

Taking into account the definition of $g_V^U(u)$, we have

$$h_U^{-1}(u, 1) = h_V^{-1}(u, g_V^U(u)(1)) = h_V^{-1}(u, \tilde{g}_V^U(u)).$$

Now, considering an arbitrary right translation and using Formula (2), we get

$$h_U^{-1}(u, f) = h_V^{-1}(u, \tilde{g}_V^U(u) f),$$

which proves the above assertion. Moreover, we see from (2) that the

2. Fiber Bundles with Structure Group

right translations act freely, preserve the local fibers, and act simple transitive on them. In other words, we have proved the following proposition.

5 Proposition *An object of $S(X, G)$ is a principal bundle with group G if and only if G acts on its total space by right translations defined by Formula (2).*

By using Theorem 3 and by the composition of the functorial isomorphism from $[Sp(X, G)]$ to $H^1(X, G_c)$ with the one from $H^1(X, G_c)$ to $[S(X, F, G)]$ we see that for every representation of G as an effective left transformation group on a topological space F there is a natural equivalence of the functors $[Sp(X, G)]$ and $[S(X, F, G)]$. In the sequel, we give a direct definition of this equivalence.

Let $p: E \to X$ be a principal bundle with group G and F a space on which G acts effectively from the left. Consider the space $F \times E$ and define on it an equivalence relation $(f, e) \sim (f', e')$ by the conditions

$$f' = g^{-1}f, \quad e' = eg, \quad e, e' \in E, f, f' \in F, g \in G,$$

where in the second relation we have the right translation D_g on E. Denote by A the corresponding quotient space and define $q([f, e]) = p(e)$ ($[f, e]$ is the class of (f, e)). We obtain a projected space $q: A \to X$.

Now, let $\{U_i, h_i\}_{i \in I}$ be an atlas of E with transition functions g_i^j. Define

$$\bar{h}_i: q^{-1}(U_i) \to U_i \times F$$

by

$$\bar{h}_i(f, e) = (p(e), gf),$$

where g is given by the relation $h_i(e) = (p(e), g)$, and

$$\bar{h}_i^{-1}: U_i \times F \to q^{-1}(U_i)$$

given by

$$\bar{h}_i^{-1}(u, f) = [f, h_i^{-1}(u, 1)].$$

It is easily seen that $\bar{h}_i$ and $\bar{h}_i^{-1}$ are continuous, well defined maps, inverse to each other. Hence, $\{U_i, \bar{h}_i\}$ defines on A a structure of fiber bundle with base space X, fiber F, and group G, and this structure is uniquely determined. It is also easy to see that the transition functions of the atlas obtained on A are again the g^i_j, which means that A represents the class of isomorphic bundles which corresponds to the class of E.

6   Definition   The fiber bundle A constructed above is called the *associated bundle* of the principal bundle E with fiber F.

Conversely, let $q: A \to X$ be an object of $S(X, F, G)$ and consider

$$E = \bigcup_{x \in X} \{l \mid l: F \approx A_x, h_x \circ l \in G\},$$

where A_x is the local fiber of A at $x \in X$, the maps l are homeomorphisms, and the $h_x: A_x \to F$ are induced by the charts of the atlas of A. We understand that $h_x \circ l \in G$ for every chart h whose coordinate domain contains x.

There is a natural projection $p: E \to X$ and if we consider the atlas $\{U_i, h_i\}$ of A, we can define

$$h_i^*: p^{-1}(U_i) \to U_i \times G$$

by

$$h_i^*(l) = (p(l), h_{i, p(l)} \circ l)$$

and

$$h_i^{*-1}: U_i \times G \to p^{-1}(U_i)$$

by

$$h_i^{*-1}(u, g)(f) = h_i^{-1}(u, gf).$$

Then, the h_i^* are bijective and define on E a structure of a principal bundle of sets with the same transition functions as A. Hence, there is a uniquely determined topology on E, which makes E an object of $Sp(X, G)$, whose isomorphism class corresponds to the isomorphism class of A.

We insist on the fact that the correspondence is only between the isomorphism classes of principal and associated bundles; if we consider the associated bundle of E constructed above we get a bundle which is isomorphic to the given bundle A, but is not A. Our previous assertion about the generation of the fiber bundles with group G by the principal bundles

2. Fiber Bundles with Structure Group

has to be understood in the sense of the above constructed natural equivalence of $[Sp(X, G)]$ to $[S(X, F, G)]$.

The existence of right translations in a principal bundle allows the following considerations. Let $p: E \to X$ be a principal bundle with group G. Then, G acts effectively from the left on E by the formula $ge = eg^{-1}$ and we can consider bundles with the structure group G and fiber E. Such fiber bundles will be called *metaprincipal*. If $q: W \to Y$ is a metaprincipal bundle, one can easily see that W has a structure of a principal bundle with group G and base space $Y \times X$.

We consider also the category $Sp(X)$ defined as the subcategory of $S(X)$ whose objects are the principal bundles with different groups G and whose morphisms are the morphisms φ of $S(X)$ such that the corresponding functions φ_U are group homomorphisms. $Sp(X)$ is strengthened by the total space functor. In $Sp(X)$ we can, for instance, consider principal subbundles of a principal bundle $p: E \to X$ with group G. Such a subbundle will be $p': E' \to X$, where $E' \subseteq E$ and the inclusion is a morphism in $Sp(X)$. G' is then a subgroup of G.

We consider an important example of a principal bundle. Let G be a topological group and G' a closed subgroup of G. Denote by $X = G/G'$ the quotient space of left classes of the elements of G modulo G' and by $p: G \to X$ the natural projection $p(g) = [g]_{G'} = \{gG'\}$. Next, denote $x_0 = p(1)$ where 1 is the unity of G; this is a distinguished base point of X. The projected space $p: G \to X$ is said to *have a local cross section* if it has a cross section over some open neighborhood U_{x_0} of the base point x_0.

7 Proposition *If $p: G \to X$ has a local cross section, it has a canonical structure of a principal fiber bundle with group G'.*

We use the previous notation. G acts as a left transformation group on X by the formula $g[\gamma] = [g\gamma]$ ($g, \gamma \in G$) and G' acts on G by right translations. Let U_{x_0} and $s_0: U_{x_0} \to G$ be respectively an open neighborhood of x_0 and a local cross section over it. For $x \in X$, choose an element $g \in p^{-1}(x)$ and consider the open neighborhood in X, $U_x = gU_{x_0}$. Then, there is a local cross section s over U_x, given by the relation

$$s(y) = gs_0([g^{-1}p^{-1}(y)]), \qquad y \in U_x.$$

This enables us to construct a homeomorphism

$$h_{U_x}: p^{-1}(U_x) \to U_x \times G'$$

by

$$h_{U_x}(\gamma) = ([\gamma], (s[\gamma])^{-1} \cdot \gamma), \qquad \gamma \in G,$$

whose inverse is given by

$$h_{U_x}^{-1}(y, g') = s(y) \cdot g', \qquad y \in U_x, g' \in G'.$$

Considering the right translation with $\bar{g}' \in G'$ we see that

$$[h_{U_x}^{-1}(y, g')]\bar{g}' = s(y) \cdot g', \bar{g}' = h_{U_x}^{-1}(y, g'\bar{g}')$$

and our proposition follows from Proposition 5.

The same result could be obtained by calculating the transition functions. We remark that if we change the neighborhood U_{x_0} and the cross section s_0 we obtain another atlas of the same structure.

Proposition 7 has an important generalization. Let $p: E \to X$ be a principal bundle with group G and let G' be a closed subgroup of G such that G has a local cross section over G/G'. Consider the quotient space of E with respect to the right translations by elements of G', E/G'. We get then the commutative diagram

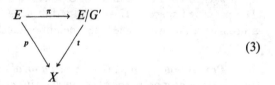

(3)

If $\{U_i\}_{i \in I}$ are coordinate domains on X for the atlas of E and $\{h_i\}_{i \in I}$ are the corresponding local charts, one sees immediately that the correspondences

$$k_i^{-1}: U_i \times G/G' \to t^{-1}(U_i)$$

given by

$$(u, [g]) \mapsto [h_i^{-1}(u, g)]_{G'}, \qquad g \in G,$$

turns the projected space $t: E/G' \to X$ into a fiber bundle over X with fiber G/G', structure group G and with the same transition functions as E.

Next, consider the covering $\{k_i^{-1}(U_i \times V_\alpha)\}$ of E/G', where $\{V_\alpha\}$ is an open covering of G/G' which is trivializing for the bundle structure of Proposition 7 and denote by s_α the cross sections over V_α introduced in the proof of that proposition. We can then define

$$l_{i\alpha}^{-1}(k_i^{-1}(u, v), g') = h_i^{-1}(u, s_\alpha(v) \cdot g'), \tag{4}$$

whose inverses are obtained in an obvious manner. It follows that

$$l_{i\alpha}: \pi^{-1}(k_i^{-1}(U_i \times V_\alpha)) \to k_i^{-1}(U_i \times V_\alpha) \times G'$$

are homeomorphisms and $\pi: E \to E/G'$ is a fiber bundle with fiber G'. Considering the right action of G' on E we see by (4) that the condition of Proposition 5 is satisfied. Hence, $\pi: E \to E/G'$ is a principal bundle with group G', which is the generalization of Proposition 7 referred to above.

We shall also establish another remarkable property of Diagram (3), namely, that the inverse image $t^*(E)$ and the bundle with fiber G associated with π (G' acting on G by left translations) are isomorphic when regarded as principal bundles with base space E/G' and group G.

In fact, the second of the two bundles considered is the quotient space A of $G \times E$ by the equivalence relation $(g, e) \sim (g'^{-1}g, eg')$ ($g' \in G'$, $g \in G$, $e \in E$), and the first is the subspace $W \subseteq E/G' \times E$ consisting of the pairs (v, e) such that $t(v) = p(e)$. A continuous map $\varphi: A \to W$ is given by

$$\varphi([g, e]) = (\pi(e), eg)$$

and one sees that in corresponding local charts φ is expressed by the identity maps. Hence, φ is the isomorphism sought.

This result also has another interpretation. Let $\xi \in H^1(X, G_c)$ be the cohomology class of E (see Theorem 3) and $\hat{\xi} \in H^1(E/G', G'_c)$ the class of E over E/G'. Let $h^*: H^1(E/G', G'_c) \to H^1(E/G', G_c)$ be the map induced by the inclusion $h: G' \subseteq G$. The previous result is then equivalent to the equation

$$h^*\hat{\xi} = t^*\xi. \tag{5}$$

To sum up, we have the following theorem.

8 Theorem *Let G be a topological group and G' a closed subgroup of G such that the projection $G \to G/G'$ has a local cross section. Then, to* *every principal bundle* $p: E \to X$ *with group* G *there is an associated commutative diagram* (3) *where*: (a) *the arrow* t *is a fiber bundle with base space* G/G' *and group* G *whose transition functions are those of* E; (b) *the arrow* π *is a principal bundle with group* G'; *and* (c) *the Equation* (5) *holds*.

An important problem related with the theory of fiber bundles is the so-called reduction of the structure group. Let $p: E \to X$ be an object of $S(X, F, G)$. We wish to know when is it possible to find an atlas of the bundle structure of E such that the transition functions $g_i^j(u)$ take values in a subgroup G' of G. If this is possible, we say that the structure group of E is *reducible* to G' and, if we actually give such an atlas, we say that there is given a *reduction* of the structure group. It should be observed that two such reductions are not necessarily compatible atlases with respect to G', though they are compatible with respect to the group G.

Clearly, if we denote by ξ the abstract bundle corresponding to E in $H^1(X, G_c)$, E has a reduction of the structure group to the subgroup $G' \subseteq G$ if and only if $\xi = h^*\eta$, where $h: G' \subseteq G$ and $h^*: H^1(X, G'_c) \to H^1(X, G_c)$ is the induced map. Such a reduction is fixed if one gives one η with the mentioned property.

Finally, another immediate remark is that giving a reduction of the structure group of a principal bundle is equivalent to giving a principal subbundle with group G' of that bundle. (In the case of arbitrary bundles, this remark can be applied after going over to their corresponding principal bundles.)

From the last remark and the notation of Diagram (3), it follows that if the principal bundle $p: E \to X$ admits a reduction of its structure group G to G', then the bundle E/G' has a global cross section. This can also be seen directly. Let the g_j^i be the transition functions of E. A reduction of the structure group is determined by functions $s_i: U_i \to G$ such that

$$g_j^i = s_j g_j'^i s_i^{-1}, \qquad g_j'^i \in G'.$$

Define

$$s(u) = [h_i^{-1}(u, s_i(u))]_{G'}.$$

As is easily seen, $s: X \to E/G'$ is a global cross section. Q.E.D.

Conversely, suppose that there is a global cross section $s: X \to E/G'$ and suppose also that $G \to G/G'$ has a local cross section, G' being a closed subgroup of G. Consider the induced map s^* and apply it to Equation (5).

2. Fiber Bundles with Structure Group

Because $ts = \text{id}$ and because, obviously, s and h^* commute, we get

$$\xi = h^*(s^*\hat{\xi}).$$

Hence, to such a cross section s there is assigned a reduction of the structure group of E to G'. Thus, we have proved the following theorem.

9   Theorem   *Let E be a principal bundle with group G and let G' be a subgroup which satisfies the above hypotheses. Then the structure group of E is reducible to G' if and only if the bundle $E/G' \to X$ has a global cross section.*

From Proposition 1.5 we see that a fiber bundle is trivial if and only if its structure group is reducible to the trivial group. Hence, using Theorem 9, we get the following corollary.

10   Corollary   *A principal bundle is trivial if and only if it has a global cross section.*

We also mention, without proof, another result related to the problem of the reducibility of the structure group. One can show that if X is Hausdorff and paracompact and if G/G' is homeomorphic to some Euclidean space R^n (R being the real field), the hypotheses of Theorem 8 being also satisfied, then the inclusion $h: G' \subseteq G$ induces a bijection $h^*: H^1(X, G'_c) \to H^1(X, G_c)$. This means that every bundle with group G has structure group reducible to G' [44].

Another important problem is the classification of the fiber bundles with structure group G, which means the determination of all these bundles by some canonical procedure. One sees easily that for such a classification it is necessary to identify isomorphic bundles. Hence, by Proposition 6, it suffices to classify the principal bundles with group G and, by Theorem 3, this is equivalent to the determination of the cohomology set $H^1(X, G_c)$. We shall briefly indicate the solution of this problem.

Let C be a class of topological spaces and consider the class

$$H^1(C) = \bigcup_{X \in C} H^1(X, G_c).$$

We make it into a category by putting

$$\mathfrak{M}(\xi, \eta) = \begin{cases} \text{a single element if } \eta = f^*(\xi) \text{ for some } f, \\ \varnothing \text{ otherwise,} \end{cases}$$

where, if $\xi \in H^1(X, G_c)$ and $\eta \in H^1(Y, G_c)$, f is a map from Y to X.

11 Definition A principal bundle corresponding to a final object of the category $H^1(C)$ is called a *universal bundle* with group G, for the class C of topological spaces (see Theorem 3). The base space of such a bundle is called a *classifying space*.

If the category $H^1(C)$ has a final object the classification problem is solved in the following manner. Let $\xi_0 \in H^1(X_0, G_c)$ be the final object, so that X_0 is the classifying space. Then, every $\xi \in H^1(X, G_c)$ is of the form $\xi = f^*(\xi_0)$ for some $f: X \to X_0$. Hence, the classification problem reduces to the construction of universal bundles.

If C is the class of paracompact Hausdorff spaces, a universal bundle has been constructed by Milnor. We indicate Milnor's construction [24].

Consider all the sequences of the form

$$\langle x, t \rangle = (t_0 x_0, t_1 x_1, \ldots, t_k x_k, \ldots),$$

where $x_i \in G$, the group for which the construction is performed, and $t_i \in [0, 1]$. It is supposed that only a finite number of the t_i are nonzero and that $\sum t_i = 1$. Two such sequences $\langle x, t \rangle$ and $\langle x', t' \rangle$ are considered as equivalent if, for all i, $t_i = t'_i$ and $x_i = x'_i$ for those i such that $t_i \neq 0$. Denote by E_G the set of classes of equivalent sequences. Next, consider the functions $t_i: E_G \to [0, 1]$ such that $\langle x, t \rangle \mapsto t_i$ (of the same sequence) and the functions $x_i: t_i^{-1}(0, 1] \to G$ given by $\langle x, t \rangle \mapsto x_i$. We now introduce on E_G the weakest topology such that the $t_i^{-1}(0, 1]$ are subspaces and the functions t_i and x_i are continuous.

Now, G acts on E_G as a right transformation group by the formula

$$\langle x, t \rangle g = \langle xg, t \rangle$$

Let X_G be the respective quotient space, and denote by $p: E_G \to X_G$ the natural projection.

One can show [24] that $p: E_G \to X_G$ is a principal universal bundle for the group G and the class of paracompact Hausdorff spaces. Hence, every principal bundle $q: E \to X$ with group G is isomorphic to a bundle of the form $f^*(E_G)$, where $f: X \to X_G$, and it can also be shown that the mapping f is determined up to a homotopy.

3 Vector Bundles

In this section, we study the fiber bundles whose fiber is a linear space and whose group is a linear group. These bundles are the source of some very important geometrical and topological theories.

1 Definition   A fiber bundle whose fiber is a Banach space and whose structure group is a subgroup of the group of linear continuous automorphisms of the fiber is called a *vector bundle*.[†]

For reasons to be seen later, we view a Banach space as one endowed with a class of equivalent norms. A fixed norm will be considered only when needed.

The vector bundles can be classified using the field of scalars of the fiber, e.g., real, complex, quaternionic vector bundles, etc. We can also distinguish between the finite dimensional and the infinite dimensional cases. Except for the cases to be specifically indicated, it is convenient to consider as structure group the group of automorphisms of the fiber with its usual norm [9].

Hence, if we denote by V the fiber and by L the structure group, we can talk of the category $S(X, V, L)$ of vector bundles on X with fiber V and group L, and we can apply to this category the results of the previous sections.

2 Proposition   *One can give the local fibers of an object of $S(X, V, L)$ the structure of a Banach space, isomorphic to V. The cross sections of such an object over an open subset $U \subseteq X$ have a module structure over the ring of continuous functions $f \colon U \to K$, where K is the field of scalars of V.*

In fact, let $p \colon E \to X$ be an object of $S(X, V, L)$ and $\{U_i, h_i\}$ be an atlas of the bundle structure. Then, for $E_u = p^{-1}(u)$, $u \in U_i$, we define

$$w_1 + w_2 = h_i^{-1}(u, v_1) + h_i^{-1}(u, v_2) = h_i^{-1}(u, v_1 + v_2), \qquad w_1, w_2 \in E_u$$
$$\alpha w = \alpha h_i^{-1}(u, v) = h_i^{-1}(u, \alpha v), \qquad \alpha \in K, w \in E_u. \tag{1}$$

[†] It is also possible to consider the more general case when the fiber is a topological vector space, but we limit ourselves to Banach spaces, which is the only case required in the sequel.

These operations are well defined because the transition functions belong to L, and this proves the first part of the proposition. Clearly, we have on E_u a family of equivalent norms induced by h_i, and the topology is the relative topology with respect to E.

It is easy to see that a vector bundle $p: E \to X$ is characterized by the following properties: (a) the local fibers are Banach spaces with respect to the induced topology; (b) there is a local trivialization whose fiber is a fixed Banach space and is such that $h_{U,u}$ are linear isomorphisms. Sometimes these conditions are easier to verify than definition 1.

The second part of Proposition 2 is obtained in just the same manner by considering in (1) cross sections w_i, w and the corresponding continuous functions v_i, v, and α.

Now, suppose that V has a basis $\{v_\alpha\}$. Then, we get over U_i the cross sections $s_\alpha = h_i^{-1}(u, v_\alpha)$, which have the property that their values at every point of U_i form a basis of the corresponding local fiber. A system of cross sections s_α having this property is called a *basis* for the sections over U_i and, if such a basis is given, it has the fiber coordinates $v_\alpha(u) = T_u v_\alpha$, where T_u is a linear isomorphism of V depending continuously on u. Hence, there is a chart of the complete atlas of the bundle, with U_i as the coordinate domain with respect to which the basis s_α is obtained from v_α as above. This is the chart given by $\bar{h}_i[h_i^{-1}(u, v)] = (u, T_u v)$. If s_α^i is a basis of the cross sections over U_i and s_α^j a basis of the cross sections over U_j, and if in $U_i \cap U_j$ we have $s_\alpha^i = T_u s_\alpha^j$, it is obvious that the T_u are just the transition functions between the charts defined by the two bases. It follows that for such vector bundles we can always use atlases whose charts are associated with bases of cross sections. For instance, this is the case for finite dimensional fibers.

From Proposition 2 it follows that, by assigning to every open subset $U \subseteq X$ the vector space over K of the cross sections of E over U, denoted by $\Gamma(U, E)$, and to every pair $U' \subseteq U$ the usual restriction of the cross sections, we have a presheaf. The corresponding sheaf of germs of cross sections, denoted by $F(E)$, is a sheaf module over the sheaf of rings of germs of K-valued continuous functions on X.

The principal bundle which is associated to a given vector bundle E can be constructed by considering the set of all the homeomorphisms l of V onto the local fibers such that (in the notation of the previous section) $h_x \circ l \in L$, i.e., $l \in L(V, E_x)$. But this set can be identified with the set of bases (when they exist) of the local fibers of E. Hence, this last set can be made into a principal bundle, and E is, up to an isomorphism, an associated bundle of this principal bundle.

3. Vector Bundles

Now, we want to obtain a category with all the vector bundles on a given basis X as objects. To obtain this we require morphisms, and we begin by considering the morphisms of the category $S(X)$ of the previous section for objects of $S(X, V, L)$ with different V. If

is such a morphism, then, with usual notations, we have

$$\varphi h_i^{-1}(u, v) = h_i'^{-1}(u, \varphi_i(u)v)$$

and it is natural to impose for morphisms of vector bundles the condition that $\varphi_i(u)$ belong to L, which is equivalent to the fact that, for every $x \in X$, $\varphi_x: E_x \to E_x'$ is linear and continuous. The category made up of the vector bundles and morphisms satisfying the previous condition will be denoted by $Sv(X)$. For such notions as monomorphism, epimorphism, etc., the category $Sv(X)$ is strengthened by the total space functor, which implies that φ is a monomorphism, epimorphism, etc., if the maps φ_x are.

We shall give an important result for the category $Sv(X)$.

3 Definition Let Φ be a functor, generally of several arguments, from the category B of Banach spaces with fixed scalar field to the same category B. Let $\varphi: X \to \mathfrak{M}$ be a continuous mapping from the topological space X to the set of morphisms of two objects of the category on which Φ is defined (with the usual topology). If for every such map φ, the map $x \mapsto \Phi(\varphi(x))$ is again continuous, Φ is called a *continuous functor*.

4 Theorem *Every continuous functor Φ on B induces, in a canonical manner, a functor with the same number of arguments and with the same covariant and contravariant characters from $Sv(X)$ to $Sv(X)$.*

It suffices to assume that Φ is a covariant functor of one argument, since the proof is the same in all the other cases.

Let $p: E \to X$ be an object of $Sv(X)$. Consider the Banach spaces $\Phi(E_x)$ ($x \in X$) and the set $\Phi(E) = \bigcup_{x \in X} \Phi(E_x)$ which has a natural projection $q: \Phi(E) \to X$. Now take an atlas $\{U_i, h_i\}$ of E, where and define

$$h_i: p^{-1}(U_i) \to U_i \times V$$

$$k_i: q^{-1}(U_i) \to U_i \times \Phi(V)$$

by

$$k_i(\alpha) = (q(\alpha), \Phi(h_{i,u})(\alpha)), \quad \alpha \in \Phi(E_u), u \in U_i.$$

By the continuity of Φ, this function makes $\Phi(E)$ into a vector bundle of sets with topological atlas. Hence $\Phi(E)$ has a uniquely determined structure of vector bundle (see Proposition 1.6.4 and its application in Sections 1 and 2). Moreover, if the $g_i^j(u)$ are the transition functions of E, the $\Phi(g_i^j(u))$ are the transition functions of the bundle $\Phi(E)$.

Finally, if $\varphi: E \to E'$ is a morphism in $Sv(X)$, it is easy to see that the sum of the maps $\Phi(\varphi_x)$ defines a corresponding morphism $\Phi(E) \to \Phi(E')$. Hence, the theorem is completely proved. The induced functor on $Sv(X)$ will also be denoted by Φ.

Note that: Φ sends trivial bundles to trivial bundles; Φ commutes with the functors induced by continuous maps $f: Y \to X$; and, if λ is a functorial morphism $\Phi \to \Psi$, where Ψ is another continuous functor on B, it induces a functorial morphism $\lambda: \Phi \to \Psi$ on $Sv(X)$.

An immediate consequence of the previous theorem is the existence of the direct sum of two objects of $Sv(X)$. Indeed, the direct sum (direct topological product) of Banach spaces, regarded as a two-argument functor, is continuous. Hence, the direct sums $E_x \oplus E'_x$ of the local fibers of two vector bundles E, E' on X define a new vector bundle, denoted by $E \oplus E'$ and called the *Whitney sum* of the two given bundles. It is easy to see that this is just the direct sum in $Sv(X)$.

The category $Sv(X)$ has a zero object, which is the product bundle of X and the trivial linear space. We can also define the sum of two morphisms between E and E' in $Sv(X)$ by taking the necessary sums at every point $x \in X$, and in this manner we get a structure of an Abelian group on $\mathfrak{M}(E, E')$; moreover, we have a structure of a vector space over K and of a module over the ring of K-valued continuous functions on X.

These remarks have the following result.

5 Proposition *$Sv(X)$ is an additive category.*

One can see that $Sv(X)$ is not an Abelian category, but there are some

3. Vector Bundles

morphisms which have kernel, cokernel, image, and coimage for which sufficient conditions are given.

6 Proposition Let $\varphi: E \to E'$ be a morphism in $Sv(X)$ such that: (a) for every $x \in X$, im φ_x is a direct summand in E'_x and ker φ_x is a direct summand in E_x; (b) if $\varphi_i(u): V \to V'$ are the functions which express φ in the atlases $\{(U_i, h_i)\}, \{(U_i, h'_i)\}$, there are, for every i, isomorphisms $k_i(u): \text{im } \varphi_i(u) \to W$, where W is a fixed Banach space, depending continuously on u. Then $\bigcup_{x \in X} \ker \varphi_x$, $\bigcup_{x \in X} \text{im } \varphi_x$, $\bigcup_{x \in X} \text{coker } \varphi_x$, $\bigcup_{x \in X} \text{coim } \varphi_x$ have natural structures of objects of $Sv(X)$ and they define ker φ, im φ, coker φ and coim φ, respectively, in the category $Sv(X)$.

Consider first im $\varphi = \bigcup_{x \in X} \text{im } \varphi_x \xrightarrow{q} X$, where q is the natural projection. Obviously, the correspondence

$$h'^{-1}_i(x, \varphi_i(x)f) \mapsto (x, k_i(x)\varphi_i(x)f)$$

defines an isomorphism $l_i: q^{-1}(U_i) \to U_i \times W$ and we have

$$l_i l_j^{-1}(x, w) = (x, k_j(x)g'^i_j(x)k_i^{-1}(x)w),$$

where the g'^i_j are the transition functions of E'. We derive from this the vector bundle structure of im φ.

From hypothesis (a), it follows that ker $\varphi_i(x)$ is a direct summand of the fiber V of E. Since ker $\varphi_i(x) = \ker (k_i(x) \varphi_i(x))$, by the identification of W with a direct summand of the fiber V' of E' (which is possible by (a)), we get isomorphisms $t_i(x): V/\ker \varphi_i(x) \approx W$, depending continuously on x. Hence, we also have isomorphisms $s_i(x): \ker \varphi_i(x) \approx V'/W$ depending continuously on x and which are defined over every U_i. Using the isomorphisms $s_i(x)$ in just the same manner as the $k_i(x)$ were previously used we obtain the vector bundle structure of ker φ.

Similar results concerning coker φ and coim φ are obtained determining first by (a) the isomorphisms

$$u_i(x): \text{coker } \varphi_i(x) \to V'/W, \qquad v_i(x): \text{coim } \varphi_i(x) \to W,$$

depending continuously on x. The last assertion of Proposition 6 is trivial. Hence, the proposition is proved and, moreover, one sees that the canonical morphism $\mu: \text{coim } \varphi \to \text{im } \varphi$ (1.4.5) is an isomorphism.

7 Definition A morphism φ of $Sv(X)$ such that the hypotheses (a) and (b) of Proposition 6 are satisfied is called a *regular morphism*.

Remark that if E and E' have finite-dimensional fibers, the condition (a) of the previous definition is always satisfied, and (b) means that rank $\varphi_x = $ constant. The composition of two regular morphisms is not always regular.

8 Definition   A sequence of the form

$$\cdots \xrightarrow{\varphi_{i-1}} E_i \xrightarrow{\varphi_i} \cdots, \qquad i \in I, \tag{2}$$

where E_i are objects in $Sv(X)$ and φ_i are regular morphisms, is called an *exact sequence of vector bundles* if, for every i, we have im $\varphi_{i-1} = $ ker φ_i.

We can consider the category of exact sequences, short exact sequences, etc. An obvious result follows.

9 Proposition   *The sequence* (2) *is exact if and only if, at every point of the base space, the corresponding sequence of local fibers is exact or, what is the same, the induced fiber sequence is exact.*

Assume a sequence of the form

$$0 \to E' \xrightarrow{\varphi} E, \tag{3}$$

and suppose that the base space X is Hausdorff and paracompact. Then, we can find a left inverse for φ. Indeed, let U_i be an open locally finite trivializing covering of X. Expressing φ locally, in the usual notation by

$$\varphi h_i'^{-1}(u, v') = h_i^{-1}(u, \varphi_i(u)v'),$$

we find the local inverses $\psi_i : E|_{U_i} \to E'|_{U_i}$ of φ given by

$$\psi_i h_i^{-1}(u, v) = h_i'^{-1}(u, \varphi_i^{-1}(u) \operatorname{pr}_{\operatorname{im} \varphi_i(u)} v).$$

Now, let $\{\chi_i\}$ be a partition of unity subordinated to the covering $\{U_i\}$. Then $\psi = \sum_i \chi_i \psi_i$ (which is a finite sum at every point $x \in X$) defines a regular morphism $\psi : E \to E'$ and $\psi \circ \varphi = 1_{E'}$ since

$$\psi\varphi(e') = (\sum_i \chi_i \psi_i) \circ \varphi_i(e') = e'.$$

Analogously, if the sequence

3. Vector Bundles

$$E \xrightarrow{f} E'' \to 0$$

is exact we can construct a right inverse of f.

Consider now the short exact sequence of vector bundles

$$0 \to E' \xrightarrow{\varphi} E \xrightarrow{f} E'' \to 0 \qquad (4)$$

over a paracompact Hausdorff basis. We get a vector bundle isomorphism $\chi: E \to E' \oplus E''$ given by

$$\chi(e) = \psi(e) \oplus f(e),$$

where ψ is the left inverse of φ constructed above. Hence, we get the following result.

10 Theorem (Whitney) *If (4) is an exact sequence of vector bundles on a paracompact Hausdorff basis, then $E \approx E' \oplus E''$ and the given exact sequence splits.*

Another interesting result can be stated as follows.

11 Proposition *If $p: E \to X$ is an object of $Sv(X)$, whose fiber is finite dimensional and whose basis is Hausdorff and compact, then there is an object E' in $Sv(X)$, with finite dimensional fiber such that $E \oplus E'$ is trivial.*

If the fiber of a vector bundle is finite dimensional, the bundle will be called *finite dimensional*. To prove the proposition consider a finite open covering $\mathfrak{U} = \{U_i\}$ ($i = 1, \ldots, k$) of X belonging to a trivialization of E and a partition of unity $\{\chi_i\}$ subordinated to $\mathfrak{U}$. To every chart $h_i^{-1}: U_i \times V \to p^{-1}(U_i)$ of E, we assign the map $\bar{h}_i: X \times V \to E$ given by

$$\bar{h}_i = \begin{cases} \chi_i h_i^{-1} & \text{in} \quad U_i \times V, \\ 0 & \text{otherwise.} \end{cases}$$

Then, we get a regular bundle morphism

$$\alpha: X \times V^k \to E, \qquad V^k = \underbrace{V \times \cdots \times V}_{k \text{ times}}$$

given by

$$\alpha(x, v_1, \ldots, v_k) = \sum_i \bar{h}_i(x, v_i),$$

which is obviously surjective on each local fiber. It follows that

$$X \times V^k \xrightarrow{\alpha} E \to 0$$

is an exact sequence and α has a right inverse whose image is the bundle E' sought.

Proposition 6 leads to the possible subbundles of a vector bundle. Thus, $p' : E' \to X$ will be considered a subbundle of $p : E \to X$ only if $E' \subseteq E$ and the inclusion is a regular morphism in $Sv(X)$. This means that the sequence $0 \to E' \xrightarrow{i} E$, where i is the inclusion, is an exact sequence. Clearly then, $E' = \operatorname{im} i$ and the bundle coker i exists. It will be denoted by $E/E' = E''$ and is called the *quotient bundle* of E with respect to E'. This bundle is characterized, up to an isomorphism, by the exactness of the sequence $0 \to E' \xrightarrow{i} E \xrightarrow{j} E'' \to 0$, where j is the canonical projection and is a regular morphism. For a triple (subbundle, bundle, quotient bundle) the fibers are in the same relations.

Now we consider the important notion of a norm on a vector bundle.

12 Definition   A *norm* on a vector bundle $p : E \to X$ is a family $\| \ \|_x$ ($x \in X$) of norms of the local fibers of E, equivalent to the norms induced by the vector bundle structure, such that if $s : U \to E$ is a cross section of E over an open subset $U \subseteq X$ then $\|s(u)\| = \|s(u)\|_u$ ($u \in U$) is a continuous real valued function on U. If the bundle E is given such a norm, it will be called a *normed bundle*.

13 Proposition   *For every vector bundle $p : E \to X$ with paracompact Hausdorff base space X, there is a norm.*

Let $\{U_i\}$ be a trivialization covering of X and $\{\chi_i\}$ a subordinated partition of unity. On $E|_{U_i}$ we can define a norm by

$$\|h_i^{-1}(u, v)\|_i = \|v\|, \tag{5}$$

where, on the right side, the norm is that of the fiber V. Now define for every $e \in E$

$$\|e\|_{p(e)} = \sum_i \chi_i(p(e)) \|e\|_i. \tag{6}$$

This is easily seen to be a norm on E.   Q.E.D.

3. Vector Bundles

Denote by Is (V) the group of linear isometries of the Banach space V, i.e., the group of linear norm preserving transformations. We then get

14 Proposition *Every reduction of the structure group of E to $\text{Is}(V)$ defines a norm on E.*

Indeed, such a norm is given by Formula (5) where h_i are charts of an atlas whose transition functions belong to $\text{Is}(V)$.

It is obvious that we also have the following result.

15 Proposition *If E is a normed bundle with an atlas whose associated functions $h_{i,x}$ are isometries, then there is a corresponding reduction of the structure group of E to $\text{Is}(V)$.*

16 Definition Let E be a vector bundle whose fiber is a Hilbert space V. If for every point $x \in X$, E_x is given a Hilbert scalar product g, compatible with the topology of E_x and such that, if s_1 and s_2 are cross sections of E over an open subset $U \subseteq X$, and $g(s_1(x), s_2(x))$ is a real-valued continuous function on U, we say that there is defined a *Hilbertian structure* on E, or that E together with the indicated structure, is a *Hilbert bundle*.

Clearly, a Hilbert bundle is a normed bundle with the norm defined by the scalar product. With the same proof as for Proposition 13 we have the following result.

17 Proposition *On every vector bundle $p: E \to X$ with paracompact Hausdorff basis and whose fiber is a Hilbert space there is a Hilbertian structure.*

18 Proposition *The definition of a Hilbertian structure on a bundle E is equivalent to the definition of a reduction of the structure group of E to $\text{Is}(V)$.*

In fact, if we have a reduction of the structure group we get a Hilbertian structure just as we got the norm in Proposition 14. Conversely, given a scalar product we get an atlas with transition functions belonging to $\text{Is}(V)$, by taking local charts associated with orthonormal bases of sections defined over the coordinate neighborhood.

From Propositions 17 and 18 we obtain the following result.

19   Corollary   *For every bundle $p: E \to X$ with paracompact Hausdorff base space and whose fiber is a Hilbert space V there is a reduction of the structure group to $\text{Is}(V)$.*

The previous results can be applied to the case of finite dimensional vector bundles. In this case, a real Hilbert bundle is called a *Riemannian bundle* and the corresponding scalar product is called a *Riemannian metric of the bundle*. In the complex case, the bundle is called *Hermitian* and the scalar product is called an *Hermitian metric*. Sometimes these names are used in the general case too. By Corollary 19, the structure group of a finite dimensional vector bundle with paracompact Hausdorff basis can always be reduced to the orthogonal group in the real case and to the unitary group in the complex case. In these cases, the scalar products are expressed, in fiber coordinates with respect to charts associated to bases of cross sections, by formulas of the type $g_{ij}\xi^i\eta^j$ with $g_{ij} = g_{ji}$ and with a positive definite form $g_{ij}\xi^i\xi^j$, in the Riemannian case, and $h_{ij}\xi^i\bar{\eta}^j$ with $h_{ij} = \bar{h}_{ji}$ and a positive definite form $h_{ij}\xi^i\bar{\xi}^j$ in the Hermitian case. In the intersection of two coordinate neighborhoods g_{ij} (respectively, h_{ij}) satisfy the classical tensor transformation law.

We now consider another important structure.

20   Definition   An *almost-complex structure* on the real vector bundle $p: E \to X$ is a field J of continuous linear automorphisms $J_x: E_x \to E_x$ ($x \in X$) with the property $J_x^2 = -1_{E_x}$ such that if $s: U \to E$ is a cross section of E over an open subset $U \subseteq X$, $J_x s(x)$ is again a cross-section over U.

In the sequel, we shall omit the index x of J and we shall denote 1_{E_x} by I. The bundle E together with the field J is called an *almost-complex bundle*. Any complex vector bundle $p: E \to X$ has an almost-complex structure, whose operator J consists of the multiplication by $\sqrt{-1}$. Generally, the local fibers of an almost-complex vector bundle can be considered as complex vector spaces if we put $\sqrt{-1}\, v = Jv$. We can now prove the following result.

21   Proposition   *For a real finite dimensional vector bundle, the definition of an almost-complex structure is equivalent to the definition of a reduction of the structure group to a complex linear gruop.*

This proposition is meaningful because of the known identification of the n-dimensional general linear group $\text{GL}(n, C)$ (C is the complex field) with a subgroup of $\text{GL}(2n, R)$ (the general real $2n$-dimensional linear group) given by

$$(\alpha_j^i) \mapsto \begin{pmatrix} a_j^i & -b_j^i \\ b_j^i & a_j^i \end{pmatrix}, \quad \alpha_j^i = a_j^i + \sqrt{-1}\, b_j^i.$$

If we have a reduction of the structure group to a complex linear group, the respective bundle has a structure of a complex vector bundle; hence, as we have already remarked, it has an almost complex structure.

Conversely, let J be an almost-complex structure on E. From $J^2 = -I$, we have $(\det J)^2 = (-1)^m$, where m is the dimension of the fiber of E. It follows that $m = 2n$. The local fibers of E have an n-dimensional complex structure, and we can find bases of local cross sections giving at each point $x \in X$ a complex basis for E_x. If s_α is such a basis, s_α and Js_α is a real basis of local cross sections, and by using their associated charts we get an atlas with transition functions in $GL(n, C)$.   Q.E.D.

Note that the transition functions of an almost-complex bundle are matrices with positive determinant because $GL(n, C) \subseteq GL^+(2n, R)$, the special linear group.

22 Definition Let $p: E \to X$ be a real finite dimensional vector bundle. If its structure group is reducible to $GL^+(m, R)$, E is called an *orientable bundle* and a reduction of the structure group is an *orientation* of E.

Hence, we have proved the following result.

23 Corollary *An almost-complex finite dimensional vector bundle is an orientable bundle.*

An important problem is that of constructing classifying spaces and universal bundles for vector bundles. We do not study this problem here; we shall only say that, for paracompact Hausdorff base spaces, the classifying space of the real (complex) vector bundles with n-dimensional fiber is the so-called Grassmann manifold. For a treatment of this matter, see, for instance, [24].

4 Operations with Vector Bundles. Characteristic Classes

In this section, we use Theorem 3.4 in order to get new operations with vector bundles. Recall that one such operation was the direct sum.

We shall denote in the category B of Banach spaces $\mathfrak{M}(V, V') = L(V, V')$. As known, $L(\cdot, \cdot)$ is a functor which is contravariant in the first argument and covariant in the second and takes values in B [9]. This functor is obviously continuous and, hence, its action extends to vector bundles. It follows that for two objects E, E' of $Sv(X)$ (the notation is, generally, that of the previous section) we can introduce the bundle $L(E, E')$, denoted also by $\mathrm{Hom}(E, E')$, with the local fibers $L(E_x, E'_x)$ (of course this bundle is not the same as $\mathfrak{M}(E, E')$ in $Sv(X)$). In particular, $\mathrm{Hom}(E, X \times K)$ is denoted by E^* and is called the *dual bundle* of E. The bundle E^{**} is the *bidual bundle* of E and, if it identifies with E, the bundle E is called *reflexive*. For example, the finite dimensional bundles and the bundles whose fibers are Hilbert spaces are reflexive.

By generalizing the previous operation, we are led to associate functorially to n bundles of $Sv(X)$, $E_1, \ldots, E_n$, the bundle $L(E_1^* \times \cdots \times E_n^*, X \times K)$ whose fibers are the respective sets of continuous multilinear mappings. This will be denoted by $E_1 \otimes^* E_2 \otimes^* \cdots \otimes^* E_n$ and is called the *tensor product* of the given bundles. Its elements are called *tensors*. This definition of tensor products suffices for differential geometry, and in the case of finite dimensional bundles it is isomorphic with the classically defined tensor product $E_1 \otimes \cdots \otimes E_n$, obtained by applying Theorem 3.4 to the usual tensor product functor for vector spaces.

If in the previous construction we have $E_1 = \cdots = E_h$, $E_{h+1} = \cdots = E_n = E_1^*$, we get a bundle whose elements are called *tensors of type* $(n - h, h)$ of the bundle E_1. Remark that there are several spaces of tensors of the same type, which are obtained by giving different places to the spaces E_1 and E_1^* in the previous product.

If we consider the functor which assigns to the Banach space V the space of the alternate multilinear continuous maps

$$\underbrace{V^* \times \cdots \times V^*}_{p \text{ times}} \to K$$

and if we use Theorem 3.4, we get for every bundle E a bundle $\wedge^{*p}E$ called the *pth exterior power of* E. In the case of finite dimensional bundles this is isomorphic to the classically defined pth exterior power $\wedge^p E$ obtained from the exterior product functors of vector spaces. The elements of $\wedge^{*p}E$ are called *skew-symmetric tensors* of E and the cross sections of this bundle are called *exterior forms*.

In the finite dimensional case, it is simple to see which are the transition functions of the bundles obtained by the operations mentioned above.

4. Operations with Vector Bundles. Characteristic Classes

First, we can always consider the fibers to be the arithmetical spaces K^n and the structure group the general linear group $GL(n, K)$. The transition functions $g_i^j(u)$ are matrices $((g_j^i)_\alpha^\beta(u))$ of order n and with non-vanishing determinant, whose elements are continuous K-valued functions on the intersections $U_i \cap U_j \neq \varnothing$ of the coordinate neighborhoods.

In order to get the needed results, we have only to use the remark, made in the proof of Theorem 3.4, that the transition functions of the bundle which is obtained by a functor Φ are $\Phi(g_i^j)$, g_i^j being the transition functions of the given bundle.

From well known facts of linear algebra it follows that if E and E' are vector bundles on X whose fibers have respectively m and n dimensions and whose transition functions are g_i^j and $g_i'^j$, then the transition functions of the Whitney sum are the matrices

$$\begin{pmatrix} g_i^j(u) & 0 \\ 0 & g_i'^j(u) \end{pmatrix} \qquad (1)$$

It follows that also the transition functions of the dual bundle E^* are ${}^t(g_i^j(u))^{-1}$, where t means the transpose of the matrix.

The transition functions of the tensor product $E \otimes E'$ are $g_i^j(u) \otimes g_i'^j(u) = ((g_i^j)_\alpha^\beta (g_i'^j)_\lambda^\mu)$, i.e., the tensor or the Kronecker product of matrices.

Finally, the transition functions of the exterior power $\wedge^p E$ are given by the matrices whose elements are the minors of order p of the matrices g_i^j (the p-compounds of g_i^j). In particular, if the fiber of E is K^n, $\wedge^n E$ has as transition functions the determinants of the matrices g_i^j and, hence, its structure group is the multiplicative group K^* of the nonzero elements of K. Such bundles are called *line bundles* and $\wedge^n E$ is called the *canonical line bundle* of E.

We make also two general remarks. First, all the properties of the operations with vector spaces which are compatible with the respective functor Φ are also valid for the corresponding operations with vector bundles. Second, the operations with vector bundles obtained by using continuous functors are compatible with bundle isomorphisms, i.e., if we replace the bundles by isomorphic bundles the results are also isomorphic. This last remark is a consequence of the fact that the transition functions of the resulting bundle depend only on the transition functions of the given bundles and, of course, on Φ. Hence, we can consider the respective operations as performed on the cohomology sets $H^1(X, GL(n, K)_c)$ with different n; the previously introduced names of the operations are preserved.

The operations introduced give different algebraic structures on $Sv(X)$, which define functors on Top or on subcategories of Top, and play an important role in geometrical and topological problems. We shall shortly consider such problems in connection with the Whitney sum and the tensor product.

In the sequel, the vector bundles are determined up to an isomorphism and, moreover, two vector bundles will be identified whose fibers are isomorphic (and the respective isomorphism commutes with the transition functions). The classes of isomorphic bundles, equivalent in the previous sense, define a set denoted by $[Sv(X)]$, and the operations with vector bundles will be considered as operations on this set.

First, we shall consider only the Whitney sum. The set $[Sv(X)]$ together with the sum is an Abelian semigroup with unit and the correspondence which assigns $[Sv(X)]$ to X is a contravariant functor from the category of topological spaces to the category of Abelian semigroups (see Proposition 2.2). Let A be another contravariant functor between the same categories. We say that A is a *subfunctor* of $[Sv]$ if, for every space X, we have $A(X) \subseteq [Sv(X)]$ as a subsemigroup and if, for $f: X \to Y$, $A(f)$ is the restriction of $f^* = [Sv](f)$. Finally, let B be another, arbitrary, contravariant functor from Top to the category of semigroups. Then, we introduce the following important notion.

1 Definition   A functorial morphism κ from a subfunctor A of $[Sv]$ to a contravariant functor B: Top $\to$ Semigroups is called a *characteristic class*.

This means that for every space X we must have a semigroup homomorphism

$$\kappa_X : A(X) \to B(X)$$

commuting with the morphisms induced by continuous maps. If $\xi \in A(X) \subseteq [Sv(X)]$, $\kappa_X(\xi) = \kappa(\xi)$ (when no confusion arises) is called the characteristic class κ of ξ, or the characteristic class $\kappa(E)$ when $\xi = [E]$, E being a vector bundle. Since κ is a semigroup homomorphism, we have

$$\kappa(\xi \oplus \eta) = \kappa(\xi) \cdot \kappa(\eta), \qquad (2)$$

where, in the second member we have the product in $B(X)$. Since κ is a functorial morphism, if $f: X \to Y$ we have

$$\kappa(f^*\xi) = B(f)(\kappa(\xi)). \qquad (3)$$

4. Operations with Vector Bundles. Characteristic Classes

Hence, (2) and (3) are the defining conditions of characteristic classes.

The most important characteristic classes are obtained if B is a cohomology functor. Now, we shall consider such classes.

Consider first the case when the subfunctor A of Definition 1 assigns to a space X the classes of finite dimensional complex vector bundles which are Whitney sums of line bundles. We call such bundles *completely reducible* and denote by $Svr(X, C)$ their category. The bundles of $Svr(X, C)$ are generated by line bundles and are characterized by the fact that their structure groups are reducible to the group of diagonal matrices. Moreover, we shall consider that X is a paracompact Hausdorff space, which means that in Definition 1, Top is replaced by a full subcategory.

Consider the exact sequence of sheaves on X

$$0 \to J \to C_c \xrightarrow{e} C_c^* \to 0, \qquad (4)$$

where C is the additive group of complex numbers, J is the constant sheaf of the additive group of the integers, C^* is the multiplicative group of nonzero complex numbers, and e is the map which sends the function f into $\exp 2\pi i f$. Writing down the corresponding exact cohomology sequence we get an isomorphism

$$c_1 : H^1(X, C_c^*) \to H^2(X, J) \qquad (5)$$

since C_c is a fine sheaf.

Now, if $\xi \in [Svr(X, C)]$ is a line bundle it can be identified, as shown by Theorem 2.3, with an element of $H^1(X, C_c^*)$, denoted again by ξ, and we assign to it the element of the cohomology ring $H^*(X, J)$ given by

$$c(\xi) = 1 + c_1(\xi). \qquad (6)$$

Finally, let $\xi = \xi_1 \oplus \cdots \oplus \xi_r$ be an arbitrary element of $[Svr(X, C)]$ where $\xi_1, \cdots, \xi_r$ are line bundles. We assign to ξ the element of $H^*(X, J)$ given by

$$c(\xi) = \prod_{h=1}^{r} (1 + c_1(\xi_h)), \qquad (7)$$

where, on the right-hand side, the operation is the cup product.

Obviously, c given by (7) defines a functorial morphism

$$c : [Svr(\cdot, C)] \to H^*(\cdot, J), \qquad (8)$$

hence, a characteristic class. The functor $H^*(\cdot, J)$ is the cohomology ring with the cup product as operation; the semigroup $H^*(X, J)$ is not Abelian but c_X takes values in an Abelian subsemigroup.

The introduced characteristic class is called the *Chern class*. More precisely, $c(\xi)$ is called the *total Chern class* of the bundle ξ. Formula (7) shows that it has the form

$$c(\xi) = \sum_{h=0}^{r} c_h(\xi), \qquad (9)$$

where $c_h(\xi) \in H^{2h}(X, J)$ and is called the *hth Chern class* of ξ. $c_0(\xi) = 1$.

These Chern classes are called *integral Chern classes* because we used the cohomology with integral coefficients. If G is a module which extends J, the inclusion $i: J \subseteq G$ induces the cohomology homomorphism i^* and i^*c is also a characteristic class: the *G-Chern class*; $i^*c_h(\xi)$ are the corresponding hth Chern classes. For instance, we have rational, real, complex, etc., Chern classes.

We mention without proof the following very important result: *there is a unique extension of the functorial morphism* (8) *to the subfunctor of* $[Sv]$ *defined by all the finite dimensional complex vector bundles* (X *is also considered to be Hausdorff and paracompact*). Hence, we can talk of Chern classes for any finite dimensional complex vector bundle. They are integral (real, etc.) cohomology classes, uniquely determined by the following axioms.

A1. *For every isomorphism class ξ of finite dimensional complex vector bundles on X, $c_h(\xi) \in H^{2h}(X, J)$ ($h = 0, 1, \ldots, r = \dim$ of the fiber of ξ) and $c_0(\xi) = 1$. We set*

$$c(\xi) = \sum_{h=0}^{r} c_h(\xi)$$

and this is the total Chern class.

A2. *For every continuous map $f: X \to Y$, $c(f^*\xi) = f^*(c(\xi))$.*

A3. $c(\xi \oplus \eta) = c(\xi)c(\eta)$ *(cup product)*.

A4. *If ξ is a line bundle, $c_1(\xi)$ is defined by the isomorphism* (5).

In the fifth chapter of this book we prove this result for the real Chern classes of differentiable bundles. A general proof can be found in [24].

Formula (7) suggests a general algorithm for calculations with Chern classes [23]. It consists in considering a formal factorization of the total Chern class

4. Operations with Vector Bundles. Characteristic Classes

$$c(\xi) = \prod_{h=1}^{r}(1 + \gamma_h), \qquad (10)$$

which means to express the Chern classes as the elementary symmetric functions on some symbols γ_h. Then, every symmetric function on the γ_h will have a definite meaning, because it can be expressed as a function of the $c_h(\xi)$. For instance, considering first the case of completly reducible bundles and using this algorithm, we get

$$\begin{aligned}
c(\xi \oplus \xi') &= \prod_{h=1}^{r}(1 + \gamma_h)\prod_{k=1}^{r'}(1 + \gamma'_h), \\
c(\xi \otimes \xi') &= \prod_{h,k=1}^{r}(1 + \gamma_h + \gamma'_k), \\
c(\xi^*) &= \prod_{h=1}^{r}(1 - \gamma_h), \\
c(\wedge^p \xi) &= \prod_{0 \leq i_1 < \cdots < i_p \leq r}(1 + \gamma_{i_1} + \cdots + \gamma_{i_p})
\end{aligned} \qquad (11)$$

The first of these formulas is obvious and the others are, in the completely reducible case, a consequence of the isomorphism (5) and of the facts that the product operation in $H^1(X, C_c^*)$, which is a group since C^* is commutative, is just the operation of bundle tensor product (see for this the expression of the transition functions of a tensor product of bundles) and that the inverse of an element of $H^1(X, C_c^*)$ is the dual line bundle (see again the transition functions).

From the third formula of (11) we get the interesting result

$$c_h(\xi^*) = (-1)^h c_h(\xi). \qquad (12)$$

Now, let α be a finite dimensional real vector bundle and $\alpha^c = \alpha \otimes (X \times C)$ (C is the complex field) its *complexification*. Since X is Hausdorff and paracompact, we can reduce the structure group of α to the orthogonal group and, by means of the corresponding transition functions, we get $(\alpha^c)^* = \alpha^c$. By applying (12) we get

$$2c_{2k+1}(\alpha^c) = 0. \qquad (13)$$

As for the classes $c_{2k}(\alpha^c)$, they are used to define the cohomology classes

$$p_k(\alpha) = (-1)^k c_{2k}(\alpha^c), \qquad k = 1, \ldots, [n/2], \qquad (14)$$

where $n = $ dim of the fiber of α. We have $p_k(\alpha) \in H^{4k}(X, J)$ and they are called the *integral Pontrjagin classes* (analogously we have rational and real Pontrjagin classes) of the bundle α. The *total Pontrjagin class* is defined by $p(\alpha) = \sum_k p_k(\alpha)$ and, in general, it is not a characteristic class in the sense of Definition 1, but it plays an important role in many geometrical and topological problems. Because in some cases it is a characteristic class, it is always called one, by an extension of Definition 1. The formal factorization algorithm, given by Formula (10), is used for calculations with Pontrjagin classes too.

Let ξ be a complex vector bundle with complex r-dimensional fiber. By the inclusion $i: \text{GL}(r, C) \subseteq \text{GL}(2r, R)$, ξ is also a real vector bundle with real $2r$-dimensional fiber. The *Euler class* of the bundle ξ is defined by

$$e(\xi) = c_r(\xi). \tag{15}$$

Since from (10) we have

$$c_r(\xi) = \prod_{h=1}^{r} \gamma_h$$

we see that $e(\xi)$ satisfies the conditions (2) and (3) and, hence, is a characteristic class in the sense of Definition 1. It is an important fact that this class can be uniquely extended to all the real orientable vector bundles [24], and in the fifth chapter we shall see this for the differentiable case. We state the following simple but important, result about the Euler class.

2 Proposition *If the bundle ξ has a trivial complex subbundle η, then $e(\xi) = 0$.*

In fact, we have then $\xi = \eta \oplus \lambda$ and $e(\xi) = e(\eta)e(\lambda)$. But, η being trivial, we have by (5) $c(\eta) = 1$, hence $c_n(\eta) = 0$, i.e., $e(\eta) = 0$. Q.E.D.

Using the extension of the Euler class to orientable bundles one gets that for the existence of a real trivial subbundle η of ξ it is again necessary to have $e(\xi) = 0$, because, for a trivial bundle η, $e(\eta) = 0$. Hence, *for ξ to have a global, nowhere zero cross section it is necessary that $e(\xi) = 0$*, since such a section is a trivial line subbundle of ξ.

The method used for introducing the Chern classes can also be used to obtain other important characteristic classes. Consider the subfunctor $[Svr(\cdot, R)]$ of $[Sv(\cdot)]$ of the real finite dimensional completely reducible vector bundles. Since the base space is paracompact and Hausdorff, they

4. Operations with Vector Bundles. Characteristic Classes

have the structure group reducible to the orthogonal group. The obvious isomorphism $\rho: O(1) \to J_2$ ($O(n)$ is the n-dimensional orthogonal group and J_2 is the group of integers modulo 2) induces an isomorphism

$$w_1: H^1(X, O(1)_c) \to H^1(X, J_2). \tag{16}$$

Now, we can assign to every object $\xi = \xi_1 \oplus \cdots \oplus \xi_r$ of $[Svr(X, R)]$ the cohomology class

$$w(\xi) = \prod_{h=1}^{r} (1 + w_1(\xi_h)) \in H^*(X, J_2), \tag{17}$$

called the *total Stiefel–Whitney (or Whitney) class of the bundle* ξ. Obviously this is a characteristic class in the sense of Definition 1. Formula (17) associates with the bundle ξ the cohomology classes

$$w_k(\xi) \in H^k(X, J_2) \tag{18}$$

which appear in the development of the product of that formula. They are called the Stiefel–Whitney classes of ξ.

As in the case of the Chern classes, the Stiefel–Whitney classes can be extended to all the real finite dimensional vector bundles over a paracompact Hausdorff basis, and they are uniquely determined by the following axioms

1. $w_0(\xi) = 1$, $w_k(\xi) \in H^k(X, J_2)$, ($k = 1, \ldots, r = \dim$ *of the fiber of* ξ).
2. *If* $f: X \to Y$, $w(f^*\xi) = f^*(w(\xi))$.
3. $w(\xi \oplus \eta) = w(\xi) w(\eta)$ *(cup product)*.
4. *If* ξ *is a line bundle,* $w_1(\xi)$ *is given by the isomorphism* (16).

We can use the formal factorization algorithm and get Formulas (11) for Stiefel–Whitney classes also. As a consequence, one can establish that if ξ is an orientable bundle $w_1(\xi) = 0$ (by using the fact that the orientability implies that the canonical line bundle of ξ is trivial).

Now, denote by Vect X the subfunctor of $[Sv(X)]$ obtained by taking only finite dimensional vector bundles. Then, besides the Whitney sum, we can consider in a consistent manner the tensor product and, with these two operations, Vect X is a semiring.

3 Definition A *character* is a functorial morphism $\sigma: A \to B$, where A is a subfunctor of Vect X and B is a contravariant functor Top $\to$ Semirings.

To obtain the most important example, suppose that the scalar field is the complex field and that X is Hausdorff and paracompact. Let $\xi \in$ Vect X and

$$c(\xi) = \prod_{1}^{r}(1 + \gamma_i)$$

be its Chern class. The *total Chern character* of ξ is defined by

$$\operatorname{ch} \xi = \sum_{1}^{r} e^{\gamma_i} \in H^*(X, J) \otimes Q, \qquad (19)$$

where Q is the rational field and the right-hand side means that we must consider the formal development of e^{γ_i} as a power series and, next, express the terms using the elementary symmetric functions on γ_i, which are the Chern classes of ξ. We get

$$\operatorname{ch} \xi = r + \sum_{k=1,2,\ldots} \operatorname{ch}_k \xi, \qquad (20)$$

where

$$\operatorname{ch}_k \xi = \frac{1}{k!} \sum_{1}^{r} \gamma_h^k \in H^{2k}(X, J) \otimes Q$$

is the *k*th Chern character of ξ.

From Formulas (11) it follows easily that ch is a character as defined in Definition 3, B being the functor $H^*(., J) \otimes Q$. By the inclusion $J \subseteq R$, we get the *real Chern character*, which is in $H^*(X, R)$.

Now, consider Vect X, over the complex field, for a compact Hausdorff space X. Since it is a semiring we can consider the ring defined as its minimal extension. This ring is the initial object of the category whose objects are the semiring homomorphisms α: Vect $X \to I$ for rings I. It is determined up to an isomorphism, and is denoted by $K(X)$, and it is called the *Grothendieck–Atiyah–Hirzebruch ring* of the space X.

There are two principal constructions of $K(X)$. The first is to consider the free Abelian group generated by Vect X and to take its quotient by

4. Operations with Vector Bundles. Characteristic Classes

the subgroup generated by the elements of the form $\xi + \eta - (\xi \oplus \eta)$. Next, the product will be induced by the tensor product.

The second is the method used in the construction of the integers beginning with the natural numbers. Take Vect X × Vect X and define on it the equivalence generated by the binary relation $(\xi, \eta) \sim (\xi', \eta')$ if $\xi \oplus \eta' = \xi' \oplus \eta$. The set of equivalence classes $\{[\xi, \eta]\}$ is made into a ring by

$$[\xi, \eta] + [\xi', \eta'] = [\xi \oplus \xi', \eta \oplus \eta'], [\xi, \eta] \cdot [\xi', \eta']$$
$$= [(\xi \otimes \xi') \oplus (\eta \otimes \eta'), (\xi \otimes \eta') \oplus (\eta \otimes \xi')]$$

and this is just the ring $K(X)$.

Now, every element of $K(X)$ can be represented as

$$[\xi, \eta] = [\xi] - [\eta], \tag{21}$$

where, if $\xi \in \text{Vect } X$, we put $[\xi] = [\xi \oplus \lambda, \lambda] = \alpha(\xi)$, where λ is arbitrary and $\alpha \colon \text{Vect } X \to K(X)$. If θ denotes an arbitrary trivial bundle, we can consider, as shown in Proposition 3.11, a bundle η such that $\eta \oplus \eta' = \theta$, and it follows that

$$[\xi, \eta] = [\xi \oplus \eta', \eta \oplus \eta'] = [\xi', \theta] = [\xi'] - [\theta]. \tag{22}$$

It is interesting to see when $[\xi] = [\eta]$. This means $(\xi \oplus \lambda, \lambda) \sim (\eta \oplus \mu, \mu)$, i.e., $\xi \oplus (\lambda \oplus \mu) = \eta \oplus (\lambda \oplus \mu)$, and by adding ν such that $(\lambda \oplus \mu) \oplus \nu = \theta$ we find $\xi \oplus \theta = \eta \oplus \theta$.

Generally, if there are two trivial bundles θ and θ' such that $\xi \oplus \theta = \eta \oplus \theta'$, the bundles ξ and η are called *stable equivalent* (one shows that this is really an equivalence relation). The Chern classes and characters are the same for two stable equivalent bundles.

Obviously K is a contravariant functor on the category of compact topological spaces. It is the starting point of a recent topological theory of Atiyah and Hirzebruch called *K-theory*, which proves to be very useful in applications. For an exposition of this theory the reader is referred to [2]. If the functor B of Definition 3 takes values in the category of rings, the respective character extends to $K(X)$. This is the case for instances with the Chern character.

Chapter 4

DIFFERENTIAL GEOMETRY

1 Differentiable Manifolds

In this chapter, we give a brief account of matters of differential geometry which are to be used later on considering that these problems are not new for the reader. We begin with a section on differentiable manifolds, which, generally, are considered as being modeled on Banach spaces (which is a "regular enough" case). The Banach spaces considered in the sequel are supposed to be real spaces. Remember that a Banach space is viewed as a space with a class of equivalent norms (2.3). References for this chapter are to [25, 27, 35, 37, 45], etc.

In Section 1.5, we considered a derivative for continuous functions $f: U \to F$, where $U \subseteq E$ is open and E, F are locally convex topological vector spaces. Now, we consider this derivative for the case when E and F are Banach spaces. Then it is easy to see that the definition given in Section 1.5 can be expressed as

1 Definition   Let E and F be Banach spaces and let $f: U \to F$ be a continuous function defined on the open subset $U \subseteq E$. If for a given point $x_0 \in U$ there is a linear mapping $\lambda: E \to F$ such that

$$\lim_{y \to 0} \frac{\|f(x_0 + y) - f(x_0) - \lambda(y)\|}{\|y\|} = 0, \tag{1}$$

f is called *differentiable* at x_0 and λ is called the *derivative* of f at the same point and is denoted by f'_{x_0}.

Note that the equivalence of Definition 1 to the definition given in Section 1.5 shows that the derivative does not change if the norms are replaced by equivalent norms. Also we see that the derivative depends only on the germ of f at x_0. An immediate consequence of (1) is that the derivative, when it exists, is unique. Using (1) again, we get easily that the derivative of a constant function is zero, the derivative of a linear continuous map is just the respective map, and f'_{x_0} is always continuous, i.e., it belongs to $L(E, F)$ in the category B of Banach spaces, introduced in Section 3.3. This last property follows from the fact that, for small enough $\|t\|$, we have

$$\|f(x_0 + t) - f(x_0)\| \leq \varepsilon/2,$$
$$\|f(x_0 + t) - f(x_0) - f'_{x_0}(t)\| \leq \varepsilon\|t\|/2,$$

where $\varepsilon > 0$ is given and hence $\|f'_{x_0}(t)\| \leq \varepsilon$.

Two other important results are easy consequences of Definition 1.

(a) *If $\alpha : E \times F \to G$ is a bilinear continuous map it has a derivative at every point (x, y) and this derivative is given by $\alpha'_{(x,y)}(s, t) = \alpha(x, t) + \alpha(s, y)$;*
(b) *a function $f : E \to F_1 \times \cdots \times F_n$ is differentiable if and only if the functions $f_i = \mathrm{pr}_{F_i} \circ f$ are and $f'_x = f'_{1x} \times \cdots \times f'_{nx}$*. One uses the fact that the norm of the product is the supremum of the norms of the projections and, for (a), that $\|\alpha(x, y)\| \leq c\|x\| \|y\|$ ($c = \text{const.}$).

Now, we establish the rule for calculating the derivative of a composite function.

2 Proposition Let E, F, G be three Banach spaces, $A \subseteq E$ and $B \subseteq F$ open subsets of E and F, $f : A \to F$, $g : B \to G$ continuous functions, $x_0 \in A$, $y_0 = f(x_0) \in B$, and suppose that the derivatives f'_{x_0} and g'_{y_0} exist. Then $(g \circ f)'_{x_0}$ exists and we have

$$(g \circ f)'_{x_0} = g'_{y_0} \circ f'_{x_0}. \tag{2}$$

From the definition of the derivative it follows that given ε such that $0 < \varepsilon < 1$, there exists $\delta > 0$ such that for $\|s\| \leq \delta$, $\|t\| \leq \delta$ we have

1. Differentiable Manifolds

$$f(x_0 + s) = f(x_0) + f'_{x_0}(s) + \varphi_1(s), \qquad \|\varphi_1(s)\| \leq \varepsilon \|s\|,$$
$$g(y_0 + t) = g(y_0) + g'_{y_0}(t) + \varphi_2(t), \qquad \|\varphi_2(t)\| \leq \varepsilon \|t\|.$$

Next, from the continuity of f'_{x_0} and g'_{y_0} we have

$$\|f'_{x_0}(s)\| \leq a\|s\|, \qquad \|g'_{y_0}(t)\| \leq b\|t\|, \qquad a, b = \text{const.};$$

hence, for $\|s\| \leq \delta$,

$$\|f'_{x_0}(s) + \varphi_1(s)\| \leq (a + 1)\|s\|.$$

Then, for $\|s\| \leq \delta/(a + 1)$ we have

$$\|\varphi_2(f'_{x_0}(s) + \varphi_1(s))\| \leq (a + 1)\varepsilon\|s\|,$$

and

$$\|g'_{y_0}(\varphi_1(s))\| \leq b\varepsilon\|s\|,$$

whence, putting $g \circ f = h$:

$$h(x_0 + s) = g(y_0) + g'_{y_0}(f'_{x_0}(s)) + \varphi_3(s),$$

$\|\varphi_3(s)\| \leq (a + b + 1)\varepsilon\|s\|$, which proves Proposition 2.

As an immediate consequence, if the derivatives of the righthand sides exist,

$$(f + g)'_{x_0} = f'_{x_0} + g'_{x_0}, \qquad (\alpha f)'_{x_0} = \alpha f'_{x_0}, \qquad \alpha = \text{const.} \qquad (3)$$

3 Proposition *Let E and F be Banach spaces, $A \subseteq E$, $B \subseteq F$ open subsets and $f: A \to F$ which maps A onto B homeomorphically. If f is differentiable at $x_0 \in A$ and if f'_{x_0} is an isomorphism, f^{-1} is differentiable at $f(x_0)$ and*

$$(f^{-1})'_{f(x_0)} = (f'_{x_0})^{-1}. \qquad (4)$$

From the hypotheses it follows that the map φ given by $\varphi(s) = f(x_0 + s) - f(x_0)$ maps a neighborhood V of the origin of E homeomorphically onto a neighborhood W of the origin of F and its inverse

ψ is given by $\psi(t) = f^{-1}(y_0 + t) - f^{-1}(y_0)$ $(y_0 = f(x_0))$. Next, $f'^{-1}_{x_0}$ exists and is continuous, hence $\|f'^{-1}_{x_0}(t)\| \leq c\|t\|$. Consider $0 < \varepsilon < \frac{1}{2c}$ and

$$\varphi(s) = f'_{x_0}(s) + \varphi_1(s), \qquad \|\varphi_1(s)\| \leq \varepsilon\|s\|$$

for $s \in V$ and $\|s\| \leq \delta$. Choose $\delta' > 0$ such that $\{t \mid \|t\| \leq \delta'\} \subseteq W$ and the image of this set by ψ is contained in the ball $\|s\| \leq \delta$. Then, $\|t\| \leq \delta'$ implies

$$t = \varphi(\psi(t)) = f'_{x_0}(\psi(t)) + \varphi_1(\psi(t)),$$

where $\|\varphi_1(\psi(t))\| \leq \varepsilon\|\psi(t)\|$. It follows that

$$f'^{-1}_{x_0}(t) = \psi(t) + g(t)$$

with

$$g(t) = f'^{-1}_{x_0}(\varphi_1(\psi(t)))$$

and

$$\|g(t)\| \leq c\varepsilon\|\psi(t)\| \leq \tfrac{1}{2}\|\psi(t)\|.$$

Hence

$$\|f'^{-1}_{x_0}(t)\| \geq \|\psi(t)\| - \tfrac{1}{2}\|\psi(t)\| = \tfrac{1}{2}\|\psi(t)\|$$

and $\|\psi(t)\| \leq 2c\|t\|$. But, this means that $\|t\| \leq \delta'$ implies

$$\psi(t) = f'^{-1}_{x_0}(t) - g(t)$$

with

$$\|g(t)\| \leq 2c^2\varepsilon\|t\|,$$

which proves the proposition.

For a function $f: U \to F$ ($U \subseteq E$), if f'_{x_0} exists for every $x_0 \in U$ and $x_0 \mapsto f'_{x_0}$ is a continuous mapping $U \to L(E, F)$, f is said to be of class C^1 on U. By iteration we can now obtain the higher-order derivatives $f^{(p)}_{x_0}$ and define functions of class C^p and C^∞. For the sake of simplicity, we shall

1. Differentiable Manifolds

always consider that differentiable means C^∞. The derivative $f_{x_0}^{(p)}$ can be also considered as a multilinear mapping

$$\underbrace{E \times \cdots \times E}_{p \text{ times}} \to F,$$

and this mapping is seen to be symmetrical. Finally, we remark that it is possible to introduce, in the usual manner, the partial derivatives [9].

4 Proposition *Let E and F be Banach spaces, $A \subseteq E$ an open subset and $f: A \to F$ a differentiable function. If at a point $x_0 \in A$, f'_{x_0} is an isomorphism, there is an open neighborhood of x_0 in A such that the restriction of f to this neighborhood is a homeomorphism with differentiable inverse.*

By replacing f with $g = (f'_{x_0})^{-1} \circ f$, we see that it suffices to prove the theorem in the case $F = E$, $f'_{x_0} = \text{id}$ and also $x_0 = f(x_0) = 0$. In this case, if x and y belong to a suitable neighborhood of 0, the map $y + x - f(x)$ (y fixed) is a contractible map of some closed subset of E. Then, the proposition follows from the classical fixed point theorem of contractible maps of complete metric spaces. For the details of this proof see, for instance, [27].

For a complete exposition of the differential calculus in Banach spaces, see [9].

From the previous considerations, the following can be derived.

5 Theorem *The open subsets of Banach spaces as objects and the differentiable partial maps as morphisms make up a pseudocategory, which we denote by B^p and which has a derivative in the category B of Banach spaces and linear continuous maps.*

In fact, the identity map is its own first derivative at every point, and its higher-order derivatives are zero, i.e., it is of class C^∞. Next, from Proposition 2 it follows that by the composition of two C^∞ functions we get again a C^∞ function. Hence B^p is a pseudocategory. The B-valued derivative functor on B^p is obtained by assigning to a pair (U, e) ($e \in U \subseteq E$) the space E and to a germ $[f]_e$ ($f: U \to V$, $V \subseteq F$) the derivative $f'_e: E \to F$. We remark that Proposition 4 offers a condition for a morphism of the local category $\mathfrak{L}(B^p)$ (Section 1.2) to have a local inverse.

From Theorem 5 it follows that for every Banach space E we can consider the full subpseudocategory E^p of B^p whose objects are the open subsets of E. E^p has the induced derivative functor. Of course, this is also true if E is a Hilbert space.

Now, we can define the differentiable manifolds modeled on E. This is a fundamental object of modern mathematics and, for our book, it is the object of study from the viewpoint of the cohomology problems.

6   Definition   A *differentiable manifold with model E* is an E^p-manifold as defined in 1.6.8. If E is a Banach space, the manifold is called a *Banach manifold*, and if E is a Hilbert space, it is a *Hilbert manifold*. Finally, if E is finite dimensional, the manifold is *finite dimensional* and its *dimension* is the dimension of E.

Obviously, if needed, the class of differentiability can be taken C^p instead of C^∞.

Hence, a differentiable manifold is a set V together with an E^p-structure on it, given by an E^p-atlas, i.e., a covering with open coordinate neighborhoods $\{U\}$ endowed with local charts $h_U \colon U \to \mathscr{A}(U) \subseteq E$, which are bijections with the property that in $U \cap U' \neq \emptyset$ the transition functions $a_{U'}^U = h_{U'} \cdot h_U^{-1}$ are differentiable. From Proposition 1.6.4 it follows that V has a uniquely determined topology which makes the atlas topological, and we could start with a topological space V. Though many problems can be treated in the general case we shall consider the following restrictive convention.

7   Convention   *All the differentiable manifolds considered in the sequel are Hausdorff spaces.*

Other topological conditions will be imposed when needed. Here we mention only that because of Riesz's classical theorem, which says that every normed linear space which is locally compact is finite dimensional [9], the only locally compact differentiable manifolds are the finite dimensional manifolds. Of course, the compact manifolds are also finite dimensional.

An important result is given by Theorem 1.6.10, which means here that, given differentiable partial maps $a^i_j \colon E \dashrightarrow E$ such that

$$a^i_i = 1_E, \; a^j_k a^i_j = a^i_k,$$

there is a differentiable manifold, uniquely determined up to an isomorphism and not necessarily Hausdorff modeled on E and having a^i_j as its transition functions.

1. Differentiable Manifolds

The classical examples of differentiable manifolds are the spheres, the projective spaces, and, in the infinite dimensional case, different function spaces [12].

As usually for mathematical objects, we shall consider a category giving the framework for the differentiable manifolds. Obviously, the E^p-manifolds are also B^p-manifolds, and we consider the full subpseudocategory of B^p-Var (1.6) whose objects are the E^p-manifolds with different models E. This is, by definition, the pseudocategory Diff^p. Its morphisms are defined as in Section 1.6 and are called *differentiable maps of manifolds*. Similarly, we have the category Diff whose objects are the E^p-manifolds (variable E) and whose morphisms are differentiable maps $\varphi: V \to V'$, defined on the whole of V. Note that in Diff we have the direct product of two manifolds defined as for the general case of $\mathfrak{P}$-manifolds (Section 1.6). The isomorphisms in Diff are called *diffeomorphisms*.

The real field R is of course an object of Diff and, hence, we can consider the R-algebra of the real differentiable functions on V. This will be denoted by $\Phi(V)$ and, correspondingly, we have the sheaf of germs of differentiable functions, denoted by $\mathfrak{F}(V)$.

We have already seen that E^p has a B-valued derivative. Hence, we can consider the tangent object at a point of a differentiable manifold. By the construction of Section 1.6, we easily see that this tangent object exists and, using the notation of Section 1.6, $T_x(V)$ is the inductive limit of the family of Banach spaces $E_U = E$, where the U are the coordinate neighborhoods containing x, and of the linear continuous maps $(a_{U'}^U)'_{h_U(x)}$, the set $\{U\}$ being ordered by the relation: $U \leqslant U'$ for every U and U'. This limit can be constructed as for sets, and consists of classes of vectors of E_U which are such that, within a class,

$$(a_{U'}^U)'_{h_U(x)} u = v, \quad u \in E_U, v \in E_{U'}. \tag{5}$$

Note that the $(a_{U'}^U)'_{h_U(x)}$ are isomorphisms, since they have the inverses $(a_U^{U'})'_{h_{U'}(x)}$. Every such class has a unique representative in every space E_U; hence, there is a bijection between $T_x(V)$ and E which gives $T_x(V)$ the structure of a Banach space isomorphic to E. On $T_x(V)$ we have a class of equivalent norms depending on the coordinate neighborhood U.

In the finite dimensional case, we can always identify the model with a real arithmetic space. The transition functions $a_{U'}^U$ are then expressed as $x_{U'}^i = f^i(x_U^j)$ where f^i are differentiable functions. $(a_{U'}^U)'$ is the linear transformation defined by the Jacobian matrix $(\partial x_{U'}^i / \partial x_U^j)$, and (5) is just the classical transformation law for the components of a contravariant vector. Hence, we recognize the classical definition of the tangent space [45].

The vectors of $T_x(V)$ are called tangent vectors of V at x. From 1.6, we know that the tangent space has functorial character, i.e., if $f: V \to W$ is a morphism in Diff we have the induced morphism $f_x^*: T_x(V) \to T_{f(x)}(W)$, called here the *differential* of f. Hence, Diff^p has a B-valued derivative functor.

Note that the tangent space of a product manifold $V \times W$ at (x, y) ($x \in V$, $y \in W$) is isomorphic to (and, hence, can be identified with) $T_x(V) \times T_y(W)$.

We can consider particular morphisms in Diff characterized by properties of their differential map. This will be done by a slight change of the corresponding definitions of 1.6.

8 Definition Let $f: V \to W$ be a morphism of Diff. Then f is called *an immersion at* $x \in V$ if f_x^* is an injection and im f_x^* is a direct summand in $T_{f(x)}(W)$; f is an *immersion* if it is an immersion at every point of V. Dually, f is a *submersion at* x if f_x^* is a surjection and ker f_x^* is a direct summand in T_x, and f is a *submersion* if it is a submersion at every point. Finally, f is an *imbedding* if it is an immersion and maps V homeomorphically onto $f(V)$ (the last with the relative topology).

If f is an immersion at x, f is also called *regular* at x, and x is called a *regular point* of f. The points which are not regular are called *singular points*. In the finite dimensional case, the conditions regarding the direct summands of the previous definitions are always satisfied. In this case, the rank of the linear map f_x^* is called the *rank* of f at x.

We define the various submanifolds of V. If $W \subseteq V$ and $i: W \to V$ is the inclusion map, W is called a *submanifold* of V if it is also a differentiable manifold and if i is a differentiable map. Generally, i may have singular points; they are called *singular points* of the submanifold. Next, W is an *immersed submanifold* of V if i is an immersion. Then, by Proposition 4, one can establish [27] that W has the following characteristic property: every point $w \in W$ has a coordinate neighborhood U in V such that $W \cap U$ is the union of subsets where the restriction of i defines charts of W with image in the model E_1 of W, E_1 being a direct summand of the model E of V. Finally, W is an *imbedded submanifold* of V, if the inclusion i is an imbedding. For instance, if W is an open subset of V, it can be made into an imbedded submanifold by taking as coordinate neighborhoods $W \cap U$, where U are the coordinate neighborhoods of V.

The set

1. Differentiable Manifolds

$$T(V) = \bigcup_{x \in U} T_x(V).$$

has a natural projection p onto V and, from the definition of the tangent space, we see that, for every coordinate neighborhood U of V, we have a bijection

$$\bar{h}_U : p^{-1}(U) \to U \times E. \tag{6}$$

These bijections are such that

$$p\bar{h}_U^{-1}(u, e) = u, \qquad \bar{h}_{U'}\bar{h}_U^{-1}(u, e) = (u, (a_{UU'}^U)'_{h_U(u)}e).$$

From Section 3.1, it follows that there is a uniquely determined topology on $T(V)$ which makes the triple $p: T(V) \to V$ into a vector bundle with fiber E and transition functions $(a_{UU'}^U)'$ (or $(\partial x_{U'}^i/\partial x_U^j)$ in the finite dimensional case). This topology is obtained by requiring that the bijections (6) be homeomorphisms.

Moreover, composing $\bar{h}_U$ of (6) with $(h_U, 1_E)$ we get bijections $p^{-1}(U) \longrightarrow E \times E$, which are easily seen to determine an $(E \times E)^p$-structure on $T(V)$. Hence, $T(V)$ is a differentiable manifold modeled on $E \times E$ and the projection p as well as the bundle transition functions

$$U \cap U' \to L(E) = L(E, E)$$

are differentiable.

Thus, we are led to give the following definition.

9 Definition A fiber bundle (with structure group) is called *differentiable* if the base space, the total space, the fiber (the structure group), the projection, and the transition functions are differentiable, and the charts of the local trivialization are diffeomorphisms (the group operations and the action of the structure group on the fiber are also differentiable).

For differentiable fiber bundles, the results of Chapter 3 apply, requiring differentiability instead of continuity whenever possible. We then get the categories of differentiable fiber bundles, fiber bundles with structure group, principal bundles, and vector bundles, and we shall use the results given in Chapter 3 without special mention. We state the convention that, if these results are used for differentiable manifolds, then the corresponding bundles are differentiable and, if no confusion arises, we use the same notation as in Chapter 3. Should there be confusion, we add the index d. For instance, Theorem 3.2.3 becomes: there is a functorial isomorphism of the functors $[S_d(X, F, G)]$ and $H^1(X, G_d)$, etc. The extension of the results of Chapter 3 to the differentiable case is left to the reader.

Special mention is to be made of *differentiable functors* on the category of Banach spaces, which are obtained by replacing the continuity conditions in Definition 3.3.3 with the corresponding differentiability conditions. These functors extend to differentiable vector bundles by Theorem 3.3.4 proved in the differentiable case. It is clear that the functor $L(E, F)$ is differentiable and, hence, we can introduce $L(W, W')$ for two differentiable vector bundles W and W' on V. This is again a differentiable vector bundle and it follows that we can consider for the differentiable case the operations with vector bundles introduced in Section 3.4.

Special mention is also made concerning the differentiability of the structure group of a bundle. This must be understood as follows: the group G is also a differentiable Banach manifold and the group operations are differentiable. Such groups are called *Lie-Banach groups* and, in the finite dimensional case, they are the classical *Lie groups*.

The results previously obtained can be applied to get the following result.

10 Theorem *If we assign to every differentiable manifold V its tangent bundle $T(V)$ and to every differentiable map $f: V \to W$ the induced map $f^*: T(V) \to T(W)$ given by the f_x^* ($x \in V$), we get a covariant functor from* Diff *to the full subcategory of $S(Top)$ whose objects are the differentiable vector bundles.*

Remark that the tangent bundle $T(W)$ of an immersed submanifold $i: W \subseteq V$ can be identified with a vector subbundle of the tangent bundle $T(V)$ of V. Then we can consider the *normal bundle* $T(V)/T(W)$ of W.

The sheaf of germs of differentiable cross sections of $T(V)$ is an $\mathfrak{F}(V)$-sheaf module.

The existence of the tangent bundle is of fundamental importance in the theory of differentiable manifolds, since it gives the possibility of *linearizing* the resulting problems.

11 Definition The characteristic classes and the characters which exist for the tangent bundle $T(V)$ (Pontrjagin, Stiefel–Whitney, Euler, etc.) are called *characteristic classes and characters of the manifold V*.

The study of the topology of differentiable manifolds is the object of

1. Differentiable Manifolds

differential topology and we shall not consider its problems here. For this one can see, for instance, the lectures of Palais [35]. But, we need some results regarding one such problem which is important here: the existence of differentiable partitions of unity on differentiable manifolds.

First, we remember the result that for paracompact Hausdorff spaces there are always continuous partitions of unity; this result is valid on paracompact differentiable manifolds because of Convention 7. But this condition does not always assure the existence of differentiable partitions of unity. A classical result can be stated as follows.

12 Theorem *A finite dimensional paracompact differentiable manifold has differentiable partitions of unity subordinated to its open locally finite coverings.*

The proof (which we only sketch here) is based on the fact that, if A is a compact subset of R^m (where R is the real field) and B a closed subset disjoint from A, there is a real differentiable function on R^m which sends A into 1 and B into 0. It follows that if V is a finite dimensional differentiable manifold, and if $C \subseteq D \subseteq V$ such that C is compact and D open, then there is a real differentiable function on V which sends C into 1 and $V - D$ into 0.

Now, let $\{U_\alpha\}$ be an open locally finite covering of V by coordinate neighborhoods and $\{V_\alpha\}$, $\overline{V}_\alpha \subseteq U_\alpha$, another open covering. Take differentiable functions ψ_α which are 1 on $\overline{V}_\alpha$ and 0 on $V - U_\alpha$, and define $\psi = \sum_\alpha \psi_\alpha$. The functions $\varphi_\alpha = \psi_\alpha/\psi$ define the differentiable partition of unity subordinated to $\{U_\alpha\}$ as required.

The results for the infinite dimensional case are given in [27] and [35]. We state them without proof.

13 Lemma *If a paracompact differentiable manifold has differentiable partitions of unity, then every open subset of it has the same property. If every point of such a manifold has an open neighborhood with differentiable partitions of unity, then the manifold has differentiable partitions of unity.*

14 Proposition *Every open subset of a paracompact differentiable manifold is paracompact.*

An immediate consequence of Lemma 13 is given by the following proposition.

15 Proposition *A paracompact differentiable manifold has differentiable partitions of unity if and only if its model has.*

This reduces the problem to the existence of differentiable partitions of unity on Banach spaces. The most important results obtained are for the case of separable spaces, which can be seen to have the property that if their norm is a differentiable function on the subset of nonzero vectors there are differentiable partitions of unity. From this, the following result can be derived.

16 Theorem *The paracompact differentiable manifolds modeled on separable Hilbert spaces have differentiable partitions of unity.*

This last result is also a consequence of a recent theorem of Henderson [22].

17 Theorem (Henderson.) *Every differentiable manifold with separable model can be imbedded as an open subset of a Hilbert space.*

Now, to obtain Theorem 16 we have only to combine Lemma 13 with Theorem 17.

A direct proof of Theorem 16 can be found in [27].

2   Vector and Tensor Fields

The existence of the tangent bundle of a differentiable manifold allows the development of the tensor calculus, which is a very important tool in the study of the manifold.

Let V be a differentiable manifold modeled on the Banach space E and let $T(V)$ be its tangent bundle.

1 Definition A differentiable global cross section of the bundle $T(V)$ is called a *vector field* on V.

Note that there is at least the zero vector field. On the other hand, not every manifold V has a nowhere-zero vector field. For instance, if V is compact and *oriented* (see Definition 10 of Section 3) such a field exists if and only if the Euler-Poincaré characteristic of V vanishes. Finally, if V

2. Vector and Tensor Fields

has differentiable partitions of unity we can use the procedure of the proof of 3.3.13 for the construction of vector fields on V; generally they will have zeros.

We denote by $\mathscr{V}(V)$ the set of vector fields on V. It has an obvious structure of a $\Phi(V)$-module, where $\Phi(V)$ is the algebra of real differentiable functions on V.

2 Definition A *derivation* of an algebra A is a map $D: A \to A$ such that: (a) D is an endomorphism for the module structure of A, (b) $D(\alpha \cdot \beta) = D\alpha \cdot \beta + \alpha \cdot D\beta$.

3 Proposition *Every vector field on V defines a derivation of the algebra $\Phi(V)$. Different vector fields give rise to different derivations.*

Let $\varphi \in \Phi(V)$, $x \in V$ and let (U, h) be a coordinate neighborhood of x and $\chi = h(x)$. We denote by $\varphi_U : E \longrightarrow R$ the coordinate expression of φ on U. Next, let $\xi : V \to T(V)$ be a vector field and u the coordinate representative on U of the vector $\xi(x) = \xi_x$, i.e., the fiber coordinate of ξ_x. If we define

$$(\xi\varphi)(x) = \varphi'_{U,\chi}(u), \tag{1}$$

it is easy to see that the real number obtained does not depend on the choice of the local chart and that it depends only on the germ of φ at x and on ξ_x. If x is variable in V, $(\xi\varphi)(x)$ given by (1) is an element of $\Phi(V)$ which will be denoted by $\xi\varphi$. Hence, (1) defines a map $\xi : \Phi(V) \to \Phi(V)$ associated with the vector field ξ. The properties (a), (b) of Definition 2 follow immediately using the rules of differential calculus given in Section 1. Hence, the first part of the proposition is proved. We observe that, since (1) is defined at every point, we can define $\xi\varphi$ over any open subset of V.

To prove the second part of Proposition 3, it suffices to prove that the derivative defined by a nonzero vector field cannot vanish. Indeed, if $\xi_x \neq 0$, consider on V a function f whose restriction to the coordinate neighborhood U of x is expressed in coordinates as the restriction to $h(U)$ of a linear function $\varphi : E \to R$ such that $\varphi(u) \neq 0$ (u represents ξ_x as before). Such a function φ exists by the Hahn–Banach theorem and, at least locally, we can construct f. Because φ is linear, $\varphi' = \varphi$ and formula (1) gives $(\xi\varphi)(x) \neq 0$.   Q.E.D.

We observe that formula (1) can be used in just the same manner to define an action of ξ on the module $\Phi(V, F)$ of F-valued differentiable functions on V, for any Banach space F.

4 Theorem $\mathscr{V}(V)$ *has an induced structure of a Lie algebra.*

Let $\xi, \eta \in \mathscr{V}(V)$, $x \in V$ and let (U, h) be a coordinate neighborhood of x, on which ξ_x and η_x have fiber coordinates u and $\tilde{u}$, respectively, and $\chi = h(x)$. When x runs over U, u and $\tilde{u}$ can be considered as differentiable maps $u, \tilde{u}: h(U) \to E$, and we can define at every point x a vector given by $\tilde{u}'(u) - u'(\tilde{u})$.

If we go over to another coordinate neighborhood (U', k), $u: h(U) \to E$ is replaced by $v = (a_{U'}^U)'u$ $a_U^{U'}: k(U') \to E$, where $a_{U'}^U$ are the transition functions, and we get by an easy computation

$$\tilde{v}'(v) - v'(\tilde{v}) = (a_{U'}^U)'(\tilde{u}'(u) - u'(\tilde{u})).$$

Hence, the indicated construction defines a tangent vector of V at x and we denote it by $[\xi, \eta]_x$. When x runs over V, we have a vector field denoted by $[\xi, \eta]$ called the *bracket* of the vector fields ξ and η.

We see that the bracket is given by the following formula of the corresponding derivation of $\Phi(V)$:

$$[\xi, \eta]\varphi = \xi(\eta\varphi) - \eta(\xi\varphi). \tag{2}$$

Indeed, considering $\eta\varphi$ as the composite mapping $\chi \mapsto (\tilde{u}_\chi, \varphi'_{U,\chi}) \mapsto \varphi'_{U,\chi}(\tilde{u}_\chi)$, and $\xi\varphi$ as the analogous composite mapping, we have, if we omit the lower indices,

$$\xi(\eta\varphi) - \eta(\xi\varphi))(x) = \varphi''(\tilde{u}, u) + \varphi'\tilde{u}'(u) - \varphi''(u, \tilde{u}) - \varphi'u'(\tilde{u}),$$

from which the result follows, since φ'' is bilinear and symmetrical.

The bracket of vector fields is obviously skew-symmetric and satisfies the Jacobi identity:

$$[\xi, [\eta, \zeta]] + [\eta, [\zeta, \xi]] + [\zeta, [\xi, \eta]] = 0, \tag{3}$$

which can be proved by expressing the first member with Formula (2) and using the second part of Proposition 3.

Defining the product of two elements of $\mathscr{V}(V)$ by their bracket the theorem is proved.

We cannot talk of a functorial character of $\mathscr{V}(V)$ because a differentiable

2. Vector and Tensor Fields

map $f: V \to \tilde{V}$ does not induce a mapping on the vector fields, except for the case when f is a diffeomorphism. However, we have the following useful notion: if $\xi \in \mathscr{V}(V)$ and $\tilde{\xi} \in \mathscr{V}(\tilde{V})$, they are called *f-related* if for every $x \in V$ we have

$$\tilde{\xi}_{f(x)} = f^*_x(\xi_x), \tag{4}$$

i.e., if

$$\xi(\varphi \circ f)(x) = (\tilde{\xi}\varphi)(f(x)), \qquad \varphi \in \Phi(\tilde{V}). \tag{5}$$

Then, we deduce by a straightforward calculation that, if $(\xi, \tilde{\xi})$ and $(\eta, \tilde{\eta})$ are f-related pairs, the brackets $[\xi, \eta]$ and $[\tilde{\xi}, \tilde{\eta}]$ are also f-related. Hence, when f is a diffeomorphism, it induces an isomorphism of the Lie algebras of vector fields of the two manifolds.

An important configuration associated with a vector field is given by the integral curves of the field. Let J be an open interval of the real line which contains the origin. A differentiable map $\omega: J \to V$ is called a (differentiable) *curve* on V, and $\omega(0)$ is called its *origin*. Now, if $t \in J$, and U is a coordinate neighborhood of $\omega(t)$ where ω is expressed by $\omega_U: J \longrightarrow E$, we have $(\omega'_U)_t: R \to E$, and $(\omega'_U)_t(1)$ is a vector in E. It is easy to see that if we consider another coordinate neighborhood the vector obtained is given by the transformation law of the tangent vectors. Hence, for every $t \in J$, we got a vector ω'_t which is tangent to V at $\omega(t)$. This is, by definition, the *tangent vector* of the curve ω at t. Now, let ξ be a vector field on V. The curve ω is called an *integral curve* of ξ if we have $\omega'_t = \xi_{\omega(t)}$ for every $t \in J$. The existence and uniqueness theorem for ordinary differential equations in Banach spaces [27] assures the local existence and uniqueness of the integral curves of a vector field if their origin is given. They depend differentiably on the origin and are given by a differentiable partial map $\omega: R \times V \longrightarrow V$ such that

$$\omega_x(t) = \omega(t, x)(t \in J \subseteq R).$$

Generally, ω extends to some maximal interval $J(x)$ which is R only if some supplementary conditions hold, e.g., if V is compact. It is easy to see that $\omega(s + t, \omega_x(0)) = \omega(s, \omega_x(t))$, when the equality is meaningful, and it follows that the $\omega_t(x) = \omega(t, x)$, which are local diffeomorphisms of V, define a *local one-parameter group* of transformations of V. Because of this property, the vector fields are also called *infinitesimal transformations*.

This local group is a group in some important cases, e.g., if V is compact. We have only sketched these results, since they will not be used later. A full exposition can be found in [27].

If V is a finite dimensional manifold and x^i is a local coordinate system (chart) at $x \in V$, Formula (1) gives the following expression for the action of a vector field on a function (repeated indices are summation indices):

$$\xi\varphi(x) = \xi_x^i \frac{\partial \varphi}{\partial x^i}\bigg|_x. \qquad (6)$$

The $\partial/\partial x^i$ are derivations of the algebra of differentiable functions on the respective coordinate neighborhood and they correspond to the basis vectors $\xi_{(j)}^i = \delta_j^i$ (the Kronecker symbols). For this reason they are identified with the basis vectors and one usually says that on the given coordinate neighborhood $\{\partial/\partial x^i\}$ is the basis of the tangent space, or the *natural basis* of the coordinate system. Then, a vector ξ_x is identified with the operator

$$\xi_x^i \frac{\partial}{\partial x^i}\bigg|_x$$

and a vector field ξ over the coordinate neighborhood is identified with $\xi^i \, \partial/\partial x^i$. Moreover, in this case one can show that every derivation of $\Phi(V)$ is given by such an operator, and that we can identify the vector fields with the derivations of $\Phi(V)$ [21].

From (6), we also obtain the corresponding expression for the bracket in the finite dimensional case. By (2), this is

$$[\xi, \eta] = \left(\xi^i \frac{\partial \eta^j}{\partial x^i} - \eta^i \frac{\partial \xi^j}{\partial x^i}\right) \frac{\partial}{\partial x^j}.$$

In the sequel, to obtain other kinds of geometric objects on a differentiable manifold V we use the bundles obtained from $T(V)$ by the operations defined in Section 3.3.

Firstly, consider the dual bundle $T^*(V)$. The elements of $T^*(V)$ are called *covectors* or *covariant vectors*, while the elements of $T(V)$ are *contravariant vectors*. The differentiable cross sections of $T^*(V)$ are called Pfaff forms on V and we denote by $\mathscr{P}(V)$ the set of all such forms, which has an obvious structure of a $\Phi(V)$-module. We can also consider Pfaff forms over an open subset of V. The Pfaff forms are an important analytical tool for geometric problems. If $T(V)$ is reflexive we have the usual duality properties.

2. Vector and Tensor Fields

For instance, let $f: V \to R$ be a differentiable function. The induced system of maps $f_x^*: T_x(V) \to R$ defines a Pfaff form which is denoted by df and is called the *differential* of the function f. The fact that we have a Pfaff form follows from the formula

$$df(\xi) = \xi f \tag{7}$$

for any vector field ξ, which is a consequence of the definitions.

Next, taking tensor products of $T(V)$ and $T^*(V)$ we get the differentiable vector bundles of *tensors* of V. Their differentiable cross sections are *tensor fields* on V of type (h, k) (h times contravariant and k times covariant). The set of tensor fields of type (h, k) has an obvious structure of a $\Phi(V)$-module and is denoted by $\mathfrak{T}_k^h(V)$. The tensor fields of type $(0, 1)$ are just the Pfaff forms, i.e., $\mathfrak{T}_1^0(V) = \mathscr{P}(V)$. The algebraic tensor calculus (which we suppose known) can be carried over in a straightforward manner to tensor fields, and we can talk of the tensor algebra $\mathfrak{T}(V) = \oplus_{h,k} \mathfrak{T}_k^h(V)$ of the differentiable manifold V (or, if required, of an open subset of V).

From the definition of tensor fields, it follows that a field t of type (h, k) defines a $\Phi(V)$-multilinear mapping from

$$\underbrace{\mathscr{P}(V) \times \cdots \times \mathscr{P}(V)}_{h \text{ times}} \times \underbrace{\mathscr{V}(V) \times \cdots \times \mathscr{V}(V)}_{k \text{ times}}$$

to $\Phi(V)$ given by

$$t(\alpha^1, \ldots, \alpha^h, \xi_1, \ldots, \xi_k)(x) = t_x(\alpha_x^1, \ldots, \alpha_x^h, \xi_{1,x}, \ldots, \xi_{k,x}),$$

where $\alpha^i \in \mathscr{P}(V)$, $\xi_i \in \mathscr{V}(V)$ and the index x means that we take the values of the respective fields at x.

A tensor field can be expressed by local fiber coordinates if the tensors t_x are expressed in this manner. Taking into account Theorem 3.3.4, the coordinate expressions are the functions

$$\underbrace{E^* \times \cdots \times E^*}_{h \text{ times}} \times \underbrace{E \times \cdots \times E}_{k \text{ times}} \to R$$

which assign to every system of arguments the value of t_x on the vectors whose fiber coordinates are these arguments. In the intersection of two coordinate neighborhoods we shall have corresponding transition relations.

In the general case the set of tensor fields is more ample than in the classical finite dimensional case. Thus, every vector field is a tensor field of type (1, 0) but the converse is false because of the nonreflexivity of the model.

In the finite dimensional case, we can consider the natural bases $\partial/\partial x^i$ which have corresponding local dual natural *cobases* in $T^*(V)$ given by dx^i. Then, a local vector field is $\xi^i\, \partial/\partial x^i$ and a local Pfaff form is $\alpha_i\, dx^i$. Hence, a function giving a local tensor field of type (h, k) has the expression

$$t^{i_1\cdots i_h}_{j_1\cdots j_k}\, \alpha^1_{i_1}\cdots \alpha^h_{i_h}\, \xi^{j_1}_1 \cdots \xi^{j_k}_k$$

in fiber coordinates, where the $t^{\cdots}_{\cdots}$ are differentiable functions, called the *components* of the tensor field. In the intersection $U \cap U' \neq \varnothing$, these components satisfy the following transformation law

$$t^{i_1\cdots i_h}_{U'\, j_1\cdots j_k} = \frac{\partial x^{i_1}_{U'}}{\partial x^{a_1}_U}\cdots \frac{\partial x^{i_h}_{U'}}{\partial x^{a_h}_U}\frac{\partial x^{b_1}_U}{\partial x^{j_1}_{U'}}\cdots \frac{\partial x^{b_k}_U}{\partial x^{j_k}_{U'}} t^{a_1\cdots a_h}_{U\, b_1\cdots b_k}$$

and we find the classical result that the tensors and tensor fields are defined by their local components and that the local bases for the tensor spaces are given by

$$\frac{\partial}{\partial x^{i_1}}\otimes \cdots \otimes \frac{\partial}{\partial x^{i_h}}\otimes dx^{j_1}\otimes \cdots \otimes dx^{j_k}.$$

The reader will reconstruct from this the classical tensor calculus. Of course, instead of the natural bases we could also consider general local bases defined by arbitrary independent local cross sections and their dual cobases.

Now, we see that every tensor field of type (1, 0) is a vector field in the finite dimensional case. In this case, one can prove that every function

$$\mathscr{P}(V) \times \cdots \times \mathscr{P}(V) \times \mathscr{V}(V) \times \cdots \times \mathscr{V}(V) \to \Phi(V)$$

which is $\Phi(V)$-multilinear defines a tensor field [21].

One cannot say very much about the behavior of an arbitrary tensor with respect to a morphism of Diff. But, there are situations when the results are important. Thus, we already know that $f: V \to W$ induces f^*_x which sends contravariant vectors to the same type of vectors. Let now α_y be a covector at $y \in W$ and $y = f(x)$, $x \in V$. We can assign to it the co-

2. Vector and Tensor Fields

vector β_x at x given by $\beta_x(\xi_x) = \alpha_y(f_x^* \xi_x)$ and one gets an induced map $f_{*y}: T_y(W) \to T_x(V)$, which defines also a map between the Pfaff forms of the two manifolds; more precisely, to every form on W there corresponds an associated form on V. The result extends easily, and we get an induced map which sends a purely contravariant tensor at x to a tensor of the same type at $f(x)$ and a purely covariant tensor at $f(x)$ to a tensor of the same type at x. For tensor fields one generalizes the f-relatedness correspondence considered above for vector fields. To a covariant tensor field α on W there corresponds an f-related field of the same type β on V which is defined by

$$\beta_x(\xi_x^1, \ldots, \xi_x^k) = \alpha_{f(x)}(f_x^* \xi_x^1, \ldots, f_x^* \xi_x^k).$$

The differentiability of the cross section β follows from the fact that its coordinate expression can be considered as a composite map

$$h(U) \xrightarrow{\alpha \circ (f^* \times \cdots \times f^*) \times f^*} L(F^k, R) \times L(E, F) \to L(E^k, R),$$

where E is the model of V, F the model of W, and the second arrow is the linear map which sends a pair (λ, φ) ($\lambda \in L(F^k, R)$, $\varphi \in L(E, F)$) into $\mu \in L(E^k, R)$ defined by $\mu(e_1, \ldots, e_k) = \lambda(\varphi(e_1), \ldots, \varphi(e_k))$.

A vector field on a differentiable manifold V defines an important operation on the tensor field algebra of V, called the *Lie derivative*.

First, if ξ is a vector field on V and $f \in \Phi(V)$, ξf is called the *Lie derivative of f with respect to ξ* and is denoted by $L_\xi f$. If η is another vector field, we put $L_\xi \eta = [\xi, \eta]$ and call it the *Lie derivative of η with respect to ξ*.

If ω is a Pfaff form on V, we consider the linear map $L_\xi \omega: \mathscr{V}(V) \to \Phi(V)$ given by

$$(L_\xi \omega)(\eta) = \xi(\omega(\eta)) - \omega([\xi, \eta]). \tag{8}$$

Expressing this formula in local coordinates, we get (in notation already used)

$$(L_\xi \omega)(\eta)_U(\chi) = (\omega_U(\eta_U))'_\chi(\xi_{U,\chi}) - \omega_{U,\chi}(\eta'_{U,\chi}(\xi_{U,\chi}) - \xi'_{U,\chi}(\eta_{U,\chi}))$$
$$= (\omega'_{U,\chi} \xi_{U,\chi})(\eta_{U,\chi}) + \omega_{U,\chi}(\xi'_{U,\chi}(\eta_{U,\chi})). \tag{9}$$

Since in the last part of (9) the derivative of η does not enter, we see that $L_\xi \omega$ is given by a system of functions $(L_\xi \omega)_x \in L(E, R)$ which depend differentiably on x, and it defines a Pfaff form on V (which is proved by the invariant expression (8)). The Pfaff form $L_\xi \omega$ is called the *Lie derivative of the form ω with respect to ξ*.

Analogously, if t is a tensor field of the type (h, k), one defines the function $L_\xi t$ given by the formula

$$(L_\xi t)(\omega^1, \ldots, \omega^h, \xi_1, \ldots, \xi_k) = \xi(t(\omega^1, \ldots, \omega^h, \xi_1, \ldots, \xi_k))$$
$$- \sum_{i=1}^{h} t(\omega^1, \ldots, L_\xi \omega^i, \ldots, \omega^h, \xi_1, \ldots, \xi_k)$$
$$- \sum_{i=1}^{k} t(\omega^1, \ldots, \omega^h, \xi_1, \ldots, L_\xi \xi_i, \ldots, \xi_k),$$

and one sees that it gives at every point the necessary function for a tensor field of type (h, k) on V. This will be denoted by $L_\xi t$ and is called the *Lie derivative of the field t with respect to ξ*.

In fiber coordinates the local expression of Formula (10) is

$$L_\xi t(\omega^1, \ldots, \omega^h, \xi_1, \ldots, \xi_k)_U(\chi)$$
$$= t'_{U,\chi}(\xi_{U,\chi})(\omega^1_{U,\chi}, \ldots, \omega^h_{U,\chi}, \xi_{1,U,\chi}, \ldots, \xi_{k,U,\chi})$$
$$- \sum_{i=1}^{h} t_{U,\chi}(\omega^1_{U,\chi}, \ldots, \omega^i_{U,\chi} \circ \xi'_{U,\chi}, \ldots, \omega^h_{U,\chi}, \xi_{1,U,\chi}, \ldots, \xi_{k,U,\chi})$$
$$+ \sum_{i=1}^{k} t^1_{U,\chi}(\omega^1_{U,\chi}, \ldots, \omega^h_{U,\chi}, \xi_{1,U,\chi}, \ldots, \xi'_{U,\chi}(\xi_{i,U,\chi}), \ldots, \xi_{k,U,\chi}).$$

In the finite dimensional case, this formula gives the well known expression for the components of the Lie derivative of a tensor field, namely,

$$(L_\xi t)^{i_1 \cdots i_h}_{j_1 \cdots j_k} = \frac{\partial t^{i_1 \cdots i_h}_{j_1 \cdots j_k}}{\partial x^s} \xi^s - \sum_{p=1}^{h} \frac{\partial \xi^{i_p}}{\partial x^s} t^{i_1 \cdots s \cdots i_h}_{j_1 \cdots j_k}$$
$$+ \sum_{p=1}^{h} \frac{\partial \xi^s}{\partial x^{j_p}} t^{i_1 \cdots i_h}_{j_1 \cdots s \cdots j_k}.$$

One can see that the Lie derivative with respect to ξ is a derivation of the tensor algebra of V. By definition, the tensor t is said to be *invariant* with respect to the infinitesimal transformation ξ if $L_\xi t = 0$.

3 Differential Forms and Integration

Another algorithm which is very important is Cartan's exterior differential calculus.

Note that the functor $L_{alt}(E^p, R)$, where E is a Banach space and R is the real field, given by the multilinear skew-symmetric functionals on p arguments of E, is differentiable ($L_{alt}(E^p, R)$ is a closed subspace of $L(E^p, R)$). Hence, we can extend this functor to the tangent bundle $T(V)$ and get the bundle $\lambda^p T(V)$ whose elements are the covariant skew-symmetric tensors of type $(0, p)$. In the finite dimensional case this bundle is the same as $\wedge^p T^*(V)$. The differentiable cross sections of $\lambda^p T(V)$ are called *differential forms of degree p on V* (or exterior differential forms). It is convenient to consider the functions of $\Phi(V)$ as forms of degree 0 and the Pfaff forms as differential forms of degree 1. The set of differential forms of degree p will be denoted by

$$A^p(V)\,(p = 0, 1, 2, \ldots, A^0(V) = \Phi(V), A^1(V) = \mathscr{P}(V))$$

and has an obvious structure of a $\Phi(V)$-module.

If $\alpha \in \lambda^p T(V)$ and $\beta \in \lambda^q T(V)$ are in the same local fiber, their *alternating product* is an element of $\lambda^{p+q} T(V)$ defined by

$$\alpha \wedge \beta(v_1, \ldots, v_{p+q}) = \frac{1}{p!\,q!} \sum_\sigma \varepsilon(\sigma)\alpha(v_{i_1}, \ldots, v_{i_p})\beta(v_{i_{p+1}}, \ldots, v_{i_{p+q}}), \quad (1)$$

where σ is the premutation $(i_1, \ldots, i_{p+q})$ of $(1, \ldots, p+q)$ and $\varepsilon(\sigma)$ is 1 for σ even and -1 for σ odd.

This operation extends to differential forms. In fact, if α and β are differential forms, the coordinate expression of $\alpha \wedge \beta$ given by (1) in a local chart (U, h) can be considered as a composite map $h(U) \to L(E^{p+q}, R) \to L_{alt}(E^{p+q}, R)$, where the last arrow is the operation of alternation, which is linear and continuous. It follows that $\alpha \wedge \beta$ is a differential form of degree $p + q$. This operation on forms is called the *exterior product* and makes

$$A(V) = \bigoplus_p A^p(V)$$

into an algebra called the *exterior* or *Grassmann algebra* of the differentiable manifold V. The exterior product is associative and skew-symmetric. Moreover,

$$\alpha \wedge \beta = (-1)^{\deg \alpha \deg \beta} \beta \wedge \alpha \tag{2}$$

1 Proposition *The correspondence which assigns to every V the Grassmann algebra $A(V)$ is a contravariant functor from* Diff *to the category of associative real algebras with skew-symmetric product (anticommutative)*.

In fact, if $f: V \to W$ is a morphism in Diff, the procedure used in the previous sections for Pfaff forms gives a map $f_*: A(W) \to A(V)$, which is easily seen to be an algebra homomorphism.

We remark that a differential form α of degree p defines a multilinear skew-symmetric function

$$\alpha: \underbrace{\mathscr{V}(V) \times \cdots \times \mathscr{V}(V)}_{p \text{ times}} \to \Phi(V).$$

We define the *exterior derivative* $d\alpha$ of an element $\alpha \in A^p(V)$. Let

$$d\alpha: \underbrace{\mathscr{V}(V) \times \cdots \times \mathscr{V}(V)}_{(p+1) \text{ times}} \to \Phi(V)$$

be the function defined by the formula

$$d\alpha(\xi_1, \ldots, \xi_{p+1}) = \sum_{i=1}^{p+1} (-1)^{i+1} \xi_i \alpha(\xi_1, \ldots, \hat{\xi}_i, \ldots, \xi_{p+1}) \tag{3}$$
$$+ \sum_{i<j} (-1)^{i+j} \alpha([\xi_i, \xi_j], \xi_1, \ldots, \hat{\xi}_i, \ldots, \hat{\xi}_j, \ldots, \xi_{p+1}),$$

where the sign "$\wedge$" indicates the absence of the respective arguments.

Applying the definition of the bracket, Formula 2(1), and the definition of the derivative, we get after a straightforward calculation [27] that the second term of (3) has the local coordinate expression

$$\sum_{i=1}^{p+1} (-1)^{i+1} \alpha'(\xi^i)(\xi_1, \ldots, \hat{\xi}_i, \ldots, \xi_{p+1}), \tag{4}$$

3. Differential Forms and Integration

where the coordinate expressions are denoted by the same letters as the respective elements, and the indices $h(U)$ and χ, used in the previous section, are omitted.

As for the Lie derivative in Section 2, we see that the function (3) has a *local character*, i.e., it defines corresponding functions at every point of V. The differentiability of the cross section $d\alpha$ of $\lambda^{p+1}T(V)$ follows from the fact that its coordinate expression (4) can be considered as a composite function $h(U) \xrightarrow{\alpha'} L(E^{p+1}, R) \to L_{alt}(E^{n+1}, R)$, where the second arrow denotes alternation. Hence (3) defines a differential form $d\alpha$ of degree $p + 1$, called the *exterior derivative* of the form α.

We mention the following properties of the exterior derivative.

2   Proposition   *For the exterior derivative the following relations hold*

(a) $d(\alpha \wedge \beta) = d\alpha \wedge \beta + (-1)^{\deg \alpha} \alpha \wedge d\beta$,
(b) $d^2\alpha = dd\alpha = 0$,
(c) $d(f_*\alpha) = f_*(d\alpha)$, $f: V \to W$ and $\alpha \in A(W)$.

The proof is immediate from the local expression (4). It follows that $A(V)$ and the operator d define a cochain complex, functorially associated with V, which is called the *de Rham complex* of the differentiable manifold V. Its cocycles, i.e., the differential forms ω such that $d\omega = 0$ are called *closed forms*, and its coboundaries $\omega = d\psi$ are called *exact forms*. The corresponding cohomology is called the *de Rham cohomology of V*.

If V is finite dimensional, we recover from the previous formulas the classical results [16]. The local expression of a differential form of degree p with respect to natural cobases is of the type

$$\alpha = \frac{1}{p!} a_{i_1 \cdots i_p} \, dx^{i_1} \wedge \cdots \wedge dx^{i_p}. \tag{5}$$

If we also have the form

$$\beta = \frac{1}{q!} b_{j_1 \cdots j_q} \, dx^{j_1} \wedge \cdots \wedge dx^{j_q},$$

then

$$\alpha \wedge \beta = \frac{1}{p!q!(p+q)!} \delta^{i_1 \cdots i_p j_1 \cdots j_q}_{h_1 \cdots h_{p+q}} a_{i_1 \cdots i_p} b_{j_1 \cdots j_q} \, dx^{h_1} \wedge \cdots \wedge dx^{h_{p+q}}, \tag{6}$$

where $\delta^{\cdots}_{\cdots}$ are the Kronecker symbols and their value is 1 if the upper indices are obtained by an even permutation of the lower indices, -1 if they are given by an odd permutation, and 0 in all the other cases.

Finally, we have

$$d\alpha = \frac{1}{p!} da_{i_1 \cdots i_p} \wedge dx^{i_1} \wedge \cdots \wedge dx^{i_p}$$

$$= \frac{1}{p!(p+1)!} \delta^{i i_1 \cdots i_p}_{j_1 \cdots j_{p+1}} \frac{\partial a_{i_1 \cdots i_p}}{\partial x^i} dx^{j_1} \wedge \cdots \wedge dx^{j_{p+1}}$$

$$= \frac{1}{p+1} \sum_{h=1}^{p+1} (-1)^{h+1} \frac{\partial a_{i_1 \cdots \hat{i}_h \cdots i_{p+1}}}{\partial x^{i_h}} dx^{i_1} \wedge \cdots \wedge dx^{i_h}$$

$$\wedge \cdots \wedge dx^{i_{p+1}}.$$

We indicate, without proof, an important application of the exterior derivative.

A *distribution* on the differentiable manifold V is a family of linear subspaces of the tangent spaces of V such that their union is a differentiable subbundle of $T(V)$. These subspaces clearly have the same dimension. If $\mathfrak{D}$ is such a distribution, we say that it is an *involutive distribution*, if, for any two vector fields ξ and η which belong to $\mathfrak{D}$, the bracket $[\xi, \eta]$ also belongs to $\mathfrak{D}$. The distribution $\mathfrak{D}$ is called *integrable* if every point of V has an open neighborhood and a submanifold immersed in this neighborhood such that the tangent bundle of the submanifold is the corresponding restriction of $\mathfrak{D}$. Then, we have the well known *Frobenius' theorem*.

3 Theorem *The distribution $\mathfrak{D}$ is integrable if and only if it is involutive.*

The necessity is obvious and the sufficiency follows from the existence theorem for ordinary differential equations with parameters. For the proof, see, for instance, [27].

Now, using the definition of the exterior derivative, we see that being involutive is equivalent to the following condition: *for every Pfaff form ω which vanishes on $\mathfrak{D}$, the form $d\omega$ also vanishes on $\mathfrak{D}$*. This is just the application of the exterior derivative which we had in mind. In the finite dimensional case, this is the classical statement of Frobenius' theorem: *the Pfaff system of equations $\omega^i = 0$ is completely integrable if and only if $d\omega^i = 0$ (modulo $\omega^i = 0$).*

3. Differential Forms and Integration

The exterior calculus can be generalized as follows. Let S be a differentiable vector bundle on V. Consider bundles of the form

$$L(\underbrace{T^*(V) \times \cdots \times T^*(V)}_{h \text{ times}} \times \underbrace{T(V) \times \cdots \times T(V)}_{k \text{ times}}, S).$$

The elements of such a bundle are called *mixed tensors* and the differentiable cross sections are *mixed tensor fields* on V.

Now, consider the bundles

$$L_{\text{alt}}(\underbrace{T(V) \times \cdots \times T(V)}_{p \text{ times}}, S).$$

The cross sections of such bundles are called *S-valued differential forms of degree p* or *vector forms*. For such forms, we define the exterior product in the following manner: let α be an S-valued p-form and β an M-valued q-form, and let $(\alpha_x, \beta_x) \mapsto \alpha_x \beta_x$ be a differentiable cross section of $L(S \times M, N)$; then $\alpha \wedge \beta$ is an N-valued $(p + q)$-form given by Formula (1), where the products $\alpha\beta$ are replaced by the previously defined bilinear map.

If $S = V \times F$, the local expression (4) is meaningful for S-valued forms α and, together with Formula (3), it defines the exterior derivative of the form α. In this case, α is also called an F-valued form.

For general S-valued forms with nontrivial S we cannot obtain in the same manner an exterior derivative.

To obtain an integration theory of differential forms on a manifold V, we have to introduce corresponding domains of integration.

4 Definition The *standard p-dimensional simplex* ($p = 0, 1, 2, \ldots$) is the subset $\Delta^p \subseteq R^{p+1}$ defined by

$$\Delta^p = \left\{ x \mid x = (x^i), x^i \geq 0, \sum_{i=1}^{p+1} x^i = 1 \right\}.$$

A *p-dimensional singular simplex* of the differentiable manifold V is a differentiable map $\sigma_{(p)} \colon \Delta^p \to V$.

Here, the differentiability of $\sigma_{(p)}$ means that $\sigma_{(p)}$ is the restriction to Δ^p of a differentiable map $U \to V$, where U is an open neighborhood of Δ^p in R^{p+1}.

5   Definition   If G is an Abelian group, a *p-dimensional singular chain of V with coefficients in G* is a formal finite linear combination $l_{(p)} = \sum_i g_i \sigma^i_{(p)}$ where the $\sigma^i_{(p)}$ are singular simplexes.

Obviously, if we denote by $C_p(V, G)$ ($p = 0, 1, 2, \ldots$) the set of p-dimensional singular chains defined above, this has a structure of an Abelian group. In particular we shall consider the groups $C_p(V, J)$ and $C_p(V, R)$ (J is the additive group of integers and R of the real numbers), the last of them being a real linear space.

If $f: V \to W$ is a morphism in Diff and $\sigma_{(p)}$ is a singular simplex of V, $f \circ \sigma_{(p)}$ is a singular simplex of W, and this correspondence extends by linearity to the induced homomorphisms

$$f_{\#p}: C_p(V, G) \to C_p(W, G). \tag{8}$$

Let now $U \subseteq R^N$ be a convex open subset of some real arithmetic space and $A_0, \ldots, A_p$ be $p + 1$ arbitrary points of U. Then, we define a singular simplex of U, denoted by $(A_0, \ldots, A_p)$, by

$$(A_0, \ldots, A_p)((x^i)) = \sum_i x^i A_i, \qquad (x^i) \in \Delta^p.$$

Such simplexes $(A_0, \ldots, A_p)$ will be called *special singular simplexes* with $A_0, \ldots, A_p$ as the *vertices* of the simplex. For instance, the identity map of Δ^p is a special singular simplex with vertices $X_j(\delta^i_j)$ ($j = 0, \ldots, p$) which we again denote by Δ^p.

If $l_{(h)} = \sum_i g_i \lambda^i_h$ is an h-dimensional singular chain of some open neighborhood of Δ^p, where the singular simplex $\sigma_{(p)}$ of V is defined, we put

$$\sigma_{(p)} \circ l = \sum_i g_i \sigma_{(p)} \circ \lambda^i_{(h)}$$

and this is a singular chain of V.

For a special singular simplex $(A_0, \ldots, A_p)$ of R^N, we define a $(p-1)$-dimensional special chain of R^N, called the *boundary* of the given simplex, by the formula

$$\partial(A_0, \ldots, A_p) = \sum_{i=0}^p (-1)^i (A_0, \ldots, \hat{A}_i, \ldots, A_p), \tag{9}$$

where "$\wedge$" means again the absence of the respective argument.

3. Differential Forms and Integration

Next, if $\sigma_{(p)}$ is a singular simplex of the manifold V, we define its *boundary* by

$$\partial \sigma_{(p)} = \sigma_{(p)} \circ \partial \Delta^p. \tag{10}$$

This is easily seen to agree with (9), i.e., if we apply (10) to $(A_0, \ldots, A_p)$ we get the same result as in (9).

Finally, ∂ extends by linearity to a homomorphism

$$\partial: C_p(V, G) \to C_{p-1}(V, G), \tag{11}$$

which, obviously, commutes with the induced homomorphisms (8).

By a straightforward calculation using the formula (9), we get $\partial^2(A_0, \ldots, A_p) = 0$ and then, by (10) and the above mentioned commutativity property we obtain the general property

$$\partial^2 = 0. \tag{12}$$

Hence, we have the following result.

6 Proposition   *There is a covariant functor from Diff to the category of chain complexes which associates to V the complex $(C_p(V, G), \partial)$.*

7 Definition   The homology groups of the complex $(C_p(V, G), \partial)$ are called *singular homology groups of V with coefficients G* and are denoted by $H_p(V, G)$.

$H_p(V, G)$ is the quotient group of the *cycles* of V, i.e., chains l such that $\partial l = 0$, by the *boundaries* $l = \partial l'$ of V and the homomorphisms (8) induce new homomorphisms

$$f_{*p}: H_p(V, G) \to H_p(W, G), \tag{13}$$

so that $H_p(\cdot, G)$ is a covariant functor. Particularly, we have the *integer homology* when $G = J$ and the *real homology* when $G = R$.

It is to be remarked that if in all previous considerations we replace differentiability by continuity we get singular homology groups of arbitrary topological spaces.

Before considering integrals, we shall point out another property of the singular chains, which will be needed in the next chapter. Let $(A_0, \ldots, A_p)$

be a special singular simplex of R^N and consider

$$(A_0, \ldots, A_p)(\Delta^p) \times I \subseteq R^{N+1}.$$

Denote the points $(A_i, 0)$ by A_i^0 and $(A_i, 1)$ by A_i^1 ($i = 0, \ldots p$). We define the singular chain of R^{N+1}:

$$P(A_0, \ldots, A_p) = \sum_{i=0}^{p} (-1)^i (A_0^0, \ldots, A_i^0, A_i^1, \ldots, A_p^1), \qquad (14)$$

and call it the *prism* of the simplex $(A_0, \ldots, A_p)$. Extending by linearity the operator P, we obtain Pl for every special singular chain of an arithmetic space. From (9) and (14), we get by a straightforward calculation the formula

$$\partial P(A_0, \ldots, A_p) + P\partial(A_0, \ldots, A_p) = (A_0^1, \ldots, A_p^1) - (A_0^0, \ldots, A_p^0),$$

which, applied to the simplex Δ^p with vertices X_i gives

$$(X_0^1, \ldots, X_p^1) - (X_0^0, \ldots, X_p^0) = \partial P \Delta^p + P \partial \Delta^p, \qquad (15)$$

where the chains are in R^{p+2}.

Two p-dimensional singular simplexes σ and σ' of the differentiable manifold V are called *homotopic* and denoted $\sigma \simeq \sigma'$ if the corresponding maps are differentiably homotopic. This means that there is a differentiable map F of an open neighborhood of $\Delta^p \times I$ into V such that

$$\sigma = F \circ (X_0^0, \ldots, X_p^0) \quad \text{and} \quad \sigma' = F \circ (X_0^1, \ldots, X_p^1).$$

If this is the case, applying $F_\#$ to the two members of Formula (15) we get

$$\sigma' - \sigma = \partial F_\#(P\Delta^p) + F_\#(P\partial\Delta^p) \qquad (16)$$

or, by putting $D = F_\# P$,

$$\sigma' - \sigma = \partial D \Delta^p + D \partial \Delta^p. \qquad (17)$$

More generally, let $l = \sum_i \alpha_i \sigma_i$ and $l' = \sum_i \alpha_i' \sigma_i'$ be two p-dimensional singular chains of V. They are called *homotopic* ($l \simeq l'$) if $\alpha_i = \alpha_i'$ and

3. Differential Forms and Integration

$\sigma_i \simeq \sigma_i'$ with homotopy F^i. Then, writing Formula (17) for each σ_i, multiplying by α_i, and adding, we get

$$l' - l = \partial(\sum_i \alpha_i D^i)\Delta^p + (\sum_i \alpha_i D^i)\partial\Delta^p,$$

where $D^i = F^i_\# P$, i.e.,

$$l' - l = \partial \mathfrak{D}\Delta^p + \mathfrak{D}\partial\Delta^p \qquad (18)$$

where $\mathfrak{D} = \sum_i \alpha_i D^i$ and the singular chains are in V.

This is just the formula which we need, and we note that it remains valid when differentiability is replaced by continuity.

If $l \simeq l'$ we have a family l_t of chains which depend differentiably on $t \in [0, 1]$ and such that $l_0 = l$, $l_1 = l'$, i.e., $l_t = \sum_i \alpha_i \sigma_{i,t}$ where

$$\sigma_{i,t}(x) = F_i(x, t), \qquad x \in \Delta^p.$$

We now return to the integration problem. The domains of integration announced at the beginning will be just the singular simplexes and the real singular chains of the manifold. Let ω be a differential form of degree p on V and $\sigma_{(p)}: \Delta^p \to V$ a p-dimensional singular simplex of V.

8 Definition The *integral* of the form ω over the simplex $\sigma_{(p)}$ is defined by

$$\int_{\sigma(p)} \omega = \int_{\Delta^p} \sigma_{(p)*}(\omega), \qquad (19)$$

where the second member is the ordinary integral.

Formula (19) extends by linearity to real p-dimensional chains and we get, for $l = \sum_i a_i \sigma^i$,

$$\int_l \omega = \sum_i a_i \int_{\sigma^i} \omega.$$

The properties of this integral are easily obtained from the corresponding properties of multiple integrals and, therefore, we do not discuss them, except for the very important *Stokes' formula*.

9 Theorem (Stokes)

$$\int_l d\omega = \int_{\partial l} \omega, \qquad \deg \omega = \dim l - 1.$$

By linearity it suffices to prove that

$$\int_{\Delta^p} d\omega = \int_{\partial \Delta^p} \omega$$

for $\omega = a(x)dx^1 \wedge \cdots \wedge dx^{p-1}$, which is a simple exercise which we leave to the reader. (See, for instance, [45].)

There is an important case where the integral can be defined by other means.

10 Definition The n-dimensional differentiable manifold V is called *orientable* if its tangent bundle $T(V)$ is orientable in the sense of Definition 3.3.22.

By Definition 3.3.22 this is the same as the existence on V of an atlas such that the Jacobians of its transition functions are always positive, or as the existence of an atlas such that, on every coordinate neighborhood, there is a differentiable form of degree n and on the intersection of two coordinate neighborhoods, the respective forms differ by a positive factor. Such an atlas is called an *orientation* of V and, if we choose that atlas, V is called an *oriented manifold*.

Now let V be a compact, n-dimensional, oriented manifold and ω a form of degree n on V. Let $\{U_i\}$ be a finite covering of V with coordinate neighborhoods from the orientation atlas of V and let $\{\varphi_i\}$ be a partition of unity subordinated to $\{U_i\}$. Denote by $\omega|_{U_i} = \omega_i$ and $\varphi_i \omega_i = \pi_i$ where ω_i is extended by 0 on $V - U_i$. Because $\omega = \sum_i \pi_i$, we are led to define

$$\int_V \omega = \sum_i \int_{h_i(U_i)} \pi_{i,U_i}, \tag{20}$$

where h_i is the chart on U_i and π_{i,U_i} the expression of π_i in this chart.

One can prove that the integral (20) is equal to the integral in the sense of Definition 8. More precisely, the hypothesis concerning V implies that there is an n-dimensional chain on V which is an integral cycle and which is characterized by the property that every integral cycle is a multiple of it, and (20) is just the integral of ω on this chain. This also proves the in-

4. Absolute Differential Calculus

dependence of the integral (20) from the chosen partition of unity.

From these considerations and Theorem 9 we get, if V is compact and oriented, *Stokes' formula*

$$\int_V d\omega = 0, \tag{21}$$

for every form ω of the degree $n - 1$.

For a more complete treatment of the integral calculus on manifolds see, for instance, [45].

4 Absolute Differential Calculus

We shall now develop the absolute differential calculus, which is determined if a supplementary structure called a connection is given.

1 Definition Let S be a differentiable vector bundle on the differentiable manifold V. A *connection* on S is a homomorphism of sheaves of linear spaces

$$D: F(S) \to F(L(T(V), S)),$$

where F denotes the sheaf of germs of differentiable cross sections of the respective vector bundles, such that the condition

$$D(fs)(t) = (tf)s + fDs(t) \tag{1}$$

is satisfied for every $s \in F(S)$, $t \in T(V)$ and $f \in \mathfrak{F}(V)$ (Section 1) with the same projection on V.

The meaning of the condition (1) becomes clear from the relation $tf = df(t)$. By Definition 1, we see that D induces a connection on $S|_U$ for every open subset $U \subseteq V$.

2 Definition For $s \in F(S)$, Ds is called the *covariant* or *absolute derivative* of s.

This definition explains the name of absolute differential calculus.

We know from Section 2.2 that D induces a homomorphism of the corresponding canonical presheaves. These presheaves are identical with the presheaves which assign to an open subset $U \subseteq V$ the space of differentiable cross sections of S and, respectively, $L(T(V), S)$ over U, the restrictions being the usual ones for cross sections. Hence, for every such open subset U we have an operator D_U on $\Gamma(S|_U)$ with values in $\Gamma(L(T(U), S|_U))$, where Γ denotes the functor assigning cross sections. This operator is a homomorphism of real linear spaces and satisfies (1) for $s \in \Gamma(S|_U)$ and $f \in \Phi(U)$; they commute with restrictions to subsets. In particular, there is such an operator D_V for the global cross sections. The converse is obvious: if we know the operators D_U, D is determined. Hence, the set of operators $\{D_U \mid U \subseteq V\}$ can be used to define a connection. Moreover, it suffices to know D_U for the coordinate neighborhoods U of an atlas V which belongs to a local trivialization of S. It is interesting to observe that in the finite dimensional case the global operator D_V determines the connection uniquely. A proof of this is given in [21]. We shall also use the terms of Definition 2 for the operators D acting on cross sections, and we shall omit the indices U, i.e., D will denote the action of the operator both on germs and on cross sections.

If $Ds = 0$, s is called a horizontal *cross section* of S and the vectors of this cross section are called *parallel*.

If $S = T(V)$, D is called a *connection on V*.

Let us find the coordinate expression of a connection D. Denote by M the fiber of S and consider the restriction of S to a trivializing neighborhood U, which is also a coordinate neighborhood on V with the local chart h. The fiber coordinate expression of $s: U \to S$ is $\sigma: h(U) \to M$, and let $\sigma'_\chi: E \to M (\chi \in h(U))$ be its derivative, E being the model of V. Ds has a local expression $D\sigma: h(U) \to L(E, M)$, which satisfies a condition of the type (1).

Define the local operator

$$\Gamma_U: \Gamma(S|_U) \to \Gamma(L(T(U), S|_U))$$

by the coordinate expression

$$(\Gamma_U \sigma)(\chi) = (D\sigma)_\chi - \sigma'_\chi.$$

It is easily seen that $\Gamma_U(fs) = f\Gamma_U s$ ($f \in \Phi(U)$).

Now, the coordinate expression of the covariant derivative is

$$D\sigma = \sigma' + \Gamma_U \sigma, \tag{2}$$

4. Absolute Differential Calculus

where Γ_U is a differentiable cross section of $L(S|_U, L(T(U), S|_U))$.

It follows that there exists always a connection on $S|_U$ for a coordinate neighborhood U and, using the technique of the proof of proposition 3.4.13, we get immediately that *if V has a differentiable partition of unity, every vector bundle S on V has a connection* (generally, of course, not a unique one).

When V is a finite dimensional differentiable manifold and $S = T(V)$, Formula (2) is just the classical formula of the covariant derivative of a contravariant vector

$$\sigma^i_{|j} = \frac{\partial \sigma^i}{\partial x^j} + \Gamma^i_{kj}\sigma^k \tag{3}$$

where on the intersection $U \cap U'$ of two coordinate neighborhoods there are the following transition relations

$$\Gamma^i_{U'jk} = \frac{\partial x^i_{U'}}{\partial x^h_U}\frac{\partial x^p_U}{\partial x^j_{U'}}\frac{\partial x^q_U}{\partial x^k_{U'}}\Gamma^h_{Upk} + \frac{\partial x^i_{U'}}{\partial x^h_U}\frac{\partial^2 x^h_U}{\partial x^j_{U'}\partial x^k_{U'}}, \tag{4}$$

the Γ^i_{jk} being the coefficients of connection.

Now, consider the more general case where V is an arbitrary manifold (which may be infinite dimensional) and S is a finite dimensional vector bundle on V (not necessarily the tangent bundle).

Let e_i be a local basis for the cross sections of S over the coordinate neighborhood U. Then, we have

$$\Gamma_U e_i = \omega^j_i e_j,$$

where the ω^j_i are Pfaff forms over U. (Einstein's summation convention is used in the sequel). Also, $e'_{i,\chi} = 0$, since e_i is the coordinate expression of a base section and, hence, it is constant. It follows that the local expression of the connection is

$$De_i = \omega^j_i e_j, \tag{5}$$

or, following [6] and using the matrix calculus with matrices whose elements are vectors or forms,

$$De = \omega e. \tag{6}$$

The ω^j_i are called *local forms of the connection* and the matrix ω is the *matrix of the connection.* When $S = T(V)$ (and, hence, V is also finite dimensional), we have simply $\omega_i^j = \Gamma_{ik}^j dx^k$.

If the local matrices of the connection are given, the connection is defined since, for $\xi = \xi^i e_i$ we get

$$D\xi = (d\xi^i + \omega_j^i \xi^j) e_j. \tag{7}$$

On the intersection $U \cap U'$ of two coordinate neighborhoods we have the transition relations $e' = ge$ and, by applying (1), we get

$$\omega' = g\omega g^{-1} + dg \cdot g^{-1}. \tag{8}$$

Relations (4) above are a particular case of the relations (8).

We now return to the general case. If $\omega: J \to V$ is a differentiable curve, Formula (2) shows that the expression $D\sigma_{\omega(t)} \circ \omega_t'$ is meaningful even if σ is defined only along the curve ω and $D\sigma_{\omega(t)} \circ \omega_t' = 0$ is an ordinary differential equation. This equation is called the *equation of the parallel translation* and, when satisfied, the vectors of the respective cross section σ are said to define a *horizontal lift* of the curve ω. The existence and uniqueness theorem for such equations shows that we can always obtain such lifts and we see that a connection as considered here gives also a connection as defined in 3.1.7. The curves of V with the property that the tangent vectors are parallel along the curve are called *autoparallel* or *geodesic curves* of the connection.

Next, we introduce the *covariant differential with respect to a given vector field* ξ. By definition and in the notation previously used this is given by

$$\nabla_\xi s = Ds(\xi) \tag{9}$$

and for a fixed s it is an S-valued 1-form on V; if ξ is also fixed, it is just a cross section of S. For the covariant differential, we have

$$\nabla_\xi (f\sigma) = df(\xi)\sigma + f\nabla_\xi \sigma. \tag{10}$$

We see from (9) that if the operators ∇_ξ are known for every open subset $U \subseteq V$, if they satisfy the given conditions and the relation (10), the operators D are defined. Hence, ∇_ξ can be used for the definition of a connection. If V is finite dimensional it suffices to know $\nabla_\xi s$ for globally defined ξ and s [21].

4. Absolute Differential Calculus

A differential form Ω of the second degree on V, with values in $L(S, S)$, is defined by the formula

$$\Omega(\xi, \eta) = \nabla_\xi \nabla_\eta - \nabla_\eta \nabla_\xi - \nabla_{[\xi,\eta]}. \tag{11}$$

In fact, if we apply the second member of (11) to a cross section s, and if we use the local coordinate expression (2), we get the coordinate expression of Ω:

$$\Omega(\xi, \eta)(\sigma) = \Gamma'(\xi)(\sigma)(\eta) - \Gamma'(\eta)(\sigma)(\xi)$$
$$+ \Gamma((\Gamma\sigma)(\eta))(\xi) - \Gamma((\Gamma\sigma)(\xi))(\eta), \tag{12}$$

where we omit the indices U and χ corresponding to the local chart and the point at which the calculation is performed. Formula (12) shows that Ω is really a differential form of the kind discussed.

Formula (11) is to be considered as a commutation rule for ∇.

3 Definition The $L(S, S)$-valued differential form Ω defined on V by Formulas (11) and (12) is called the *curvature form* of the connection D.

When $S = T(V)$, there is also another important notion. Considering

$$T(\xi, \eta) = \nabla_\xi \eta - \nabla_\eta \xi - [\xi, \eta] \tag{13}$$

we get a $T(V)$-valued differential form of the second degree, globally defined on V, whose local coordinate expression is

$$T(\xi, \eta) = \Gamma(\eta)(\xi) - \Gamma(\xi)(\eta).$$

This form is called the *torsion form* of the connection. Observe that (13) is also a kind of commutation rule.

A *curvature tensor* R of type $(1,3)$ is now defined by

$$R(\alpha, \zeta, \xi, \eta) = \alpha(\Omega(\eta, \xi)\zeta). \tag{14}$$

The fact that this formula defines a tensor field follows from Formula (12). The field R is skew-symmetric in the last two arguments.

Similarly

$$T(\alpha, \xi, \eta) = \alpha(T(\eta, \xi))$$

defines a tensor field of type (1,2) which is skew-symmetric in the arguments ξ and η and is called the *torsion tensor* of the connection.

Let V and S be finite dimensional and express the connection by the local formulas (6) where $\omega_i^j = \Gamma_{i\alpha}^j \, dx^\alpha$ (the italic indices run from 1 to the dimension of the fiber of S and the Greek indices from 1 to dim V). Then, the curvature form Ω is locally determined by the forms Ω_i^j of degree two defined by the relations

$$\Omega(\xi, \eta)e_i = \Omega_i^j(\xi, \eta)e_j,$$

e being the basis used in (6). Now, by (12), we get

$$\Omega_i^j = -\tfrac{1}{2} R_{i\alpha\beta}^j \, dx^\alpha \wedge dx^\beta, \tag{15}$$

where

$$R_{i\alpha\beta}^j = \frac{\partial \Gamma_{i\alpha}^j}{\partial x^\beta} - \frac{\partial \Gamma_{i\beta}^j}{\partial x^\alpha} + \Gamma_{i\alpha}^h \Gamma_{h\beta}^j - \Gamma_{i\beta}^h \Gamma_{h\alpha}^j. \tag{16}$$

The forms Ω_i^j are called the *local curvature forms*, and the matrix $\Omega = (\Omega_i^j)$ is the *curvature matrix*.

Next, consider the more general case when only S is assumed to be finite dimensional (V can be infinite or finite dimensional). Just as above, the curvature forms Ω_i^j can also be defined.

By straightforward calculations we get the following expression of the curvature forms

$$\Omega_i^j = d\omega_i^j - \omega_i^h \wedge \omega_h^j \tag{17}$$

or, using matrices

$$\Omega = d\omega - \omega \wedge \omega. \tag{18}$$

If U and U' are two coordinate neighborhoods and $e' = ge$ on $U \cap U'$, then by multiplying (8) at the right by the matrix g and taking the exterior derivative we get the transition relations

$$\Omega' = g\Omega g^{-1}. \tag{19}$$

Suppose $S = T(V)$ and both are finite dimensional; the italic and Greek

4. Absolute Differential Calculus

indices now run over the same values. Then, (14) shows that (16) are just the components of the curvature tensor, the right-hand side being the classical expression for these components. Similarly we find for the components of the torsion tensor the classical expression

$$T^i_{jk} = \Gamma^i_{jk} - \Gamma^i_{kj}. \qquad (20)$$

We can also introduce the torsion forms T^i given by

$$T(\xi, \eta) = T^i(\xi, \eta) \frac{\partial}{\partial x^i}$$

and we find

$$T^i = -\tfrac{1}{2} T^i_{jk}\, dx^j \wedge dx^k = d(dx^i) - dx^j \wedge \omega^i_j. \qquad (21)$$

This last expression remains valid if the natural cobases dx^i are replaced by general cobases θ^i, which means that we have

$$T^i = d\theta^i - \theta^j \wedge \omega^i_j. \qquad (22)$$

If θ is the dual cobasis of e, one finds easily, on $U \cap U'$, $\theta = \theta' g$, where θ is a matrix with one row, from which it follows that

$$T = T'g. \qquad (23)$$

Now, consider the expression $\Omega(\xi, \eta)\zeta$ with fixed ξ and ζ and denote by $\Omega_{\xi\zeta}$ the corresponding operator on η. In the finite dimensional case, we can introduce

$$\rho(\zeta, \xi) = \operatorname{Tr} \Omega_{\xi\zeta}, \qquad \operatorname{Tr} = \text{trace},$$

and this defines on V a twofold covariant tensor field called the *Ricci tensor*. The associated quadratic form is called the *Ricci form* of the connection. In fact, with the help of the previous formulas, one sees that ρ is the tensor with components $\rho_{ij} = R^h_{ijh}$, which is the classical expression for the Ricci tensor.

A connection on the vector bundle S defines a connection on the dual bundle S^* given by

$$(D\omega)(\xi)(s) = \xi(\omega(s)) - \omega(Ds(\xi)), \qquad (24)$$

where $\omega \in F(S^*)$, $s \in S$, $\xi \in T(V)$ and all have the same projection on V. We can see that the derivative of s does not enter in the right-hand side of (24) by considering the corresponding local expression which can be obtained by means of (2).

The covariant differential of a form α with respect to this connection is

$$(\nabla_\xi \alpha)(s) = \xi(\alpha(s)) - \alpha(\nabla_\xi s) \qquad (25)$$

and we shall say that this is the *covariant differential* of α with respect to the connection of S.

In a similar manner we can get a covariant differential with respect to D for any tensor of S by the formula (with obvious notation)

$$(\nabla_\xi t)(\alpha^1, \ldots, \alpha^h, s_1, \ldots, s_k) = \xi(t(\alpha^1, \ldots, \alpha^h, s_1, \ldots, s_k))$$
$$- \sum_{i=1}^{h} t(\alpha^1, \ldots, \nabla_\xi \alpha^i, \ldots, \alpha^h, s_1, \ldots, s_k) \qquad (26)$$
$$- \sum_{i=1}^{k} t(\alpha^1, \ldots, \alpha^h, s_1, \ldots, \nabla_\xi s_i, \ldots, s_k).$$

This formula is again justified by the corresponding local coordinate expression, which we leave to the reader.

In the finite dimensional case this gives the classical formula

$$t^{i_1 \cdots i_h}_{j_1 \cdots j_k / l} = \frac{\partial t^{i_1 \cdots i_h}_{j_1 \cdots j_k}}{\partial x^l} + \sum_{p=1}^{h} \Gamma^{i_p}_{sl} t^{i_1 \cdots s \cdots i_h}_{j_1 \cdots j_k}$$
$$- \sum_{p=1}^{k} \Gamma^{s}_{j_p l} t^{i_1 \cdots i_h}_{j_1 \cdots s \cdots j_k} \qquad (27)$$

Using the expression (16) for the components of the curvature tensor, one gets by a straightforward calculation the known commutation rule for covariant derivatives

$$t^{i_1 \cdots i_h}_{j_1 \cdots j_k / pq} - t^{i_1 \cdots i_h}_{j_1 \cdots j_k / qp} = \sum_{u=1}^{h} R^{i_u}_{spq} t^{i_1 \cdots s \cdots i_h}_{j_1 \cdots j_k}$$
$$- \sum_{u=1}^{k} R^{s}_{j_u pq} t^{i_1 \cdots i_h}_{j_1 \cdots s \cdots j_k} - t^{i_1 \cdots i_h}_{j_1 \cdots j_k / s} T^{s}_{pq}. \qquad (28)$$

There are also induced covariant derivatives on other vector bundles

4. Absolute Differential Calculus

associated with S. For instance, it is important to note that on $L(S, S)$ we have the connection given by the formula

$$D\lambda(\xi)(s) = D(\lambda(s))(\xi) - \lambda(Ds(\xi)), \qquad (29)$$

where $\lambda \in F(L(S, S))$, $\xi \in T(V)$, $s \in S$ and they have the same projection on V. For this connection,

$$\nabla_\xi \lambda(s) = \nabla_{\xi_i}(\lambda(s)) - \lambda \nabla_\xi(s). \qquad (30)$$

If we have a connection on the vector bundle S, we can define an *exterior covariant derivative* of S-valued vector forms (Section 3) by a formula similar to 3(3)

$$D\alpha(\xi_1, \ldots, \xi_{p+1}) = \sum_{i=1}^{p+1} (-1)^{i+1} \nabla_{\xi_i} \alpha(\xi_1, \ldots, \hat{\xi}_i, \ldots, \xi_{p+1}) \qquad (31)$$
$$+ \sum_{i<j} (-1)^{i+j} \alpha([\xi_i, \xi_j], \xi_1, \ldots, \hat{\xi}_i, \ldots, \hat{\xi}_j, \ldots, \xi_{p+1}).$$

This formula has to be justified by the local coordinate expressions, which are left to the reader.

If S is finite dimensional and if we again use the notation of Formula (5), then the operator D applied to a form Φ of degree p with values in $L(S, S)$ can be expressed locally by a simple formula [6]. Indeed, in this case Φ is locally represented by

$$\Phi(e_i) = \Phi_i^j e_j,$$

where Φ_i^j are local scalar forms of degree p. By (31) and using (30) and (5) we get the desired formula

$$D\Phi = d\Phi - [\omega, \Phi], \qquad (32)$$

where we used matrix notation, and for two matrices λ and μ, whose elements are forms of degrees p and q, respectively, we introduced the operation

$$[\lambda, \mu] = \lambda \wedge \mu - (-1)^{pq} \mu \wedge \lambda \qquad (33)$$

By a twofold application of Formula (32), and using (18), we get

$$D^2\Phi = [\Phi, \Omega] \tag{34}$$

Applying the operator of (31) to $L(S, S)$ one obtains by straightforward calculation the important *Bianchi's identity*

$$D\Omega = 0. \tag{35}$$

When $S = T(V)$, using (31) for $T(V)$ we get the second Bianchi identity

$$DT(\xi_1, \xi_2, \xi_3) = \Omega(\xi_1, \xi_2)\xi_3 + \Omega(\xi_2, \xi_3)\xi_1 + \Omega(\xi_3, \xi_1)\xi_2. \tag{36}$$

From these formulas the reader will easily derive the known formulas for $S = T(V)$ and when both are finite dimensional:

$$\begin{aligned}
R^i_{jkh/s} + R^i_{jhs/k} &+ R^i_{jsk/h} + R^i_{juk}T^u_{sh} + R^i_{jus}T^u_{hk} \\
&+ R^i_{juh}T^u_{ks} = 0, \\
R^i_{jkh} + R^i_{khj} &+ R^i_{hjk} = T^i_{jk/h} + T^i_{kh/j} + T^i_{hj/k} + T^i_{uk}T^u_{jh} \\
&+ T^i_{uj}T^u_{hk} + T^i_{uh}T^u_{kj}.
\end{aligned} \tag{37}$$

5 Riemannian and Foliated Riemannian Manifolds

When the structure group of a differentiable vector bundle is reduced to a subgroup G one can derive more special properties. If this is the case for the tangent bundle of a differentiable manifold, the manifold is said to have a *G-structure*. In this section two such important structures are considered, and another one will be the subject of the next section.

1 Definition If V is a differentiable manifold with a real Hilbertian model H, a *Riemannian structure* on V is a Hilbertian structure g on the tangent bundle $T(V)$ (Definition 3.3.16), which is defined by a twofold differentiable covariant tensor field g.

Of course, not every field of type (0, 2) defines a Hilbertian structure. If one is defined, the respective Riemannian metric on $T(V)$ (Section 3.3) is called a *Riemannian metric on V*, and the pair (V, g), where g is such a

5. Riemannian and Foliated Riemannian Manifolds

metric, is called a *Riemannian manifold* or a *Riemann space*.

In the finite dimensional case, the local coordinate expression of the metric is denoted by

$$ds^2 = g_{ij}dx^i dx^j, \qquad i,j = 1,\ldots,\dim V, \tag{1}$$

where the product on the right-hand side is tensorial, $g_{ij} = g_{ji}$, the quadratic form (1) is positive definite, and the functions g_{ij} are differentiable.

On a differentiable manifold with Hilbertian model having differentiable partitions of unity there exists always a Riemannian metric which can be obtained as in Proposition 3.3.17. This is particularly true for finite dimensional paracompact differentiable manifolds and for infinite dimensional manifolds with separable Hilbertian model.

The existence of the metric allows one to define the scalar product of two tangent vectors at the same point by

$$(v, w) = g(v, w) \tag{2}$$

and the length of a vector by

$$|v| = \sqrt{(v, v)}. \tag{3}$$

We can also consider the usual formula for the angle between two vectors:

$$\cos \varphi = \frac{(v, w)}{|v|\,|w|}. \tag{4}$$

Let $\omega: J \to V$ be a curve in V defined on the open interval J and $[t_0, t_1] \subseteq J$. Then $\omega([t_0, t_1])$ is, by definition, a curve with initial point $\omega(t_0)$ and endpoint $\omega(t_1)$, and is denoted again by ω. If ω'_t is the tangent vector of ω at t, the length of this curve is defined by

$$l(\omega) = \int_{t_0}^{t_1} |\omega'_t|\, dt. \tag{5}$$

Now, we can obtain, in the obvious manner, the definition of the length of piecewise differentiable curves.

Let ξ be a tangent vector of V at x. Then $g(\xi, \cdot)$ defines a tangent covector at the same point and, by a classical theorem of Riesz for Hilbert spaces, it is known that in this manner we get an isomorphism of $T_x(V)$

onto $T_x^*(V)$. This gives the so-called operation of raising and lowering indices which associates with $g(\xi, \cdot)$ the vector ξ and conversely. The operation naturally extends to all types of tensors. In the finite dimensional case, raising and lowering indices are given componentwise by

$$t^{\cdots\cdots}_{\cdots h\cdots} = g_{hi}t^{\cdots i\cdots}_{\cdots\cdots}, \quad t^{\cdots h\cdots}_{\cdots\cdots} = g^{hi}t^{\cdots\cdots}_{\cdots i\cdots},$$

where (g^{hi}) is the inverse matrix of (g_{hi}), whence we have $g_{ih}g^{hk} = \delta_i^k$. The matrix (g^{hi}) is also symmetric and nondegenerate. We remark that in order to perform the indicated operations we must preserve the positions of the indices. We also remark that these operations extend to tensor fields because g is a tensor field.

It is important to remember that by proposition 3.3.18 the existence of the Riemannian metric allows a reduction of the structure group of the tangent bundle of V to $\text{Is}(H)$, i.e., to the orthogonal group in the finite dimensional case. It follows that we can consider orthonormal bases in the tangent spaces and the principal bundle of such bases on V.

We can obtain a canonical connection on the tangent bundle of a Riemannian manifold, which enables us to use the absolute differential calculus. This is done by the following theorem.

2   Theorem   *If (V, g) is a Riemannian manifold, the tangent bundle $T(V)$ has a uniquely defined connection such that (a) $\nabla_\xi g = 0$ for every vector field ξ, (b) the torsion of the connection vanishes.*

In fact, the hypotheses imply that, for every triple ξ, η, ζ of vector fields on V,

$$\nabla_\xi g(\eta, \zeta) = \xi g(\eta, \zeta) - g(\nabla_\xi \eta, \zeta) - g(\eta, \nabla_\xi \zeta) = 0,$$
$$\nabla_\xi \eta - \nabla_\eta \xi - [\xi, \eta] = 0. \tag{6}$$

If in the first relation we cyclically permute ξ, η, ζ we get two other relations which, together with the first one, define a system of relations $R_i = 0$ ($i = 1, 2, 3$). The relation $R_1 - R_2 + R_3 = 0$ becomes, after using the second condition (6),

$$2g(\eta, \nabla_\xi \zeta) = \xi g(\eta, \zeta) - \eta g(\zeta, \xi) + \zeta g(\xi, \eta)$$
$$- g([\xi, \eta], \zeta) - g(\eta, [\zeta, \xi]) + g([\eta, \zeta], \xi) \tag{7}$$

and this defines $\nabla_\xi \zeta$.

5. Riemannian and Foliated Riemannian Manifolds

By using the local coordinate expressions one sees that this is the covariant differential with respect to a connection and that it satisfies the conditions (6). Hence, the theorem is proved.

Related to Theorem 2, we remark that condition (a) is also meaningful for a Hilbertian structure defined by a differentiable tensor field g of an arbitrary vector bundle on V. The connections of the bundle which satisfy this condition are called *Euclidean* or *metrical connections*, and (a) means just that the parallel translation preserves scalar products. The connection of Theorem 2 is the unique Euclidean connection with vanishing torsion on $T(V)$.

3 Definition The connection of Theorem 2 is called the *Levi–Civita connection* of the Riemannian manifold.

In the finite dimensional case, writing (6) in local coordinates, we get that the connection coefficients of the Levi–Civita connection are the *Christoffel symbols*

$$\begin{Bmatrix} m \\ k\ i \end{Bmatrix} = \tfrac{1}{2} g^{mj}\left(\frac{\partial g_{ij}}{\partial x^k} + \frac{\partial g_{jk}}{\partial x^i} - \frac{\partial g_{ki}}{\partial x^j}\right). \tag{8}$$

For Riemannian manifolds, the covariant derivatives will always be taken with respect to the Levi–Civita connection. Also, by definition, the geodesics of a Riemannian manifold are those of the Levi–Civita connection of the given metric. If the geodesic curves can be always extended to the real line (as defining interval J) the Riemannian manifold is called *complete* and this definition extends to arbitrary connections on $T(V)$.

4 Definition The *curvature* of a Riemannian manifold (or metric) is the curvature of its Levi–Civita connection.

From Formula 4.(14) defining the curvature tensor

$$R(\omega, \zeta, \xi, \eta) = \omega(\Omega(\eta, \xi)\zeta), \tag{9}$$

we get *Riemann's curvature tensor*

$$R(\tau, \zeta, \xi, \eta) = g(\tau, \Omega(\eta, \xi)\zeta). \tag{10}$$

By definition, this tensor is skew-symmetric in the last two arguments.

It is also skew-symmetric in the first two arguments as may be seen from (10) and (6).

To obtain other properties of the curvature tensor, we consider the Bianchi identity 4(36) and condition (b) of Theorem 2. From them we get the first *Bianchi identity* for Riemannian manifolds

$$R(\tau, \zeta, \xi, \eta) + R(\tau, \xi, \eta, \zeta) + R(\tau, \eta, \zeta, \xi) = 0. \tag{11}$$

The second Bianchi identity is obtained from 4(35). In the finite dimensional case, we have 4(37), where T vanishes and the index i is lowered to the first position.

If in (11) we cyclically permutate the arguments of the first term, we get four relations and, by adding them, we obtain the interesting property that R is symmetric with respect to the pairs of arguments (τ, ζ) and (ξ, η), i.e.,

$$R(\tau, \zeta, \xi, \eta) = R(\xi, \eta, \tau, \zeta). \tag{12}$$

Let $x \in V$ and P be the 2-plane of the space $T_x(V)$ defined by an orthonormal basis (ξ_1, ξ_2).

5   Definition   The *curvature* of the manifold V at P (the *sectional curvature*) is defined by

$$K(x, P) = -R(\xi_1, \xi_2, \xi_1, \xi_2). \tag{13}$$

It depends (as is easily seen) only on x and P.

In the finite dimensional case, we have

$$K(x, P) = -R_{ijkh}\xi_1^i \xi_2^j \xi_1^k \xi_2^h, \tag{14}$$

or, if P is defined by two arbitrary vectors ξ and η,

$$K(x, P) = \frac{R_{ijkh}\xi^i \eta^j \xi^k \eta^h}{(g_{jk}g_{ih} - g_{jh}g_{ik})\xi^i \eta^j \xi^k \eta^h}. \tag{15}$$

Recall some important results for the finite dimensional case. $K(x, P)$ does not depend on P if and only if it is constant (Schur's theorem [25]). If V is a complete Riemannian manifold which is simply connected and has nonpositive sectional curvature, V is diffeomorphic to a Euclidean space of the same dimension (the Hadamard–Cartan theorem [21]). There are generalizations of these results to the infinite dimensional case.

There is also the following generalization of the sectional curvature [47]. Let $p \leq \dim V$ be a positive even integer and P a p-dimensional plane of $T_x(V)$ defined by the orthonormal basis $\xi_1, \ldots, \xi_p$. The *p-sectional curvature* is defined by

$$K_p(x, P) = \frac{(-1)^{p/2}}{2^{p/2} p!} \sum \delta^{j_1 \cdots j_p}_{i_1 \cdots i_p} R(\xi_{i_1}, \xi_{i_2}, \xi_{j_1}, \xi_{j_2})$$
$$\cdots R(\xi_{i_{p-1}}, \xi_{i_p}, \xi_{j_{p-1}}, \xi_{j_p}). \tag{16}$$

This quantity does not change when the given orthonormal basis is changed since it is multiplied by the square of the determinant of the transformation, which is ± 1. Hence the sectional curvature is an invariant of P. When n is even, $K_n(x, T_x) = K(x)$ is a differentiable function on V called the *Lipschitz–Killing curvature* of V. For a surface, this is just Gauss' total curvature.

In the finite dimensional case, there is also another important tensor field on V, the *Ricci curvature tensor*, which is the Ricci tensor of the Levi–Civita connection of V. It will be denoted as in Section 4. An interesting class of Riemannian spaces is the class of spaces such that the Ricci tensor ρ is proportional to the metric tensor g. These spaces are called *Einstein spaces*. The Ricci tensor defines a linear operator on the tangent space of V and the trace of it, denoted by κ, is called the *scalar curvature* of the manifold V.

If $\varphi: V \to W$ is a differentiable map and g is a Riemannian metric on W, $\varphi_* g$ is a Riemannian metric on V. A diffeomorphism of two Riemannian manifolds is an *isometry* if it preserves the metric and it is a *conformal transformation* if it reproduces the metric up to a scalar factor. An infinitesimal transformation ξ on V is called *isometric* if it has the property

$$L_\xi g = 0 \tag{17}$$

and ξ is said to be a *Killing vector field* on V. An infinitesimal transformation is called *conformal* if

$$L_\xi g = \rho g. \tag{18}$$

With these we conclude our general information about Riemannian manifolds and consider in the sequel another structure.

6　Definition　A pair $(V, \mathfrak{D})$, where V is a differentiable manifold and $\mathfrak{D}$ a regular distribution on it, is called an *almost foliated manifold*. When $\mathfrak{D}$ is integrable we have a *foliated manifold*.

We know from Section 3 that the vectors of $\mathfrak{D}$ form a subbundle D of the tangent bundlte $T(V)$, whose fiber F is a direct summand of the model E of V. It follows that the structure group of $T(V)$ is reducible to the subgroup of $L(E, E)$ which leaves F invariant. In the finite dimensional case, denoting by $m + n$ the dimension of V and by m the dimension of $\mathfrak{D}$, we say that n is the *codimension* of $\mathfrak{D}$. Then, using bases of local cross sections with the last m vectors in $\mathfrak{D}$, we see that the structure group of the tangent bundle of V is reducible to the group of matrices of the form

$$\left(\begin{array}{c|c} \cdot & 0 \\ \hline \cdot & \cdot \end{array}\right)_m^n. \tag{19}$$

(More precisely, these are the matrices which give the transformations of the components of a vector.)

Returning to the general case, suppose that V is also given a Riemannian metric g (hence the model of V is a Hilbertian space H). Then, V is called a *Riemannian almost foliated manifold*. In this case, we have also the distribution $\mathfrak{D}^\perp$ of the subspaces of $T_x(V)$ ($x \in V$) which are orthogonal to those of $\mathfrak{D}$ and we call $\mathfrak{D}$ the *structural distribution* and $\mathfrak{D}^\perp$ the *transversal distribution* of V. The corresponding vector bundles will be: D, the *structural bundle*; and $D^\perp$, the *transversal bundle*. The relation

$$T(B) = D \oplus D^\perp \tag{20}$$

holds, which means that V has an *almost product structure*, and the structure group of $T(V)$ is reducible to the subgroup of $\mathrm{Is}(H)$ which leaves F invariant (and leaves invariant the subspace orthogonal to F as well); in the finite dimensional case this is the group of orthogonal matrices of the form

$$\left(\begin{array}{c|c} \cdot & 0 \\ \hline 0 & \cdot \end{array}\right)_m^n. \tag{21}$$

A differential form ω on a Riemannian almost foliated manifold is said to be of type (p, q) if it has degree $p + q$ and, for vectors ξ_i in D or in $D^\perp$, $\omega(\xi_1, \ldots, \xi_{p+q}) = 0$ except for at most the case when p of the arguments ξ_i are in $D^\perp$ and q of them are in D. Obviously, every form of degree r has a unique representation as a sum of forms of types (p, q) with

5. Riemannian and Foliated Riemannian Manifolds

$p + q = r$. Indeed, the *component* of the type (p, q) of ω is the form ω_{pq}, which is equal to ω for all cases when p arguments are in $D^\perp$ and q arguments are in D, and is equal to 0 in all other cases. We remark that this is true for every almost product structure.

For a Riemannian almost foliated manifold V, we can consider the Levi–Civita connection of the metric g on $T(V)$ and this will be called the *first connection* of V. We also introduce another connection canonically defined by the structure of V.

For an arbitrary connection on $T(V)$ and with the notation of Section 4, the torsion has the decomposition

$$T(\xi, \eta) = \tau(\xi, \eta) + \tau^\perp(\xi, \eta), \qquad (22)$$

where $\tau(\xi, \eta) \in \mathfrak{D}$ and $\tau^\perp(\xi, \eta) \in \mathfrak{D}^\perp$. There is also the decomposition

$$[\xi, \eta] = [\xi, \eta]_1 + [\xi, \eta]_2, \qquad (23)$$

where the two terms are in $\mathfrak{D}$ and $\mathfrak{D}^\perp$, respectively

7 Theorem [49] *On a Riemannian almost foliated manifold V there is a connection (on $T(V)$) uniquely defined by the conditions*: (a) *if $\eta \in \mathfrak{D}$ (respectively $\in \mathfrak{D}^\perp$), then $\nabla_\xi \eta \in \mathfrak{D}$ (respectively $\mathfrak{D}^\perp$) for every ξ;* (b) *if $\xi, \eta, \zeta \in \mathfrak{D}$ ($\mathfrak{D}^\perp$) then $(\nabla_\xi g)(\eta, \zeta) = 0$;* (c) *$\tau(\xi, \eta) = 0$ if at least one of the arguments is in $\mathfrak{D}$ and $\tau^\perp(\xi, \eta) = 0$ if at least one of the arguments is in $\mathfrak{D}^\perp$.*

Because of the linearity of $\nabla_\xi \eta$ in its two arguments, it suffices to obtain the covariant derivative only when ξ and η belong either to $\mathfrak{D}$ or to $\mathfrak{D}^\perp$ or $\xi \in \mathfrak{D}, \eta \in \mathfrak{D}^\perp$, or, finally, $\xi \in \mathfrak{D}^\perp, \eta \in \mathfrak{D}$.

Let $\xi \in \mathfrak{D}, \eta \in \mathfrak{D}^\perp$. We have

$$T(\xi, \eta) = \nabla_\xi \eta - \nabla_\eta \xi - [\xi, \eta]_1 - [\xi, \eta]_2,$$

whence, with (a) and (c),

$$\tau^\perp(\xi, \eta) = \nabla_\xi \eta - [\xi, \eta]_2 = 0,$$

which defines $\nabla_\xi \eta$ in this case.

Similarly, if $\xi \in \mathfrak{D}^\perp$ and $\eta \in \mathfrak{D}$, we have

$$\tau(\eta, \xi) = -\nabla_\xi \eta - [\eta, \xi]_1 = 0$$

and we again obtain the value of $\nabla_\xi \eta$.

Next, let $\xi, \zeta \in \mathfrak{D}$ and consider also $\eta \in \mathfrak{D}$. Condition (b) gives

$$(\nabla_\xi g)(\eta, \zeta) = \xi g(\eta, \zeta) - g(\nabla_\xi \eta, \zeta) - g(\eta, \nabla_\xi \zeta) = 0.$$

By the same procedure as in Theorem 2 and using (c) we get

$$2g(\eta, \nabla_\xi \zeta) = \xi g(\eta, \zeta) - \eta g(\zeta, \xi) + \zeta g(\xi, \eta)$$
$$- g([\xi, \eta]_1, \zeta) - g(\eta, [\zeta, \xi]_1) + g([\eta, \zeta]_1, \xi), \qquad (24)$$

whence, taking into account the condition (a), $\nabla_\xi \zeta$ ($\xi, \zeta \in \mathfrak{D}$) is defined.

In the same way, we get $\nabla_\xi \zeta$ for $\xi, \zeta \in \mathfrak{D}^\perp$, hence the given conditions define the connection uniquely. By a straightforward computation one sees that condition 4(10) and the linearity properties for a connection are satisfied, and by using the usual local coordinate expressions, we verify the differentiability properties, so that we really have a connection. This completes the proof of Theorem 7.

The geometrical meaning of the condition (a) is that $\mathfrak{D}$ and $\mathfrak{D}^\perp$ are invariant by any parallel translation. The meaning of (b) is that the length of the vectors of $\mathfrak{D}(\mathfrak{D}^\perp)$ is preserved by parallel translations along curves tangent to $\mathfrak{D}(\mathfrak{D}^\perp)$.

8 Definition The connection of Theorem 7 is called the *second connection* of the Riemannian almost foliated manifold V.

Generally, the elements of the first connection will be called the *first* elements and those of the second connection the *second* elements of V. We speak of the first curvature, the second curvature, etc.

We remark that it is possible to consider a more general *almost multifoliated* structure on a manifold which is defined by a *regular lattice of distributions* on V, i.e., a family of distributions closed under intersections and unions (defined by the intersections and unions of the respective subspaces at each point of V), such that this lattice is finite and distributive, contains the zero and the tangent distributions of V, and such that the corresponding lattices of subspaces at every point of V are isomorphic. A finite dimensional almost multifoliated manifold V is *Riemannian* if it is given a Riemannian metric *compatible* with the almost multifoliate structure in the following sense: at every point of V any two planes of the lattice of distributions are *weakly orthogonal*, i.e., if the planes are of dimensions s and t, respectively, and if their intersection is of dimension i then, the first plane contains $s - i$ independent vectors orthogonal to the second plane and the second plane contains $t - i$ independent vectors orthogonal to the first plane. It is shown in [48] that Theorem 7 has a generalization for such Riemannian manifolds. If the distributions of the given lattice are integrable, we get a *multifoliated manifold*, as studied in [26].

We now consider foliated manifolds V, i.e., manifolds together with integrable distributions $\mathfrak{D}$. We can see [8, 49] that these are $\mathfrak{P}$-manifolds for a convenient pseudocategory $\mathfrak{P}$. To obtain this, consider two product spaces $E \times F$ and $E' \times F'$ (i.e., direct topological sums). An arbitrary continuous linear map of $L(E \times F, E' \times F')$ can be written as a second-order matrix

$$\begin{pmatrix} a & b \\ c & d \end{pmatrix}$$

where $a \in L(E, E')$, $b \in L(E, F')$, $c \in L(F, E')$, and $d \in L(F, F')$. If $b = 0$, the linear map will be called *foliated*.

Now, we can introduce the pseudocategory $\mathfrak{P}f$ whose objects are open subsets of the product spaces $E \times F$ and whose morphisms are differentiable maps between such subsets such that their derivative at every point is a foliated linear map.

9 Proposition   *A foliated manifold is a $\mathfrak{P}f$-manifold such that the local charts take values in the same product space $E \times F$, and conversely.*

This is a straightforward consequence of the definition of an integrable distribution and of some properties of such distributions given in [27]. We omit the details of the proof.

Proposition 9 is important because it enables us to apply in the case of foliated manifolds the general theory of $\mathfrak{P}$-manifolds. Thus we can talk of $\mathfrak{P}f$-mappings and we shall call them *foliated mappings*. We can also consider the tangent space but, as is easily seen, this is the same as the tangent space defined by the differentiable structure of the manifold.

A differentiable manifold is a special case of a foliated manifold where the dimension of $\mathfrak{D}$ is zero. Hence we can consider foliated mappings of a foliated manifold into an arbitrary differentiable manifold and, in particular, foliated real functions on V. The last make up an algebra $\Psi(V)$ of foliated functions and the corresponding sheaf of germs will be denoted by $\mathfrak{G}(V)$.

10 Definition   The maximal connected integral manifolds of $\mathfrak{D}$ are called the *leaves* of the foliated manifold V.

The study of topological and differential properties of the leaves and of the space of leaves is an essential problem in the theory of foliated manifolds, but lies beyond the scope of this book.

It follows from the definition of foliated functions that they are just the differentiable functions on V which remain constant on the leaves.

The tangent bundle $T(V)$ of the foliated manifold V also has a foliated structure. In fact, we know (Section 1) that it has a differentiable structure with the model $(E \times F) \times (E \times F)$, and it follows that for every $\mathfrak{P}f$-atlas on V whose coordinates take values in $E \times F$ there is a corresponding $\mathfrak{P}f$-atlas on $T(V)$ with coordinates taking values in $(E \times E) \times (F \times F)$, which justifies our assertion. The cross sections of $T(V)$ which are foliated maps $V \to T(V)$ are called *foliated vector fields* or *foliated infinitesimal transformations* on V, and the corresponding local one-parametric groups send leaves to leaves.

Next, we shall consider *Riemannian foliated manifolds*. The first important fact about these manifolds is that we can introduce a foliated exterior derivative [49]. Let α be a differential form of type (p, q), and consider $d\alpha$. From Formula 3(3) we see that the only possible nonzero components of $d\alpha$ are those of types $(p, q+1)$, $(p+1, q)$, and $(p+2, q-1)$, i.e., we have a decomposition

$$d = d_{01} + d_{10} + d_{2,-1},$$

where the indices denote the type of the respective operators. The *foliated exterior derivative* is defined as $d_f = d_{01}$ and it extends by linearity to all the differential forms on V. From 3(3)

$$d_f\alpha(\xi_1, \ldots, \xi_p, \xi_{p+1}, \ldots, \xi_{p+q+1})$$

$$= \sum_{i=1}^{q+1}(-1)^{i+p+1}\xi^{p+i}\alpha(\xi_1, \ldots, \xi_p, \xi_{p+1}, \ldots, \hat{\xi}_{p+i}, \ldots, \xi_{p+q+1})$$

$$+ \sum_{i<j=1}^{q+1}(-1)^{i+j}\alpha(\xi_1, \ldots, \xi_p, [\xi_{p+i}, \xi_{p+j}],$$

$$\xi_{p+1}, \ldots, \hat{\xi}_{p+i}, \ldots, \hat{\xi}_{p+j}, \ldots, \xi_{p+q+1})$$

$$+ \sum_{i=1}^{p}\sum_{j=1}^{q}(-1)^{i+j+p}\alpha([\xi_i, \xi_{p+j}]_2, \xi_1, \ldots, \hat{\xi}_i, \ldots,$$

$$\xi_p, \xi_{p+1}, \ldots, \hat{\xi}_{p+j}, \ldots, \xi_{p+q+1}), \qquad (25)$$

when $\xi_1, \ldots, \xi_p \in \mathfrak{D}^\perp$, $\xi_{p+1}, \ldots, \xi_{p+q+1} \in \mathfrak{D}$ (hence, because of Fro-

5. Riemannian and Foliated Riemannian Manifolds

benius' theorem, $[\xi_{p+i}, \xi_{p+j}]$ are also in $\mathfrak{D}$), and

$$d_f\alpha(\xi_1, \ldots, \xi_{p+q+1}) = 0$$

when the number of arguments in $\mathfrak{D}^\perp$ is not equal to p, all the arguments being in $\mathfrak{D}$ or in $\mathfrak{D}^\perp$. The value of $d_f\alpha$ is then extended by linearity and skew-symmetry to all the other possible arguments.

In this manner, we get a differential form $d_f\alpha$ of type $(p, q + 1)$, since the local coordinate expression of Formula (25) is, by 3(4) (omitting the chart and the point indices),

$$d_f\alpha(\xi_1, \ldots, \xi_p, \xi_{p+1}, \ldots, \xi_{p+q+1}) \tag{26}$$

$$= \sum_{i=1}^{q+1} (-1)^{p+i+1} \alpha'(\xi_{p+i})(\xi_1, \ldots, \xi_p, \xi_{p+1}, \ldots, \hat{\xi}_{p+i}, \ldots, \xi_{p+q+1}).$$

Hence, the differentiability of $d_f\alpha$ follows just as the differentiability of the exterior derivative in Section 3.

11 Proposition *The foliated exterior derivative satisfies the following relations:*
(a) $d_f(\alpha \wedge \beta) = d_f\alpha \wedge \beta + (1)^{\deg \alpha} \alpha \wedge d_f\beta$,
(b) $d_f^2\alpha = 0$,
(c) $d_f(\varphi_*\alpha) = \varphi_*(d_f\alpha)$
for $\varphi: V \to W$ foliated, and the metric on V being induced by φ from the metric on W.

All these properties are easily deducible from the local coordinate expression (26).

12 Definition A differential form α of degree p on V is called a *foliated form* if is of type $(p, 0)$ and $d_f\alpha = 0$.

When $p = 0$ these are just the foliated functions.

For the Riemannian foliated manifolds we can use the second connection defined by Theorem 7. We only remark that, in the notation of this theorem, we have $[\xi, \eta]_1 = [\xi, \eta]$ if $\xi, \eta \in \mathfrak{D}$, and Formula (24) shows that on every leaf the covariant derivative with respect to the second connection is just the covariant derivative with respect to the Levi–Civita connection of the metric induced on the leaf by the metric of V.

An important class of Riemannian foliated manifolds was introduced by B. Reinhart in [38]. We shall define these manifolds, as in [49], by

13 Definition A Riemannian foliated manifold is called a *Reinhart space* if and only if

$$(\nabla_\xi g)(\eta, \zeta) = 0 \quad \text{for} \quad \xi \in \mathfrak{D}, \eta, \zeta \in \mathfrak{D}^\perp, \tag{27}$$

where the covariant derivative is taken with respect to the second connection of the manifold.

Geometrically, the condition (27) means that the length of a vector of $\mathfrak{D}^\perp$ is preserved by parallel translation with respect to the second connection along a curve which is tangent to $\mathfrak{D}$.

Some of the previous results extend to the case of manifolds V modelled on Banach spaces by considering a foliation and a supplementary distribution of it. This case was treated in [8].

We give the coordinate expressions of the introduced elements in the finite dimensional case, following [49].

Suppose that the indices take the following values: $a, b, \ldots = 1, \ldots, n$; $u, v, \ldots = n+1, \ldots, n+m$, where $n + m = \dim V$ and n is the codimension of the foliation. The expressions will be considered for charts of a $\mathfrak{P}f$-atlas, which are local charts (x^a, x^u) on V such that the structural distribution $\mathfrak{D}$ has the equation

$$dx^a = 0. \tag{28}$$

In fact, the coordinate transformations between such charts are

$$\tilde{x}^a = \tilde{x}^a(x^a, x^u), \quad \tilde{x}^u = \tilde{x}^u(x^a, x^u)$$

with

$$\frac{\partial \tilde{x}^a}{\partial x^u} = 0. \tag{29}$$

Considering the equations of $\mathfrak{D}^\perp$ in the form

$$\theta^u = dx^u + t_a^u\, dx^a = 0, \tag{30}$$

we have in $T_x^*(V)$ the cobases $\{dx^a, \theta^u\}$, and it follows that the dual bases are

$$X_a = \frac{\partial}{\partial x^a} - t_a^u \frac{\partial}{\partial x^u}, \quad X_u = \frac{\partial}{\partial x^u}, \tag{31}$$

5. Riemannian and Foliated Riemannian Manifolds

such that $\mathfrak{D}$ is given by $\{X_u\}$ and $\mathfrak{D}^\perp$ by $\{X_a\}$. The components of all the tensors will be considered with respect to these bases and cobases.

The coordinate expression of the metric g of the manifold is

$$ds^2 = h_{ab}\,dx^a\,dx^b + h_{uv}\theta^u\theta^v, \tag{32}$$

and we shall denote by $(\tilde{h}^{ab})$ and $(\tilde{h}^{uv})$ the inverse matrices of (h_{ab}) and (h_{uv}), respectively. h_{uv} is the metric induced on the leaves.

If W is another Riemannian foliated manifold of dimension $p + q$ and codimension p, and if $\varphi: V \to W$ is a map with coordinate expression

$$y^\alpha = y^\alpha(x^a, x^u), \qquad y^\sigma = y^\sigma(x^a, x^u),$$

$\alpha = 1, \ldots, p$; $\sigma = p + 1, \ldots, p + q$, where (y^α, y^σ) is a chart of a $\mathfrak{P}f$-atlas of W, then φ is foliated if and only if

$$\frac{\partial y^\alpha}{\partial x^u} = 0. \tag{33}$$

In particular, a differentiable function φ on V is foliated if and only if

$$\frac{\partial \varphi}{\partial x^u} = 0 \tag{34}$$

and a vector field (ξ^a, ξ^u) is foliated if and only if

$$\frac{\partial \xi^a}{\partial x^u} = 0. \tag{35}$$

A differential form α of the type (p, q) has the local expression

$$\alpha = \frac{1}{p!\,q!}\alpha_{a_1\cdots a_p u_1\cdots u_q}\,dx^{a_1} \wedge \cdots \wedge dx^{a_p} \wedge \theta^{u_1} \wedge \cdots \wedge \theta^{u_q}, \tag{36}$$

with differentiable coefficients which are skew-symmetric separately in the indices a and in the indices u.

The operator d_f is given by

$$d_f\alpha = \frac{(-1)^p}{p!\,q!\,(q+1)!}\delta^{uv_1\cdots v_q}_{u_1\cdots u_{q+1}}\frac{\partial \alpha_{a_1\cdots a_p v_1\cdots v_q}}{\partial x^u}\,dx^{a_1} \wedge \cdots$$
$$\wedge dx^{a_p} \wedge \theta^{u_1} \wedge \cdots \wedge \theta^{u_{q+1}}. \tag{37}$$

Imposing the conditions of Theorem 7, one gets the following expressions for the coefficients of the second connection of V:

$$\Gamma^a_{ub} = \Gamma^a_{uv} = \Gamma^u_{ab} = \Gamma^u_{av} = \Gamma^a_{bu} = 0, \qquad \Gamma^u_{va} = X_v t^u_a, \tag{38}$$

$$\Gamma^a_{bc} = \tfrac{1}{2}\tilde{h}^{ad}(X_c h_{bd} + X_b h_{cd} - X_d h_{bc}),$$
$$\Gamma^u_{vw} = \tfrac{1}{2}\tilde{h}^{ut}(X_w h_{vt} + X_v h_{wt} - X_t h_{vw}). \tag{39}$$

Now, the relation (35) becomes

$$\zeta^a_{/u} = 0 \tag{40}$$

and the relation (27) which defines the Reinhart spaces and which means

$$h_{ab/u} = 0 \tag{41}$$

is the same as

$$\frac{\partial h_{ab}}{\partial x^u} = 0. \tag{42}$$

In these formulas the symbol / denotes the covariant derivative with respect to the second connection of the manifold.

6 Complex and Almost Complex Manifolds

In this section we shall consider a new type of manifold, the complex manifolds. First, we introduce them independently and then we consider their connection with differentiable manifolds.

Consider the pseudocategory (which is similar to B^p used for differentiable manifolds) whose objects are open subsets of complex Banach spaces and whose morphisms are partial complex differentiable mappings of such subsets. We denote this pseudocategory by B^p_c.

1 Definition A *complex manifold* is a B^p_c-manifold (in the sense of the theory of $\mathfrak{P}$-manifolds (Section 1.6)) whose local charts take values in a fixed complex Banach space, called the *model*.

6. Complex and Almost Complex Manifolds

Hence we can use the theory of $\mathfrak{P}$-manifolds. Thus the B_c^p-mappings are called *holomorphic mappings*, and the functions $\varphi: V \longrightarrow C$, which are holomorphic mappings of the complex manifold V to the complex field C, with the natural B_c^p-structure, are called *holomorphic functions* on V. The set of holomorphic functions defines a C-algebra denoted $O(V)$ and the corresponding sheaf of germs will be denoted by $\Omega(V)$. Next, just as for differentiable manifolds, we get the tangent space $T_x(V)$ ($x \in V$) and the tangent bundle $T(V) = \bigcup_{x \in V} T_x(V)$, which is a complex vector bundle whose fiber is the model of V.

For the complex manifolds we obviously have a topology and we accept again Convention 1.7, i.e., that the manifold is a Hausdorff space. We can introduce in the usual manner the complex submanifolds and the product of two complex manifolds.

Since the complex Banach spaces are also real Banach spaces and complex differentiability implies real differentiability, it follows that every complex manifold V has a subordinate structure of a differentiable manifold, which we shall always have in mind when talking of differentiable elements on V. The tangent bundle of the differentiable structure of V is the tangent bundle of the complex structure with its natural structure of a real vector bundle.

Denoting by Hol (form holomorphic) the category of complex manifolds and holomorphic maps our previous assertions mean that there is a covariant functor Hol → Diff, which we denote by F_d. The converse is not true: not every differentiable structure is subordinated to a complex structure.

In the remaining part of this section, we limit our discussion to the finite dimensional case, the one used in the next chapter. Then, the complex Banach spaces used are C^n for different n and, in order to characterize the complex differentiable maps between them, it suffices to know what a complex differentiable function $f: C^n \longrightarrow C$ is, on an open subset of C^n.

Denote the complex coordinates on C^n by

$$z^\alpha = x^\alpha + iy^\alpha, \qquad \alpha = 1, \ldots, n; i = \sqrt{-1} \tag{1}$$

and consider the class of complex-valued real differentiable functions on x^α, y^α. Such a function can be expressed as $f(z^\alpha, \bar{z}^\alpha)$ where $\bar{z}^\alpha$ are the complex conjugates of z^α.

Introduce on $C^n = R^{2n}$ (R is the real field) the complex-valued Pfaff forms

$$dz^\alpha = dx^\alpha + i\,dy^\alpha, \qquad d\bar{z}^\alpha = dx^\alpha - i\,dy^\alpha \tag{2}$$

and express df using dz^α, $d\bar{z}^\alpha$. The coefficients of this expression are

$$\frac{\partial f}{\partial z^\alpha} = \frac{1}{2}\left(\frac{\partial f}{\partial x^\alpha} - i\frac{\partial f}{\partial y^\alpha}\right), \qquad \frac{\partial f}{\partial \bar{z}^\alpha} = \frac{1}{2}\left(\frac{\partial f}{\partial x^\alpha} + i\frac{\partial f}{\partial y^\alpha}\right). \qquad (3)$$

Thus, we have a kind of partial differentiation which can be easily seen to satisfy all the usual rules of the differential calculus.

It is a classical result in the theory of functions of a complex variable that the function $f: C^n \longrightarrow C$ is complex differentiable if and only if f is a real differentiable function on (x^α, y^α) and satisfies the *Cauchy–Riemann conditions* [9],

$$\frac{\partial f}{\partial \bar{z}^\alpha} = 0. \qquad (4)$$

Then, f has a Taylor series development in the neighborhood of every point (z^α) of the domain where it is differentiable and we say that f is an *analytic function*; this is a name which will sometimes replace the name holomorphic function (in the finite dimensional case). Moreover, in this case the $\partial f/\partial z^\alpha$ of (3) are just the complex partial derivatives of the function f and the derivative of f as described in Definition 1.1 is the linear map

$$(\xi^\alpha) \mapsto \left(\frac{\partial f}{\partial z^\alpha}\xi^\alpha\right).$$

It follows that on a complex n-dimensional manifold we have local complex coordinates z^α and on the intersection $U \cap U'$ of coordinate neighborhoods the functions $z_{U'}^\alpha(z_U^\beta)$ ($\alpha, \beta = 1, \ldots, n$) are analytic. The subordinate real differentiable structure is defined by the real local coordinates (x^α, y^α) introduced in (1). As local bases for the tangent bundle we can consider $\partial/\partial z^\alpha$ and the dual cobases are dz^α. But, if we wish to identify the vector fields with derivations of the algebra of differentiable functions, we must express them locally by

$$a^\alpha \frac{\partial f}{\partial x^\alpha} + b^\alpha \frac{\partial f}{\partial y^\alpha}$$

which, because of (3) is

$$\xi^\alpha \frac{\partial f}{\partial z^\alpha} + \bar{\xi}^\alpha \frac{\partial f}{\partial \bar{z}^\alpha}.$$

6. Complex and Almost Complex Manifolds

In this form the multiplication by a scalar λ gives

$$(\lambda \xi^\alpha) \frac{\partial f}{\partial z^\alpha} + (\overline{\lambda \xi^\alpha}) \frac{\partial f}{\partial \bar{z}^\alpha}.$$

2 Proposition *If the differentiable structure of a manifold V is subordinated to a complex finite dimensional structure, the real dimension of V is even and V is orientable,*

The first assertion is trivial and the second follows from the fact that on every coordinate neighborhood we can consider the differential form

$$\bigwedge_\alpha (dx^\alpha \wedge dy^\alpha) = \left(\frac{i}{2}\right)^n \bigwedge_\alpha (dz^\alpha \wedge d\bar{z}^\alpha),$$

and on the intersection of two coordinate neighborhoods these forms differ only by the positive factor

$$\left|\frac{\partial z_{U'}^\alpha}{\partial z_U^\beta}\right| \cdot \left|\overline{\frac{\partial z_{U'}^\alpha}{\partial z_U^\beta}}\right|.$$

But the conditions of Proposition 2 do not assure us that the differentiable structure of V is subordinated to a complex structure. The problem of finding the range of the functor F_d is important, and we shall consider it in the finite dimensional case.

If V^{2n} is a $2n$-dimensional differentiable manifold whose structure is subordinated to a complex structure, it is obvious that the structure group of its tangent bundle is reducible to $GL(n, C)$; i.e., because of proposition 3.3.21, the tangent bundle is almost complex. This leads to the following definition.

3 Definition A differentiable manifold whose tangent bundle has a reduction of the structure group to the complex general linear group is called an *almost complex manifold* and a reduction of the structure group is an *almost complex structure* of the manifold. For an almost complex manifold, we shall consider a fixed almost complex structure. An almost complex structure is called *integrable* if it is defined by a complex structure of the manifold.

Hence, the range of F_d is the class of manifolds with almost complex integrable structure. But it is important to note that there are also nonintegrable almost complex structures. Any almost complex manifold satisfies the conditions of Proposition 2, as follows by corollary 3.3.23. These conditions are necessary but they are not sufficient for a manifold to have an almost complex structure. We shall not study the problem of finding sufficient conditions for the existence of an almost complex structure on V, but we mention a condition for a given almost complex structure to be integrable.

4   Proposition   *An almost complex structure on V^{2n} is equivalent to each of the two following configurations*:
   (a) *A differentiable field J of endomorphisms of the tangent spaces with $J^2 = -\text{Id}$*.
   (b) *A decomposition $T^c(V) = T_1 \oplus \overline{T}_1$ where $T^c(V)$ is the complexification of $T(V)$, T_1 is a complex subbundle with complex n-dimensional fiber of $T^c(V)$ and $\overline{T}_1$ is the set of vectors which are complex conjugates of the vectors of T_1*.

The first part can be proved in a manner similar to the proof of Proposition 3.3.21, the only difference being that J is differentiable.

For the second part, extend J by linearity to $T^c(V)$. On each local fiber the decomposition indicated is the decomposition of the space as the direct sum of the eigenspaces of the operator J. It is easy to see that these spaces are

$$T_{1x} = (J_x + iI)T_x(V), \qquad \overline{T}_{1,x} = (J_x - iI)T_x(V),$$

where I is the identity operator and the corresponding eigenvalues are $\pm i$.

Conversely, if the decomposition (b) is given we obviously obtain from it the field J as follows. For a real tangent vector v we have a unique decomposition $v = w + \overline{w}$ with $w \in T_{1,x}$ and we put $Jv = i(w - \overline{w})$.

5   Theorem   *An almost complex structure on the differentiable manifold V^{2n} is integrable if and only if*

$$N(\xi, \eta) = [J\xi, J\eta] - J[J\xi, \eta] - J[\xi, J\eta] - [\xi, \eta] = 0 \qquad (5)$$

for every pair of vector fields (ξ, η).

The corresponding tensor field N of type (1, 2) is called the *Nijenhuis tensor* of J.

In order to prove the theorem, we see first that the almost complex structure defined by J is integrable if and only if we can introduce local

6. Complex and Almost Complex Manifolds

complex coordinates z^α ($\alpha = 1, \ldots, n$) such that the dz^α generate locally the space $(J_x + iI)T_x^*(V)$. If J has (local) components F_β^α, this means that we must have the complex complete integrability of the Pfaff system

$$\theta^\alpha = (F_\beta^\alpha + i\delta_\beta^\alpha)dx^\beta = 0.$$

When V and J are of class C^ω, i.e., real analytic, the condition for the complete integrability of this system is given by the theorem of Frobenius (Section 3), $d\theta^\alpha = 0 \pmod{\theta^\alpha}$, which after the necessary calculations gives (5). A complete proof is given for instance in [25]. When V and J are differentiable the theorem is also true, but the proof is much more complicated, it was first given by Newlander and Nirenberg [34].

The almost complex manifolds can be introduced in the infinite dimensional case too, by the condition that the operator J be given by a differentiable section of $L(T(V), T(V))$, but we do not study such manifolds here.

On the almost complex and complex manifolds it is important to consider the tensor and exterior calculus associated with the bundle $T^c(V)$. It has a similar development with the calculus of Sections 2, 3, and 4, and we consider here only some specific aspects.

Since we are in the finite dimensional case a differential form on V of degree r and with complex values can be identified with a complex linear skew-symmetric function

$$\underbrace{\Gamma(T^c(V)) \times \cdots \times \Gamma(T^c(V))}_{r \text{ times}} \to \Phi^c(V)$$

where $\Phi^c(V)$ is the algebra of complex-valued differentiable functions on V and Γ denotes as usual the set of global cross sections.

From Proposition 4(b) we have on $T^c(V)$ a so-called *complex almost product structure*, we can define the forms of type (p, q) just as for the Riemannian almost foliated manifolds in Section 5, and an arbitrary form of degree r has a unique decomposition into a sum of forms of types (p, q) with $p + q = r$.

The exterior product of complex differential forms is calculated as in Section 3. We can introduce the complex conjugate of a form and the form will be a real one if and only if it is equal to its complex conjugate form. For the exterior derivative we shall give the definition by Formula 3(3), where the action of a vector field on a function and the bracket operation extend by linearity to the complexification of the tangent bundle.

In the case of a complex manifold, if we consider the decomposition of Proposition 4(b), T_1 has locally the basis $(\partial/\partial z^\alpha)$ and $\overline{T}_1$ has the basis $(\partial/\partial \bar{z}^\alpha)$, whence we obtain that the bracket of two vector fields of T_1 is in T_1 and that the same is true for $\overline{T}_1$. It follows that we can introduce an exterior derivative analogous with the foliated derivative given by 5(25), having the type $(0, 1)$, which will be denoted by $d_{\bar{z}}$. But we can also define a similar derivative d_z of type $(1, 0)$ by working on the first p arguments. The local coordinate expressions of these operators are obtained in the following way. As we already know, $T^c(V)$ has the local bases $(\partial/\partial z^\alpha, \partial/\partial \bar{z}^\alpha)$ and their dual cobases are $(dz^\alpha, d\bar{z}^\alpha)$, given by (2). It follows that a differential form α of type (p, q) has the local expression

$$\alpha = \frac{1}{p!q!} a_{\alpha_1\cdots\alpha_p\beta_1\cdots\beta_q} dz^{\alpha_1} \wedge \cdots \wedge dz^{\alpha_p} \wedge d\bar{z}^{\beta_1} \wedge \cdots \wedge d\bar{z}^{\beta_q}, \qquad (6)$$

where the $a \ldots$ are complex-valued differentiable functions on the respective coordinate neighborhood and they are skew-symmetric in $\alpha_1, \ldots, \alpha_p$ and $\beta_1, \ldots, \beta_q$. The operators d_z and $d_{\bar{z}}$ are obtained by considering the parts of type $(p + 1, q)$ and $(p, q + 1)$, respectively, of the exterior derivative $d\alpha$. This means that we have

$$d = d_z + d_{\bar{z}} = d_{10} + d_{01}. \qquad (7)$$

From $d^2 = 0$, we get

$$d_z^2 = d_{\bar{z}}^2 = 0, \qquad d_z d_{\bar{z}} + d_{\bar{z}} d_z = 0. \qquad (8)$$

The derivatives of an exterior product are given by the same rule as for d.

6 Definition A differential form α of degree p on the complex manifold V is called a *holomorphic or analytic form* if it is of type $(p, 0)$ and $d_{\bar{z}}\alpha = 0$.

We shall denote by $O^p(V)$ the space of analytic forms of degree p and by $\Omega^p(V)$ the corresponding sheaf of germs. For $p = 0$ we obtain the analytic functions on V.

Next, note that the theory of connections and the absolute differential calculus can be transposed without change to the case of a differentiable complex vector bundle S on a differentiable manifold V. It suffices to consider that in Formula 4(1) f is a germ of a complex valued differentiable

6. Complex and Almost Complex Manifolds

function on V, and to take this into account for the remainder of the development.

Now let V be an almost complex manifold and $S = T(V)$, which we know has a structure of a complex bundle. A complex connection on $T(V)$ is called an *almost complex connection on* V. Obviously, every such connection is at the same time a real connection on $T(V)$, but the converse is false. We obtain the condition for a real connection $T(V)$ to be complex by requiring 4(1) to be valid for $f = i$. Then, in the notation of formula 4(1), the condition is

$$D(Js)(t) = JDs(t), \qquad (9)$$

or, by using the covariant differential

$$\nabla_\xi(J\eta) = J(\nabla_\xi \eta), \qquad (10)$$

for any two vector fields ξ and η on V. We recall that, because we are in the finite dimensional case, $\nabla_\xi \eta$ for global vector fields defines the connection.

For an almost complex connection on V, we can again use the local expressions 4(5) or 4(6), where, now, the connection matrix ω consists of complex Pfaff forms. Correspondingly, the curvature forms of the connection will be given by 4(18) and the torsion forms by 4(22). By asking the forms θ^i in 4(22) to be of type $(1, 0)$ we see that *the existence of an almost complex connection with vanishing torsion is a condition equivalent to the integrability of the almost complex structure of* V, since it is equivalent to the complete integrability of the Pfaff system $\theta^i = 0$.

When V is a complex manifold, its almost complex connections are called *complex connections*.

7 Definition An *almost Hermitian manifold* is an almost complex manifold together with a differentiable Hermitian metric on its tangent bundle as defined in Section 3.3 (with continuity replaced by differentiability). If, moreover, the almost complex structure is integrable, the manifold is called *Hermitian*.

It is easily seen [6] that the real part of a Hermitian metric defines a Euclidean metric, whence we have a Riemannian metric on every almost Hermitian manifold. Conversely, a Riemannian metric on an almost complex manifold is the real part of a Hermitian metric if and only if [6]

$$(v, w) = (Jv, Jw). \tag{11}$$

The imaginary part of a Hermitian metric defines a bilinear skew-symmetric form and, hence, a corresponding exterior quadratic form. When this last form is closed the manifold V is called *almost Kählerian* or, in the integrable case, a *Kählerian manifold*.

The previous statements are easy consequences of the characteristic Hermitian property $(w, v) = \overline{(v, w)}$, and the proofs are left to the reader.

In the case of a complex manifold, a Hermitian metric will be expressed in local coordinates in the form

$$ds^2 = h_{\alpha\beta} dz^\alpha d\bar{z}^\beta, \tag{12}$$

the product on the right-hand side being tensorial and

$$h_{\alpha\beta} = \bar{h}_{\beta\alpha}. \tag{13}$$

On the intersection $U \cap U'$ of two coordinate neighborhoods we have

$$h_{\alpha\beta} = \frac{\partial z_U^\lambda}{\partial z_{U'}^\alpha} \frac{\partial \bar{z}_U^\mu}{\partial \bar{z}_{U'}^\beta} h_{\lambda\mu}. \tag{14}$$

The exterior quadratic form associated with the metric is, up to a constant factor, the form of type (1, 1) given by

$$\gamma = \frac{i}{2} h_{\alpha\beta} dz^\alpha \wedge d\bar{z}^\beta, \tag{15}$$

where the constant factor was introduced in order to have a real form γ. This is the *fundamental form* of the Hermitian metric and

$$d\gamma = 0 \tag{16}$$

characterizes the Kählerian manifolds.

If $(h^{\alpha\beta})$ is the inverse matrix of $(h_{\alpha\beta})$, we have

$$h_{\alpha\beta} h^{\beta\gamma} = \delta_\alpha^\gamma, \qquad h^{\alpha\beta} = \bar{h}^{\beta\alpha} \tag{17}$$

and on $U \cap U'$

6. Complex and Almost Complex Manifolds

$$h_{U'}^{\alpha\beta} = \frac{\partial \bar{z}_{U'}^{\alpha}}{\partial \bar{z}_{U}^{\lambda}} \frac{\partial z_{U'}^{\beta}}{\partial z_{U}^{\mu}} h_{U}^{\lambda\mu}. \tag{18}$$

With $h_{\alpha\beta}$ and $h^{\alpha\beta}$ we can raise and lower indices as in the case of the Riemannian metrics.

We shall now derive an almost complex connection canonically associated with an almost Hermitian manifold (see, for instance, [25]).

8 Theorem[†] *Let V^{2n} be an almost Hermitian manifold with almost complex structure J and Riemannian metric g of the Hermitian structure. Then, on V there is a unique connection such that:*
(a) $(\nabla_\xi g)(\eta, \zeta) = 0$,
(b) $\nabla_\xi (J\eta) = J(\nabla_\xi \eta)$,
(c) $T(J\xi, \eta) = T(\xi, J\eta)$,
where ξ, η, ζ are arbitrary vector fields on V and T is the torsion of the connection.

Condition (b) is the condition for an almost complex connection; we shall call it the *Hermitian connection* of V.

We begin with the remark that, if ∇ is a connection on V, we can define its *complexification* ∇^c as an operator given by

$$\nabla^c_\xi \eta = \nabla_\xi \eta, \qquad \nabla^c_{(i\xi)} \eta = \nabla^c_\xi (i\eta) = i \nabla_\xi \eta,$$

where ξ and η are real vector fields, and which is extended by linearity to arbitrary complex vector fields on V (i.e., differentiable cross sections of $T^c(V)$). ∇^c is complex linear with respect to ξ and η, and satisfies the conditions

$$\nabla^c_{f\xi} \eta = f \nabla^c_\xi \eta, \qquad \nabla^c_\xi (f\eta) = f \nabla^c_\xi \eta + (\xi f)\eta, \qquad f \in \Phi^c(V). \tag{19}$$

We can also consider the *complexification* g^c of g as the Hermitian metric of $T^c(V)$ given by

$$g^c(i\xi, \eta) = ig(\xi, \eta), \qquad g^c(\xi, i\eta) = -ig(\xi, \eta),$$

for real vectors ξ and η, and extended by linearity to complex vectors.

By straightforward calculations one sees that the conditions (a), (b),

---

[†] For a generalization of this theorem see our note: *Connexions remarquables sur les variétés hor-ehresmanniennes.* C. R. Acad. Sci. Paris, Ser. A, **273**, 1253–1256 (1971).

and (c) of the theorem are equivalent to the fact that ∇^c satisfies the conditions

(a') $\zeta g^c(\xi, \eta) - g^c(\nabla^c_\zeta \xi, \eta) - g^c(\nabla^c_\zeta \bar{\eta}, \bar{\xi}) = 0$
(the bar means the complex conjugate),

(b') $\nabla^c_\xi(J\eta) = J(\nabla^c_\xi \eta)$,

(c') $T^c(\xi, \eta) = \nabla^c_\xi \eta - \nabla^c_\eta \xi - [\xi, \eta] = 0$

if $\xi \in T_1$ and $\eta \in \bar{T}_1$, which are the spaces of Proposition 4(b).

These conditions define the complex linear operator ∇^c such that (19) is satisfied. Indeed, it suffices to get $\nabla^c_\xi \eta$ for the cases

(1) $\xi \in T_1$, $\eta \in \bar{T}_1$,
(2) $\xi \in \bar{T}_1$, $\eta \in T_1$,
(3) $\xi, \eta \in T_1$,
(4) $\xi, \eta \in \bar{T}_1$.

(1) and (b') above and the expressions previously given for $T_{1,x}$, $\bar{T}_{1,x}$ show that $\nabla^c_\xi \eta \in T_1$, and it is obtained by separating in (c') the terms in T_1 and the terms in $\bar{T}_1$. A similar procedure determines $\nabla^c_\xi \eta \in \bar{T}_1$ in (2). In the cases (3) and (4) the result is obtained from (a') with ξ, η, ζ in T_1 and $\bar{T}_1$, respectively, and from the results of the cases (1) and (2). The verification of the conditions (19) is by straightforward calculation.

The restriction of the operator ∇^c above to real vector fields gives just the connection sought, and the theorem is proved.

We can give an interesting local expression of the conditions of Theorem 8. To do this, we consider the connection given by Equations 4(5),

$$De_\alpha = \omega^\beta_\alpha e_\beta$$

where e_α is a local complex basis of the tangent bundle and ω^β_α are complex Pfaff forms and observe that ∇^c can be obtained from D, just as ∇ was obtained, by formula 4(9), if we put $D\bar{e}_\alpha = \bar{\omega}^\beta_\alpha \bar{e}_\beta$. Condition (b') is implicitly contained in the equations of the connection because we started with an almost complex connection. Taking in (a') an arbitrary ζ with $\xi = e_\alpha$, $\eta = e_\beta$ we see it is equivalent to

$$dh_{\alpha\beta} - h_{\sigma\beta}\omega^\sigma_\alpha - h_{\alpha\sigma}\bar{\omega}^\sigma_\beta = 0, \qquad (20)$$

where $h_{\alpha\beta} = (e_\alpha, e_\beta)$ (the Hermitian scalar product).

Finally, (c') is easily seen to be equivalent to the fact that the torsion forms 4(22) are sums of forms of type (2, 0) and (0, 2), i.e., they do not have terms of type (1, 1). Hence this property and (20) are the expressions sought. With these expressions, the Hermitian connection could more

6. Complex and Almost Complex Manifolds

easily be obtained, but we prefer the proof of Theorem 8 given above because it generalizes to the infinite dimensional case, which we do not study here.

For a Hermitian manifold, we can take $e_\alpha = \partial/\partial z^\alpha$ and we get, from 4(22),

$$T^\alpha = -dz^\sigma \wedge \omega^\alpha_\sigma,$$

whence it follows that T^α does not have components of type (1, 1) and is of type (2, 0), if and only if the forms ω^β_α are of type (1, 0). This last condition has an invariant meaning because in Formula 4(8) we have $g = (\partial z^\alpha_{U'}/\partial z^\beta_U)$, and this is a matrix of analytic functions, hence $d_{\bar{z}} g = 0$ and dg is of type (1, 0). We thus obtain the following result.

9 Theorem *On a Hermitian manifold there is a unique complex connection satisfying* (20) *such that the local forms of the connection are of the type* (1, 0).

This is called the *Hermitian connection* of the manifold. It has torsion forms of type (2, 0) and curvature forms of type (1, 1).

The Hermitian connection of Theorem 8 is, of course, a Euclidean connection on $T(V)$ (condition (a)) but generally it does not coincide with Levi-Civita's connection of the metric g. If they coincide, the torsion of the Hermitian connection vanishes and V is a Hermitian manifold. Moreover, by straightforward calculations one sees that the two connections coincide if and only if V is a Kählerian manifold. If this is the case, the sectional curvatures of 2-planes which are one dimensional complex subspaces of the tangent space are called *holomorphic sectional curvatures*. Similarly, one can consider higher-order holomorphic sectional curvatures in the sense of Formula 5(16). This notion can be generalized by using the curvature of the Hermitian connection. Such questions are treated, for instance, in [6, 25, 47].

Condition (20) means geometrically that the Hermitian scalar product is preserved by parallel translation and it is obvious that this condition is meaningful for an arbitrary Hermitian finite dimensional complex vector bundle on a differentiable manifold V. Every connection of such a bundle which satisfies (20) is called a *general Hermitian connection of the bundle*.

The example of the tangent bundle of a complex manifold, which is itself a complex manifold and which has holomorphic projection and transition functions, leads to the definition of a new class of vector bundles where the topological spaces are complex manifolds and the projection, the transition functions and the charts are holomorphic. The structure group will also be a complex manifold, and the group operations will be holomorphic, i.e., the group will be a *complex Lie group*. The bundles of this class are called *analytic bundles*.

If S is an arbitrary analytic Hermitian vector bundle on the complex manifold V, the condition that the local forms of a connection on S be of type $(1, 0)$ is meaningful just as for the tangent bundle. If (20) is asked we get the following result.

10   Theorem   *On every Hermitian analytic vector bundle S there is a uniquely defined Hermitian connection having connection forms of type $(1, 0)$.*

For a detailed treatment of the differential geometry of complex and almost complex manifolds the reader is referred to [25] (and the literature cited therein).

Chapter 5

COHOMOLOGY CLASSES AND DIFFERENTIAL FORMS

1 The Theorems of de Rham and Allendoerfer–Eells

In the second chapter we studied the cohomology of topological spaces with coefficients in a sheaf. For the general case, there is no simple analytic expression for the cohomology classes. But for differentiable manifolds there is such an expression, using differential forms whose connection with the cohomology can be obtained from the abstract de Rham's Theorem 2.6.7. This chapter is devoted to the study of these expressions in some important cases.

We begin with the constant sheaf R on a differentiable manifold V, whose model is a Banach space with differentiable partitions of unity. This sheaf defines the classical real cohomology of the manifold.

1 Theorem (Poincaré's Lemma.) *Let E be a Banach space and U a convex open neighborhood of its origin. If ω is a differential form of degree $p \geqslant 1$ on U such that $d\omega = 0$, there is a differential form ψ on U such that $\omega = d\psi$.*

This is the same as saying that the de Rham complex of differential forms on U is acyclic, and the proof is obtained by constructing a complex homotopy between the identity and zero maps of this complex (Definition 2.4.3). We consider the same construction as in [27].

Given the form ω, we define on U a form $k\omega$ of degree $p - 1$ by

$$(k\omega)_x(\xi_1, \ldots, \xi_{p-1}) = \int_0^1 t^{p-1}\omega_{tx}(x, \xi_1, \ldots, \xi_{p-1})dt, \tag{1}$$

which is meaningful because if $x \in U$ and $0 \leqslant t \leqslant 1$, $tx \in U$ since U is convex.

Now, from the computations:

$$(dk\omega)_x(\xi_1, \ldots, \xi_p) = \sum_{i=1}^{p}(-1)^{i+1}(k\omega)'_x(\xi_i)(\xi_1, \ldots, \hat{\xi}_i, \ldots, \xi_p)$$

$$= \sum_{i=1}^{p}(-1)^{i+1}\int_0^1 t^{p-1}\omega_{tx}(\xi_i, \xi_1, \ldots, \hat{\xi}_i, \ldots, \xi_p)dt$$

$$+ \sum_{i=1}^{p}(-1)^{i+1}\int_0^1 t^p \omega'_{tx}(\xi_i)(x, \xi_1, \ldots, \hat{\xi}_i, \ldots, \xi_p)dt,$$

$$(kd\omega)_x(\xi_1, \ldots, \xi_p) = \int_0^1 t^p d\omega_{tx}(x, \xi_1, \ldots, \xi_p)dt$$

$$= \int_0^1 t^p \omega'_{tx}(x)(\xi_1, \ldots, \xi_p)dt$$

$$+ \sum_{i=1}^{p}(-1)^i \int_0^1 t^p \omega'_{tx}(\xi_i)(x, \xi_1, \ldots, \hat{\xi}_i, \ldots, \xi_p)dt,$$

it follows that

$$(dk\omega + kd\omega)_x(\xi_1, \ldots, \xi_p) = \int_0^1 pt^{p-1}\omega_{tx}(\xi_1, \ldots, \xi_p)dt$$

$$+ \int_0^1 t^p \omega'_{tx}(x)(\xi_1, \ldots, \xi_p)dt$$

$$= \int_0^1 \frac{d}{dt}(t^p \omega_{tx}(\xi_1, \ldots, \xi_p))dt = \omega_x(\xi_1, \ldots, \xi_p),$$

i.e.,

$$dk + kd = \text{id}. \tag{2}$$

1. The Theorems of de Rham and Allendoerfer-Eells

Hence k is the homotopy sought and Poincaré's Lemma is proved. We remark that the lemma is true, with the same proof, and the weaker condition that U be *starshaped* from the origin, i.e., if $x \in U$, $tx \in U$ for $0 \leqslant t \leqslant 1$. In the finite dimensional case there is also a simple proof by induction [13].

Now let V be a differentiable manifold modeled on the Banach space E and having differentiable partitions of unity. Consider the sequence of sheaves on V

$$0 \to R \xrightarrow{i} \mathfrak{F}(V) \xrightarrow{d} \mathscr{A}^1(V) \xrightarrow{d} \cdots \xrightarrow{d} \mathscr{A}^p(V) \xrightarrow{d} \cdots \quad (3)$$

where, as in Chapter 4, R is the constant real sheaf, $\mathfrak{F}(V)$ is the sheaf of germs of differentiable functions, $\mathscr{A}^p(V)$ is the sheaf of germs of differential forms of degree p, d is the exterior derivative, and i is the natural inclusion. The condition that differentiable partitions of unity exist implies that $\mathfrak{F}(V)$ and $\mathscr{A}^p(V)$ are fine sheaves (Definition 2.6.9) and Poincaré's Lemma ensures that the sequence (3) is exact. Therefore, (3) is a fine resolution of R on V. Denoting, as in Section 4.3, by $A^p(V)$ the space of forms of degree p on V and by $Z^p(V)$ the subspace of closed differential forms of degree p, it follows from the abstract de Rham's theorem for sheaves (2.6.7) that we have the result below.

2 Theorem (de Rham.) *If V is a differentiable manifold modeled on the Banach space E with differentiable partitions of unity, the real cohomology groups of V are given by the formula*

$$H^p(V, R) = Z^p(V)/dA^{p-1}(V). \quad (4)$$

The right-hand side of Formula (4) is just the de Rham cohomology of V, as defined in Section 4.3. Following Theorem 2, it coincides with the real cohomology of V. This fundamental result obtained by de Rham in the finite dimensional case was, via the paper of A. Weil [52], the starting point of sheaf theory. For the infinite dimensional case the result was given by Eells [12].

We now give some simple consequences of Theorem 2.

3 Corollary *The p-dimensional Betti number $b^p(V)$ (Section 2.6), when it exists, is equal to the dimension of the real linear space written in the right-hand side of the formula (4).*

4   Corollary   *If V has finite dimension n, $H^p(V, R) = 0$ for $p > n$.*

Note that if V is finite dimensional it suffices to take V paracompact, because of theorem 4.1.12.

5   Remark   In Section 2.6, we saw that the cup product can also be obtained by a resolution. Applying this result, it follows *that the isomorphism of Formula* (4) *represents cup products on the exterior products of the corresponding closed differential forms.* Indeed, the resolution (3) satisfies, with respect to the exterior product, the required conditions given in Section 2.6.

6   Remark   If in all the previous conditions the real field R is replaced by the complex field C we obtain, in the same manner,

$$H^p(V, C) = Z_c^p(V)/dA_c^{p-1}(V), \tag{5}$$

where the index c denotes complex valued forms.

Theorem 2 can be generalized as follows. Let $p: S \to V$ be a differentiable vector bundle on V, whose fiber is the Banach space F. We say that S is *locally constant* if there is a trivializing atlas such that its transition functions are locally constant on the intersections of coordinate neighborhoods. Then, we can define the exterior derivative of an S-valued form on V, since such a form is expressed locally by an F-valued form whose exterior derivative exists, and on the intersection of two trivializing neighborhoods the derivatives are equal (because the transition functions are locally constant and, hence, have vanishing derivatives). In particular, we have the derivative of a cross section of S and the cross sections with vanishing derivative are locally constant. Poincaré's Lemma holds for S-valued differential forms with the same proof as in Theorem 1 but with an F-valued integral in (1).

Let $\mathscr{C}(S)$ be the sheaf of germs of locally constant cross sections of S and $\mathscr{A}^p(S)$ be the sheaf of germs of S-valued differential forms of degree p. Also, let $A^p(S)$ and $Z^p(S)$ be the spaces of forms and closed forms, respectively, of degree p on V with values in S. It follows clearly that

$$0 \to \mathscr{C}(S) \xrightarrow{i} \mathscr{A}^0(S) \xrightarrow{d} \mathscr{A}^1(S) \xrightarrow{d} \cdots \tag{6}$$

is a fine resolution of $\mathscr{C}(S)$, and we have the following result.

1. The Theorems of de Rham and Allendoerfer-Eells

7 Theorem *If V has differentiable partitions of unity,*

$$H^p(V, \mathscr{C}(S)) = Z^p(S)/dA^{p-1}(S).$$

In the remainder of this chapter, the differentiable manifolds will be assumed to have differentiable partitions of unity or, in the finite dimensional case, to be paracompact.

Let $f: V \to W$ be a differentiable map, T a locally constant vector bundle on W, and $S = f^*(T)$. Then S is also locally constant, and $f^{-1}(\mathscr{C}(T)) = \mathscr{C}(S)$. We know that there are induced homomorphisms 2.6(9),

$$f^*: H^p(W, \mathscr{C}(T)) \to H^p(V, \mathscr{C}(S)). \tag{7}$$

On the other hand, the T-valued differential forms on W are sent by f to S-valued forms on V, which gives a homomorphism of exact sequences of sheaves between the resolution (6) of $\mathscr{C}(T)$ and the corresponding resolution of $\mathscr{C}(S)$

$$
\begin{array}{ccccccccc}
0 & \longrightarrow & \mathscr{C}(T) & \xrightarrow{i} & \mathscr{A}^0(T) & \xrightarrow{d} & \mathscr{A}^1(T) & \xrightarrow{d} & \cdots \\
& & \downarrow f_* & & \downarrow f_* & & \downarrow f_* & & \\
0 & \longrightarrow & \mathscr{C}(S) & \xrightarrow{i} & \mathscr{A}^0(S) & \xrightarrow{d} & \mathscr{A}^1(S) & \xrightarrow{d} & \cdots
\end{array}
$$

Because of the functorial character of the isomorphisms of de Rham's abstract theorem (Section 2.6), it follows that the homomorphisms (7) are represented in the de Rham cohomology (Theorem 7) by the map induced by f between the differential forms. If f is a diffeomorphism, the induced homomorphisms f^* are isomorphisms since, if $f(f^{-1}(T))$ and T are identified, we can construct the inverse homomorphisms.

8 Theorem *Let $f, g: V \to W$ be homotopic differentiable maps. Then, we have*

$$f^* = g^*: H^p(W, R) \to H^p(V, R).$$

Of course, the homomorphisms f^*, g^* of this theorem are the homomorphisms (7) for $T = W \times R$ and $S = V \times R$, respectively. In the calculations below, we follow [20].

If V is a differentiable manifold modeled on E, we can consider the manifold $V \times I'$, where I' is an open neighborhood of $[0, 1]$ on the real line R, whose model is $E \times R$ and whose tangent space at

$$(x, t) \quad (x \in V, t \in I')$$

is $T_x \oplus R$. On this manifold, there is the vector field η tangent to the curves (x, t) with fixed x, which acts on a function $f(x, t)$ by the formula

$$\eta f(x, t) = \partial f/\partial t$$

and which has in the charts $U \times R$ (U is a chart of V) the fiber coordinates $(0, 1)$.

Now, every vector field ξ on V defines a unique vector field $\tilde{\xi}$ on $V \times R$ by the formula $\tilde{\xi}_{(x,t)} = (\xi_x, 0)$.

Define the operator

$$h: A^p(V \times I') \to A^{p-1}(V), \quad p \geq 1,$$

by

$$(h\omega)_x(\xi_1, \ldots, \xi_{p-1}) = \int_0^1 \omega_{(x,t)}(\eta, \tilde{\xi}_1, \ldots, \tilde{\xi}_{p-1}) dt$$

and $h = 0$ for $p = 0$.

It is easy to see that

$$[\eta, \tilde{\xi}] = 0 \quad \text{for every } \xi,$$
$$\widetilde{[\xi_1, \xi_2]} = [\tilde{\xi}_1, \tilde{\xi}_2],$$
$$\xi_x(h\omega(\xi_1, \ldots, \xi_{p-1})) = \int_0^1 \tilde{\xi}\omega_{(x,t)}(\eta, \tilde{\xi}_1, \ldots, \tilde{\xi}_{p-1}) dt,$$

whence, by using 4.3(3), we get

$$[(dh + hd)\omega]_x(\xi_1, \ldots, \xi_p) = \int_0^1 \eta\omega(\tilde{\xi}_1, \ldots, \tilde{\xi}_p) dt \qquad (8)$$
$$= \omega_{(x,1)}(\tilde{\xi}_1, \ldots, \tilde{\xi}_p) - \omega_{(x,0)}(\tilde{\xi}_1, \ldots, \tilde{\xi}_p).$$

Now let $F: V \times I \to W$ be a differentiable homotopy between f and g.

1. The Theorems of de Rham and Allendoerfer-Eells

We suppose that F is the restriction of a differentiable map $F: V \times I' \to W$ such that $F(x, 0) = f(x)$, $F(x, 1) = g(x)$. Then, introducing

$$k: A^p(W) \to A^{p-1}(V) \qquad (9)$$

by $k = h \circ F_*$, we get from (8)

$$dk + kd = g^\# - f^\#, \qquad (10)$$

where $f^\#$ and $g^\#$ are the homomorphisms induced by f and g between the de Rham complexes of W and V.

Hence k is a homotopy of complexes between $f^\#$ and $g^\#$ and the theorem follows.

The calculations of the proof of Theorem 8 can be used for another result too [20]. Suppose that we have the differential commutative diagram

$$\begin{array}{ccc} S \times I' & \xrightarrow{F} & T \\ {\scriptstyle p \times 1_{I'}} \downarrow & & \downarrow {\scriptstyle p'} \\ V \times I' & \xrightarrow{F} & W \end{array}$$

where S and T are locally constant vector bundles, the maps $\tilde{F}_t(\cdot) = \tilde{F}(\cdot, t)$, $F_t(\cdot) = F(\cdot, t)$ are diffeomorphisms for every $t \in I'$, and they are also morphisms in $S(\mathrm{Top})$. Then, $F_t^*(T)$ can be identified with S (Section 2.2) and we can talk of the isomorphisms $F_t^* : H^p(V, \mathscr{C}(S)) \to H^p(V, \mathscr{C}(T))$. Considering that the previous calculations were performed with S-valued and T-valued forms, respectively, we get $F_0^* = F_1^*$.

It is important to remark that Theorem 8 extends to the case when V and W are topological spaces and f, g are continuous homotopic maps, which shows that the cohomology groups are invariant by homotopy equivalences and depend only on the homotopy type of the space.

Returning to Theorem 7, we remark that it could be used to express the spectral sequence of a differentiable map $p: M \to V$ in the sense of Theorem 2.6.13. When p is the projection of a differentiable fiber bundle and M, V are finite dimensional manifolds with compact fibers this was done by A. Hattori [20], who showed that the corresponding spectral sequence can be obtained from the following filtration of the complex of differential forms on M with values in a locally constant, finite dimensional vector bundle $\pi: S \to M$. Define

$$A^{pq} = \{\omega \mid \omega \in A^{p+q}(M, S) \text{ and } \omega(\cdots) = 0 \text{ if more than } q$$
$$\text{arguments are fields tangent to the fibers}\}, \quad \text{for} \quad p > 0, \quad (11)$$
$$A^{pq} = A^q(M, S), \quad \text{for} \quad p \leq 0.$$

Then

$$A_p = \sum_{q \geq 0} A^{pq} \qquad (12)$$

is a filtration of the above-mentioned complex (definition 2.5.4). If we note that $d(A^{pq}) \subseteq A^{p,q+1}$ and use the notation of Section 2.5 for spectral sequences, we find

$$E_0^{pq} = A^{pq}/A^{p+1,q-1}. \qquad (13)$$

Then, using connections on M, as defined in 3.1.7 but with a special construction (horizontal fields), it can be shown that, for the same spectral sequence,

$$E_2^{pq} = H^p(V, \mathscr{C}(\mathscr{H}^q(F, S))), \qquad (14)$$

where F is the fiber of M and $\mathscr{H}^q(F,S)$ is a locally constant vector bundle on V obtained by the union of the q-dimensional cohomology spaces of the local fibers of p with coefficients $\mathscr{C}(S|_{p^{-1}(x)})$.

The differential forms on a manifold V can also be used in the calculation of the groups $H^p(V, A)$ where A is a subring of R, but in a more complicated manner. We sketch here this calculation following the original paper of Allendoerfer and Eells [1], and we refer the reader to this paper for complete proofs. In the remaining part of this section, V is a finite dimensional paracompact differentiable manifold.

9 Definition An *incompletely defined differential form* ω on V, of degree $p \geq -1$, is (a) the constant zero function on V if $p = -1$; (b) a differential form of degree p defined on V except, eventually, for a closed nowhere dense subset $e(\omega)$ of V, if $p \geq 0$.

Recall that a closed nowhere dense subset of a topological space is a subset such that the closure of its complement is the whole space.

10 Definition A pair (θ, ω) of incompletely defined forms of degrees $p + 1$ and p, respectively, $p \geq -1$, is called an *admissible pair of degree*

1. The Theorems of de Rham and Allendoerfer-Eells

$p + 1$ if: (a) for $p = -1$, $\omega = 0$ and θ is a differentiable function on $V - e(\theta)$; (b) for $p \geq 0$, ω and θ are incompletely defined forms, $e(\theta) \subseteq e(\omega)$, and θ is an extension of $d\omega$ to $V - e(\theta)$ (necessarily unique because $V - e(\omega)$ is dense in $V - e(\theta)$).

In the sequel, we shall use the singular simplexes and real singular chains of V as defined in Section 4.3. If $\sigma: \Delta^p \to V$ is a singular simplex of V, $|\sigma| = \sigma(\Delta^p)$ is called the *carrier* of σ and for a real chain $l = \sum_i a_i \sigma^i$ ($a_i \in R$), where $a_i \neq 0$ and the σ^i are all different, the *carrier* is defined by $|l| = \bigcup_i |\sigma^i|$. The following relations are immediate

$$|l_1 + l_2| \subseteq |l_1| \cup |l_2|,$$
$$|al| \subseteq |l|, \qquad a \in R, \tag{15}$$
$$|\partial l| \subseteq |l|,$$

where ∂ is the boundary operator.

Let (θ, ω) be an admissible pair of degree $p + 1$ and l a real $(p + 1)$-dimensional singular chain of V, l is said to be *admissible* for the given pair if

$$|l| \cap e(\theta) = |\partial l| \cap e(\omega) = \emptyset. \tag{16}$$

If this is the case, we define the *residue* of (θ, ω) on l by

$$R[(\theta, \omega), l] = \int_l \theta - \int_{\partial l} \omega. \tag{17}$$

This is zero if $e(\omega) = \emptyset$ because of Stokes' Theorem (4.3.9), but is not zero in the general case.

The residue is bilinear with respect to (θ, ω) and l if the necessary terms exist. It has also another important property: if l and l' are admissible *homotopic* chains in the sense that the chains l_t of a homotopy of l and l', as defined in Section 4.3. are admissible for every t, then the pair (θ, ω) has the same residue on l and l'. As in Section 4.3, we can use the operator $\mathfrak{D}$ and Formula 4.3(18) and, by the definition of the carrier one gets $|\mathfrak{D}\partial\Delta^p| \cap e(\omega) = \emptyset$, whence the residue of $\mathfrak{D}\partial\Delta^p$ exists and vanishes. Also, $|\mathfrak{D}\Delta^p| \cap e(\theta) = \emptyset$, and these lead to

$$R[(\theta, \omega), l_2 - l_1] = \int_{\partial \mathfrak{D}\Delta^p} \theta = \int_{\mathfrak{D}\Delta^p} d\theta = 0,$$

where we used Stokes' formula. The assertion is therefore proved.

We shall denote by n the dimension of V and by A a subring of R.

11   Definition   An admissible pair (θ, ω) of degree $p \geqslant 0$ is called an (A, p)-*pair* on V if the following conditions hold:

(a) $e(\theta) \subseteq \bigcup_s |s_\alpha|, \quad e(\omega) \subseteq \bigcup_\sigma |\sigma_\beta|,$

where s_α and σ_β are singular simplexes of V of dimensions $\alpha \leqslant n - p - 1$, $\beta \leqslant n - p$, and $\{|s_\alpha|\}, \{|\sigma_\beta|\}$ are locally finite families of subsets of V;

(b) the residue $R[(\theta, \omega), l] \in A$, when it exists.

12   Definition   The (A, p)-pairs (θ, ω) and (θ', ω') are called *equivalent* if they have the same residue for every chain l which is admissible for both.

The sign used to denote this equivalence is $\sim$. The reflexivity and symmetry of this relation are obvious. If $(\theta, \omega) \sim (\theta', \omega')$ and $(\theta', \omega') \sim (\theta'', \omega'')$, let l be an admissible chain for (θ, ω) and (θ'', ω''). Since $|l|$ is compact and $\{|s_\alpha|\}$ is locally finite, $|l|$ intersects only a finite number of the sets $|s'_\alpha|$ (the prime refers to the pair (θ', ω') and because l has dimension p and s'_α dimension at most $n - p - 1$, the sum of the dimensions is $< n$ and we can find a chain l' homotopic to l such that $|l'| \cap e(\theta') = \varnothing$ and $|\partial l'| \cap e(\omega') = \varnothing$. The deformation will be performed so as to preserve the admissibility of l for (θ, ω) and (θ'', ω''). Then, l' will be admissible for the three pairs and we have

$$R[(\theta, \omega), l] = R[(\theta, \omega), l']$$
$$= R[(\theta', \omega'), l']$$
$$= R[(\theta'', \omega''), l']$$
$$= R[(\theta'', \omega''), l].$$

Hence $\sim$ is transitive.

The equivalence classes of $\sim$ will be denoted by square brackets, and we can talk of the residue of a class.

The set $\{[\theta, \omega]\}$ with fixed p has an obvious structure of an A-module, whose operations are given by

1. The Theorems of de Rham and Allendoerfer-Eells

The operator c extends by linearity to chains of U'. By a straightforward calculation, we get for a special simplex $(A_0, \ldots, A_p)$

$$c(A_0, \ldots, A_p) = (x_0, A_0, \ldots, A_p),$$

whence

$$\partial c \Delta^p = \Delta^p - c \partial \Delta^p$$

and by applying $\tilde{\sigma}_\sharp$

$$\partial c \sigma = \sigma - c \partial \sigma.$$

Extending this by linearity, we get

$$\partial c l = l - c \partial l \qquad (22)$$

for any chain l of U'.

Now, let l be an admissible chain for the pair (η, ξ) defined above. Then cl is admissible for (θ, ω). Using the coordinate expression of $k\theta$, we see that

$$\int_l k\theta = \int_{cl} \theta$$

(transformation of a multiple integral into an iterated one!) and (for the same reason)

$$\int_{\partial l} k\omega = \int_{c \partial l} \omega.$$

From these relations and (22),

$$R[(\eta, \xi), l] = -R[(\theta, \omega), cl]$$

follows, which proves that (η, ξ) is an $(A, r-1)$-pair in U.

Moreover,

$$R[(0, \eta), l] = R[(\theta, \omega), l] - \int_{\partial cl} \theta \qquad (23)$$

if l is admissible for both $(0, \eta)$ and (θ, ω).

Suppose that $d[\theta, \omega] = [0, \theta] = 0$. Then the last integral in (23) vanishes and we get

$$d[\eta, \xi] = [\theta, \omega], \qquad (24)$$

which is the required Poincaré lemma and implies the exactness of the sequence (20).

Hence, to establish that (20) is a resolution of A we have still to show that $H^p(V, C^r) = 0$ for $r \geq 0$, $p \geq 1$. The idea of the proof is the same as in 2.6.8, but now the C^r are not $\mathfrak{F}(V)$-sheaf modules, which makes necessary a more complicated use of partitions of unity on open subsets of V. This difficulty can be overcome, as shown in [1], thereby ending the proof of the fact that (20) is a resolution of A.

The next theorem follows immediately.

13 Theorem (Allendoerfer-Eells [1].) *If $Z^p(V, A)$ is the module of the elements of Q^p, closed with respect to the operator d, then*

$$H^p(V, A) = Z^p(V, A)/dQ^{p-1}.$$

For $A = J$, the ring of integers, we have the integral cohomology of V. Applying Theorem 13 for J and for the ring mJ ($m \in J$), and using the exact cohomology sequence of the pair (J, mJ), one gets

$$H^p(V, J_m) = \text{the cohomology of the complex } Q^*(V, J)/Q^*(V, mJ),$$

where J_m is the ring of integers modulo m. This means that $H^p(V, J_m)$ can be expressed by (J, r)-pairs whose equivalence means that the residues are congruent modulo m.

A more detailed analysis of these results [1] shows that a cohomology class of $H^r(X, A)$ always has a representative $[\theta, \omega]$, where θ is defined (and closed) on the whole manifold V. It is possible to obtain from this the answer to the following important question: when does a closed, globally defined differential form on V represent a real cohomology class, which is the image of an integer cohomology class by the homomorphism induced by the inclusion $J \subseteq R$? The answer is that this is true if the integral of the form on any integer singular cycle of V is an integer [1]. These integrals are called the *periods* of the given form. For this and other connected questions, the reader is also referred to [52].

2 Theorems of de Rham Type for Foliated and Complex Manifolds

The foliated and complex manifolds have again important sheaves of germs of differentiable functions, the foliated and the analytic functions, respectively. The cohomology of the manifold with these coefficient sheaves can be calculated with de Rham type theorems, and this is the subject of the present section.

Consider first a Riemannian foliated manifold V, for which we use the notation of Section 4.5. Observe that the metric has only auxiliary role, the essential thing being the foliated structure. The sheaf of germs of foliated functions on V is denoted by $\mathfrak{G}(V)$ and we shall construct a resolution of this sheaf. More generally, consider the sheaf $\mathfrak{G}^p(V)$ of germs of foliated differential forms of degree p on V. First, we have the following theorem.

1 Theorem [49, 8] *Let U be an open convex neighborhood of the origin of the product $E \times F$ of the Banach spaces E and F and let α be a differential form of type (p, q) $(q \geq 1)$ on U, such that $d_f \alpha = 0$. Then, there is a form β of type $(p, q - 1)$ on U such that $\alpha = d_f \beta$.*

Here, we consider the types with respect to the given product structure and that the structural distribution $\mathfrak{D}$ is defined by F, $\mathfrak{D}^\perp$ (the transversal distribution) being defined by E. (The terminology is that of Riemannian foliated manifolds of Section 4.5.)

The proof of the theorem is similar to that of Poincaré's Lemma 1.1. For $\xi_1, \ldots, \xi_p \in \mathfrak{D}^\perp$ and $\xi_{p+1}, \ldots, \xi_{p+q+1} \in \mathfrak{D}$, define

$$(k_f \alpha)_x(\xi_1, \ldots, \xi_p, \xi_{p+1}, \ldots, \xi_{p+q-1})$$
$$= (-1)^p \int_0^1 t^{p-1} \alpha_{tx}(\xi_1, \ldots, \xi_p, x, \xi_{p+1}, \ldots, \xi_{p+q-1}) \, dt$$

and put $k_f \alpha = 0$ for any other number of arguments in $\mathfrak{D}$ and $\mathfrak{D}^\perp$. Extending this by linearity to arbitrary arguments, we get a differential form $k_f \alpha$ on U of type $(p, q - 1)$.

By the same calculation as in the proof of Theorem 1.1, but replacing d by d_f and using Formula 4.5(26), we get

$$d_f k_f \alpha + k_f d_f \alpha = \text{id},$$

whence, since $d_f \alpha = 0$, we obtain the theorem by taking $\beta = k_f \alpha$. Q.E.D.

A proof by induction for the finite dimensional case is given in [49].

Next, denote by A^{pq} the space of forms of type (p, q) on the Riemannian foliated manifold V and by $\mathscr{A}^{p,q}$ the corresponding sheaf of germs. From Theorem 1, we see that the sequence

$$0 \to \mathfrak{G}^p(V) \xrightarrow{i} \mathscr{A}^{p0} \xrightarrow{d_f} \mathscr{A}^{p1} \xrightarrow{d_f} \mathscr{A}^{p2} \xrightarrow{d_f} \cdots, \tag{1}$$

where i is the natural inclusion, is a fine resolution of the sheaf $\mathfrak{G}^p(V)$.

Let Z^{pq} be the space of forms of type (p, q) closed with respect to d_f. From the previous result we obtain the following theorem.

2 Theorem [49]   *The cohomology of V with coefficients in $\mathfrak{G}^p(V)$ is given by the formula*

$$H^q(V, \mathfrak{G}^p(V)) = Z^{pq}/d_f A^{p,q-1} = H^{pq}(V) \tag{2}$$

(*where the last equality is a definition*).

The extension of this theorem to the infinite dimensional case was given in [8]. For $p = 0$, we obtain just the cohomology sought, with coefficients $\mathfrak{G}(V)$. Another important observation follows.

3 Corollary   *If V has $n + m$ dimensions and $\mathfrak{D}$ has m dimensions, $H^q(V, \mathfrak{G}^p(V)) = 0$ for $q > m$.*

It is possible to give a more general form of Theorem 2. Let S be a differentiable vector bundle on the Riemannian foliated manifold V. We can define the type for S-valued forms just as for the usual scalar forms. By definition, S is called a *foliated bundle* if it has an atlas whose transition functions are foliated. d_f acts on S-valued forms on V, if S is foliated, and we can define *S-valued foliated forms*.

Denote by $\mathfrak{G}^p(V, S)$ the space of S-valued foliated forms on V of degree p, by $A^{pq}(S)$ the space of S-valued forms of type (p, q), and by $\mathscr{A}^{pq}(S)$ the corresponding sheaf of germs. Theorem 1 can obviously be derived for S-valued forms also. Then, we have a fine resolution of the

2. Theorems of de Rham for Foliated and Complex Manifolds

sheaf $\mathfrak{G}^p(V, S)$ if we give all the sheaves of the sequence (1) the argument S, and we get the following result.

4 Theorem $H^q(V, \mathfrak{G}^p(V, S)) = H^{pq}(V, S) = Z^{pq}(S)/d_f A^{p,q-1}(S)$.

An important example of a foliated bundle on V is given by the transversal bundle $D^\perp(V)$ of $\mathfrak{D}^\perp(V)$ (Section 4.5). It is easy to see that its transition functions are foliated. Hence, we can apply Theorem 4 and we get

$$H^q(V, \mathfrak{G}^0(V, D^\perp)) = Z^{0q}(D^\perp)/d_f A^{0,q-1}(D^\perp), \tag{3}$$

$\mathfrak{G}^0(V, D^\perp)$ being the sheaf of germs of foliated cross sections of $D^\perp$. This result is important in the theory of deformations of foliated structures [26, 49].

We now go over to the corresponding results for finite dimensional complex manifolds. If V is such a manifold, let $\Omega(V)$ be the sheaf of germs of holomorphic functions on V and $\Omega^p(V)$ the sheaf of germs of holomorphic forms of degree p. The notation will be that of Section 4.6. We also introduce $A^{pq}(V)$, $Z^{pq}(V)$, $H^{pq}(V)$ by definitions similar to those for the foliated manifolds, but with meaning corresponding to the complex manifolds and for differentiable complex valued forms. We begin with the *lemma of Grothendieck-Dolbeault*.

5 Theorem   *Let α be a form of type (p, q), with $q \geq 1$, defined on a neighborhood $|z^h| < \rho^h$ of the origin of the space $C^n = \{(z^1, \ldots, z^n)/z^h \in C\}$ (C being the complex field) and suppose $d_{\bar{z}}\alpha = 0$. Then, there is a neighborhood $|z^h| < \rho'^h < \rho^h$ of the origin of C^n and a form of type $(p, q-1)$ on it, such that $d_{\bar{z}}\beta = \alpha$.*

We give Chern's proof [6], and consider first the following preliminary result.

6 Lemma   *Let $f(z, \bar{z})$ be a complex valued differentiable function on the complex variables z and $\bar{z}$, defined for $|z| < \rho$. Then, for $|z| < \rho' < \rho$, with some real number $\rho' > 0$, there is a differentiable function $g(z, \bar{z})$ such that $\partial g/\partial \bar{z} = f$.*

To obtain this, we establish a kind of generalized Cauchy integral formula. Let $\varphi(z, \bar{z})$ be a differentiable function on a neighborhood D of the origin of the complex plane. Let $D' \subseteq D$ be another neighborhood and $\zeta \in D'$. Consider a circle with (fixed) center z and radius ε, interior to D', and denote it by Δ_ε. Consider the Pfaff form

$$\omega = \frac{\varphi(\zeta, \bar{\zeta})d\zeta}{\zeta - z}$$

defined on $D' - \Delta_\varepsilon$. Its exterior derivative is

$$d\omega = \frac{\varphi_{\bar{\zeta}}}{\zeta - z} d\bar{\zeta} \wedge d\zeta.$$

Here and in the sequel, if a variable is taken as an index, this means that we have the partial derivative with respect to it. Applying Stokes' formula (Theorem 4.3.9) we get

$$\int_{\partial D'} \omega - \int_{\partial \Delta_\varepsilon} \omega = \int_{D' - \Delta_\varepsilon} d\omega \tag{4}$$

where ∂ denotes, as usual, the boundary.

We take here the limit as $\varepsilon \to 0$ and for this purpose we note that

$$\left| \int_{\partial \Delta_\varepsilon} \frac{\varphi(\zeta, \bar{\zeta})d\zeta}{\zeta - z} - \int_{\partial \Delta_\varepsilon} \frac{\varphi(z, \bar{z})d\zeta}{\zeta - z} \right| \leq 2\pi\delta,$$

where

$$|\varphi(\zeta, \bar{\zeta}) - \varphi(z, \bar{z})| < \delta \quad \text{for} \quad |\zeta - z| < \varepsilon$$

(by the continuity of the differentiable function φ). Then,

$$\lim_{\varepsilon \to 0} \int_{\partial \Delta_\varepsilon} \omega = \lim_{\varepsilon \to 0} \int_{\partial \Delta_\varepsilon} \frac{\varphi(z, \bar{z})d\zeta}{\zeta - z} = 2\pi i \varphi(z, \bar{z})$$

and, taking the limit in (4), we get

$$2\pi i \varphi(z, \bar{z}) = \int_{\partial D'} \frac{\varphi(\zeta, \bar{\zeta})d\zeta}{\zeta - z} + \int_{D'} \frac{\varphi_{\bar{\zeta}} d\bar{\zeta} \wedge d\zeta}{\zeta - z}, \tag{5}$$

which is the formula looked for. For an analytic function, $\varphi_{\bar{\zeta}} = 0$ and we have just the classical Cauchy integral formula.

Taking in (5) the complex conjugates and replacing $\bar{\varphi}$ by ψ we obtain another formula, which holds for every differentiable function ψ:

$$-2\pi i \psi(z, \bar{z}) = \int_{\partial D'} \psi \frac{d\zeta}{\zeta - z} - \int_{D'} \psi_{\bar{\zeta}} \frac{d\zeta \wedge d\bar{\zeta}}{\zeta - z}. \tag{6}$$

Obviously, the integrals on D' which enter in Formulas (5) and (6) are improper convergent integrals.

Returning to Lemma 6, if we replace the function φ in Formula (5) by the function g we see that, if g exists, it is necessarily given by

$$2\pi i g = \int_{D'} f \frac{d\zeta \wedge d\bar{\zeta}}{\zeta - z} + h(z), \tag{7}$$

where $h(z)$ is an analytic function on z.

Now, to prove the existence of g, we first remark that the integral in the right-hand side of (7) is convergent because, if we put $\zeta = x + iy$, $z = a + ib$, the modulus of the integrand is majorized by a quantity of the form const.$/[(x - a)^2 + (y - b)^2]^{1/2}$. We must now show that $\partial g / \partial \bar{z} = f$, and to obtain this we must see how the right side of (7) depends on $\bar{z}$.

Consider on $D' - \Delta_\varepsilon$ the Pfaff form

$$\varpi = f(\zeta, \bar{\zeta}) \ln |\zeta - z|^2 \, d\zeta,$$

whose exterior derivative is

$$d\varpi = f_{\bar{\zeta}} \ln |\zeta - z|^2 \, d\zeta \wedge d\bar{\zeta} + \frac{f}{\zeta - z} d\zeta \wedge d\bar{\zeta}.$$

Using Stokes' formula we have

$$\int_{\partial D'} \varpi - \int_{\partial \Delta_\varepsilon} \varpi = \int_{D' - \Delta_\varepsilon} d\varpi. \tag{8}$$

Note that

$$\left| \int_{\partial \Delta'} f \ln |\zeta - z|^2 \, d\zeta \right| \leq 4\pi M \varepsilon \ln \varepsilon,$$

where $|f(\zeta)| \leq M$. Hence, by taking in (8) the limit as $\varepsilon \to 0$, we get

$$\int_{\partial D'} f \ln |\zeta - z|^2 \, d\zeta = \int_{D'} f_{\bar{\zeta}} \ln |\zeta - z|^2 \, d\zeta \wedge d\bar{\zeta} + \int_{D'} \frac{f}{\zeta - z} d\zeta \wedge d\bar{\zeta}.$$

Here, $|\zeta - z|^2 = (\zeta - z)(\bar\zeta - \bar z)$ and taking the derivative on both sides with respect to $\bar z$ we get

$$-\int_{\partial D'} f \frac{d\zeta}{\bar\zeta - \bar z} = -\int_{D'} f_\zeta \frac{d\zeta \wedge d\bar\zeta}{\bar\zeta - \bar z} + 2\pi i \frac{\partial g}{\partial \bar z}.$$

Comparing this relation with Formula (6) for $\psi = f$, we get $\partial g/\partial \bar z = f$, which ends the proof of the lemma. It is important to note from Formula (7) that if the function f depends analytically on a number of complex parameters, then the function g has the same property.

We now give the proof of Theorem 5. Denote by H_j the hypothesis that the form α does not contain $d\bar z^{j+1}, \ldots, d\bar z^n$. Theorem 5 is true if it is true for α satisfying any one of the hypotheses H_j, $j = 0, \ldots, n$ and, to verify this, we use induction on j. Indeed, the hypothesis H_0 implies $\alpha = 0$, since $q \geq 1$, and the theorem is obviously true. Suppose that the theorem is true in the hypothesis H_{j-1} and consider a form α satisfying H_j; α can be written as

$$\alpha = d\bar z^j \wedge \lambda + \mu,$$

where λ and μ satisfy H_{j-1} and are of types $(p, q-1)$ and (p, q), respectively.

From the condition $d_{\bar z} \alpha = 0$ and the Cauchy–Riemann conditions 4.6(4) it follows immediately that the coefficients of the forms λ and μ depend analytically on the variables $z^{j+1}, \ldots, z^n$. Then, by Lemma 6, we can find a form λ' of type $(p, q-1)$, such that the coefficients of λ are the derivatives with respect to $\bar z^j$ of the corresponding coefficients (with the same indices) of λ'. This will be written symbolically as

$$\frac{\partial \lambda'}{\partial \bar z^j} = \lambda.$$

λ' is defined on some neighborhood $|z^h| < \rho'^h < \rho^h$ and its coefficients are analytic functions on $z^{j+1}, \ldots, z^m$.

From these observations it follows that the form

$$v = d_{\bar z} \lambda' - d\bar z^j \wedge \lambda$$

satisfies the hypothesis H_{j-1} and

$$\alpha = d_{\bar z} \lambda' + \mu - v.$$

2. Theorems of de Rham for Foliated and Complex Manifolds

Using again $d_{\bar{z}}\alpha = 0$, we get $d_{\bar{z}}(\mu - \nu) = 0$, and since μ satisfies H_{j-1}, we have from the inductive assumption $\mu - \nu = d_{\bar{z}}\gamma$ on $|z^h| < \rho'^h$. Hence, on the same domain, we have

$$\alpha = d_{\bar{z}}(\lambda' + \gamma) = d_{\bar{z}}\beta$$

and Theorem 5 is proved.

Now, considering the sequence of sheaves

$$0 \longrightarrow \Omega^p(V) \xrightarrow{i} \mathscr{A}^{p0}(V) \xrightarrow{d_{\bar{z}}} \mathscr{A}^{p1}(V) \xrightarrow{d_{\bar{z}}} \cdots, \tag{9}$$

where i is the natural inclusion, we see that, because of Theorem 5, this is a fine resolution of the sheaf $\Omega^p(V)$ and, consequently, we get

7   Theorem   (Dolbeault.)

$$H^q(V, \Omega^p(V)) = H^{pq}(V) = Z^{pq}(V)|d_{\bar{z}}A^{p,q-1}(V).$$

8   Corollary $H^q(V, \Omega^p(V)) = 0$ for $q > n$, where n is the complex dimension of the manifold V.

Now, let S be an analytic finite dimensional complex vector bundle on V. Then the transition functions of S can be taken as analytic functions, and the action of the operator $d_{\bar{z}}$ extends to forms of type (p, q) with values in S, so that it is meaningful to talk of S-valued analytic forms. From Theorem 5, one derives immediately a similar theorem for S-valued forms, using fiber coordinates and, consequently, if the sheaves of the sequence (9) are given the argument S, and if this is understood as a reference to S-valued forms, we get a fine resolution of the sheaf $\Omega^p(V, S)$ of germs of analytic S-valued forms of degree p on V. The following theorem results.

9   Theorem[†]   (Dolbeault–Serre.)

$$H^q(V, \Omega^p(V, S)) = H^{pq}(V, S)$$
$$= Z^{pq}(V, S)|d_{\bar{z}}A^{p,q-1}(S).$$

Applying Theorems 7 and 9 for the case $p = 0$, we obtain the co-

---

[†] For a generalization of this theorem see our note: *Sur la cohomologie des variétés analytiques complexes feuilletées.* C. R. Acad. Sci. Paris, Ser. A, **273**, 1067–1070 (1971).

homology groups of V with coefficients in the sheaf of germs of analytic functions on V and of analytic cross sections of the bundle S, respectively.

We remark that an important example of an analytic vector bundle on V is offered by the tangent bundle $T(V)$, hence Theorem 9 applies to this bundle.

By using de Rham's theorem of the previous section and theorems 2 and 7 above, we obtain relations between the real cohomology of a foliated or complex manifold V and the cohomology with coefficients $\mathfrak{G}(V)$ or $\Omega(V)$, respectively. To get this, we remark that the real(complex)-valued differential forms on a Riemannian foliated (complex finite dimensional) manifold form a double semipositive (positive) cochain complex, as defined in Section 2.5.

We follow our paper [50]. Let

$$C = \bigoplus_{r=0,1,2,\ldots} C^r$$

be a double semipositive complex, where

$$C^r = \bigoplus_{p+q=r} C^{pq}, \quad p, q \text{ nonnegative integers,}$$

whose differential $d = C^r \to C^{r+1}$ ($d^2 = 0$) when acting on an element of type (p, q) has the decomposition

$$d = \sum_{h+k=1} d_{hk}, \quad d_{hk} = 0 \text{ for } h < 0,$$

with

$$d_{hk} \colon C^{pq} \to C^{p+h,q+k}.$$

Then, if

$$\varphi = \sum_{p+q=r} \varphi_{pq},$$

we have

$$d\varphi = \sum_{p=0}^{r+1} \sum_{h=0}^{p} d_{h,1-h} \varphi_{p-h,r-p+h} \tag{10}$$

and

2. Theorems of de Rham for Foliated and Complex Manifolds

$$d^2\varphi = \sum_{k=0}^{r+1}\sum_{p=0}^{r+1}\sum_{h=0}^{p} d_{k-p,1+p-k}\, d_{h,1-h}\, \varphi_{p-h,r-p+h} = 0.$$

This last relation is equivalent to

$$\sum_{p=0}^{k}\sum_{h=0}^{p} d_{k-p,1+p-k}\, d_{h,1-h}\, \varphi_{p-h,r-p+h} = 0,$$
$$k = 0, \ldots, r + 2. \tag{11}$$

Hence, the relations (11) hold for every system of elements $\varphi_{p-h,r-p+h}$. If all these elements except one, φ_{uv}, vanish, the system (11) gives

$$\sum_{p=u}^{k} d_{k-p,1+p-k}\, d_{p-u,1-p+u}\, \varphi_{uv} = 0, \qquad k = u, \ldots, r + 2. \tag{12}$$

Putting $k = 0$ in (12), we get

$$d_{01}^2 = 0, \tag{13}$$

and we see that the modules

$$A^u = \bigoplus_v C^{uv}, \qquad u = 0, 1, 2, \ldots,$$

together with the operator d_{01}, define, for every u, a cochain complex.

10 Definition The complex $A = A^0$ is called the *subordinated complex* of C.

If we define the homomorphisms $f: C \to C'$ of two double semipositive complexes by systems of module homomorphisms $f_{pq}: C^{pq} \to C'^{pq}$ commuting with the operators d_{hk}, then A^u and A are covariant functors.

Let φ be a cocycle of C^r. Then $d\varphi = 0$ and, from Formula (10), this is equivalent to

$$\sum_{h=0}^{p} d_{h,1-h}\, \varphi_{p-h,r-p+h} = 0, \qquad p = 0, \ldots, r + 1, \tag{14}$$

whence, taking $p = 0$, it follows that $d_{01}\varphi_{0r} = 0$.

Moreover, replacing φ by $\varphi + d\lambda$ ($\lambda \in C^{r-1}$) we get, from Formula (10), that the new component of type $(0, r)$ is $\varphi_{0r} + d_{01}\lambda_{0,r-1}$. We deduce that the correspondence which assigns to an element $\varphi \in C^r$ its $(0, r)$-component φ_{0r}, induces a homomorphism

$$p^r: H^r(C) \to H^r(A), \quad r = 0, 1, 2, \ldots, \tag{15}$$

We call these homomorphisms the *cohomology projections of the complex C*.

11 Theorem [50] *For a double semipositive complex C we have:*
(a) p^0 *is a monomorphism,* (b) *if* $H^{r-p}(A^p) = 0$ *for* $p = 1, \ldots, r$ $(r \geq 1)$, p^r *is a monomorphism;* (c) *if* $H^{r-p+1}(A^p) = 0$ *for* $p = 1, \ldots, r+1$ $(r \geq 0)$, p^r *is an epimorphism.*

The assertion (a) is trivial. To obtain (b), let $\Phi \in H^r(C)$ and $p^r(\Phi) = 0$. Then Φ is represented by an r-cocycle φ such that

$$\varphi_{0r} = d_{01}\lambda_{0, r-1}.$$

Suppose that there are elements λ of the necessary type, such that

$$\varphi_{u, r-u} = \sum_{h=0}^{u} d_{h, 1-h}\lambda_{u-h, r-u+h-1}, \quad u = 0, \ldots, p-1. \tag{16}$$

Write the relation (14) in the form

$$d_{01}\varphi_{p, r-p} + \sum_{h=1}^{p} d_{h, 1-h}\varphi_{p-h, r-p+h} = 0. \tag{17}$$

Next, replacing, in (11), $k - p$ by h, k by p, h by k and r by $r - 1$, we get

$$\sum_{h=0}^{p} \sum_{k=0}^{p-h} d_{h, 1-h} d_{k, 1-k} \lambda_{p-h-k, r-p+h+k-1} = 0 \tag{18}$$

(there is no $\lambda_{p, r-p-1}$ because of (13)). In view of (16), (18), and (13), Formula (17) becomes

$$d_{01}(\varphi_{p, r-p} - \sum_{k=1}^{p} d_{k, 1-k}\lambda_{p-k, r-p+k-1}) = 0. \tag{19}$$

Hence, if $H^{r-p}(A^p) = 0$, there is an element $\lambda_{p, r-p-1}$ such that the relation (16) also holds for $u = p$.

2. Theorems of de Rham for Foliated and Complex Manifolds

It follows that, starting with $\lambda_{0,r-1}$ and using the hypotheses of assertion (b), we can obtain (16) up to $u = r$, i.e., $d\lambda = \varphi$. Hence the cocycle φ is a coboundary, Φ vanishes, and part (b) of Theorem 11 is proved.

Now, consider the last part of the theorem. To prove it we shall look for the conditions which assure that the class $\Psi \in H^r(A)$ is in im p^r or, equivalently, that an element $\varphi_{0r} \in C^{0r}$, closed with respect to d_{01} (i.e., $d_{01}\varphi_{0r} = 0$) is the $(0, r)$-component of an element $\varphi \in C^r$, closed with respect to d ($d\varphi = 0$). This can be viewed as an extension problem: *to extend a d_{01}-closed element into a d-closed element*. The solution is given by looking for the required elements $\varphi_{p,r-p}$, step by step.

Suppose that we obtained the elements $\varphi_{p,r-p}$ for $p = 0, \ldots, s-1$ such that the conditions (14) for $p = 0, \ldots, s-1$ are satisfied. We must look for an element $\varphi_{s,r-s}$ such that the following relation holds:

$$d_{01}\varphi_{s,r-s} = -\sum_{h=1}^{s} d_{h,1-h}\varphi_{s-h,r-s+h}. \tag{20}$$

For $k = s$, the relation (11) gives

$$d_{01}\left(\sum_{h=1}^{s} d_{h,1-h}\varphi_{s-h,r-s+h}\right) + \sum_{p=0}^{s-1} d_{s-p,1+p-s}\left(\sum_{h=0}^{p} d_{h,1-h}\varphi_{p-h,r-p+h}\right) = 0,$$

and, in view of (14) with $p = 0, \ldots, s-1$, the first term here vanishes. It follows that the right-hand side of (20) is an $(r-s+1)$-cocycle of the complex A^s. Its cohomology class will be denoted by

$$\omega_s(\varphi_{0r}) \in H^{r-s+1}(A^s) \tag{21}$$

and called the *sth obstruction* of φ_{0r} in the given extension problem.

From the expression for the obstruction cocycle, we see that the obstruction depends linearly on the system of elements

$$\varphi_{p,r-p} \ (p = 0, \ldots, s-1).$$

If this system is of the type (16) with u replaced by p and p by s, conditions (14) are satisfied up to the index $s-1$, because of (11), and the same calculation which gave (19) shows that the corresponding obstruction vanishes. The possible modifications of the choice of the elements $\varphi_{p,r-p}$ at the kth stage ($1 \leq k \leq s-1$) consist just in adding some system of the type (16). It follows that ω_s depends only on the cohomology class of φ_{0r}.

Returning to Equation (20), we see that it has solutions for $\varphi_{s,r-s}$ if and only if $\omega_s(\varphi_{0r}) = 0$.

It follows that the extension problem has a solution if and only if all the possible $r + 1$ obstructions of φ_{0r} vanish. More precisely, the extension problem is the problem of extending a cohomology class of $H^r(A)$ to a cohomology class of $H^r(C)$, and this problem has a solution for the elements of im p^r; im p^r is just the submodule of $H^r(A)$ of classes with vanishing obstructions.

The hypotheses of Theorem 11 (c) ensure that every element of $H^r(A)$ has only vanishing obstructions, which implies that p^r is an epimorphism. Q.E.D.

In Theorem 11, the conditions for p^r to be an epimorphism and for p^{r+1} to be a monomorphism are the same. To explain this, consider the subcomplex C_1 of C defined by

$$C_1 = \bigoplus_{i \geq 1} C^{ij}$$

and the same differential d. Note that we have the exact sequence of complexes

$$0 \to C_1 \xrightarrow{i} C \xrightarrow{p} A \to 0, \tag{22}$$

where i is the inclusion and p is the projection which assigns to an element $\varphi \in C^r$ its $(0, r)$-component. Recall that the differential of A is d_{01}.

The exact cohomology sequence corresponding to (22) is

$$\cdots \to H^r(C_1) \xrightarrow{i^r} H^r(C) \xrightarrow{p^r} H^r(A) \xrightarrow{\delta} H^{r+1}(C_1) \to \cdots \tag{23}$$

and we call it the *exact cohomology sequence of the double semipositive complex C*.

From the proof of part (b) of Theorem 11, we see that these hypotheses imply $H^r(C_1) = 0$, hence, that p^r is a monomorphism. But from (23), $H^r(C_1) = 0$ implies p^{r-1} is an epimorphism.

Now let V be a Riemannian foliated manifold, for which we use the ordinary notation of Section 4.5. Consider the double semipositive complex

$$C = \bigoplus_{p,q} A^{pq}(V)$$

which, as we know, has the nonvanishing operators $d_{01} = d_f$, d_{10}, and

2. Theorems of de Rham for Foliated and Complex Manifolds

$d_{2,-1}$. The subordinated complex is the complex of forms of type $(0, q)$, and A^u is the complex of forms of type (u, q).

From de Rham's Theorem and Theorem 2 of this section we get

$$H^r(C) = H^r(V, R),$$
$$H^r(A^u) = H^r(V, \mathfrak{G}^u(V)) = H^{ur}(V), \qquad (24)$$
$$H^r(A) = H^r(V, \mathfrak{G}(V)).$$

Hence from the previous algebraic theory we obtain the following result.

12   Theorem   *For a foliated Riemannian manifold V there are canonical projection homomorphisms*

$$p^r: H^r(V, R) \to H^r(V, \mathfrak{G}(V)),$$

and p_0 is a monomorphism; if

$$H^{p,r-p}(V) = 0, \qquad p = 1, \ldots, r \geqslant 1$$

p^r is a monomorphism; if

$$H^{p,r-p+1}(V) = 0, \qquad p = 1, \ldots, r+1 \geqslant 1,$$

p^r is an epimorphism.

We remark that the homomorphisms p^r are just the homomorphisms induced by the inclusion of the constant sheaf R in the sheaf $\mathfrak{G}(V)$, since the resolution (1) of $\mathfrak{G}(V)$ ($p = 0$) is obtained from de Rham's resolution 1(3) by putting into the latter $dx^a = 0$, and since the isomorphisms of de Rham's abstract theorem have a functorial character.

13   Corollary   *If for a Riemannian foliated manifold V of dimension $n + m$ and codimension n we have $H^{pr}(V) = 0$ for $0 < r - m \leqslant p \leqslant n$, then $H^r(V, R) = 0$.*

In fact, the imposed conditions imply that $H^r(V, \mathfrak{G}(V)) = 0$, and that p^r is a monomorphism.

Let us now give an interpretation of the cohomology of the subcomplex C_1. To do this, let $\Psi(V)$ be the sheaf of germs of differential forms of type $(1, 0)$ on V, closed with respect to d and $\mathfrak{C}_1^r$ the sheaf of germs of the forms of the spaces

$$C_1^r = \bigoplus_{\substack{i+j=r \\ i \geq 1}} A^{ij}(V).$$

Consider the sequence of sheaves

$$0 \to \Psi(V) \xrightarrow{i} \mathfrak{C}_1^1 \xrightarrow{d} \mathfrak{C}_1^2 \xrightarrow{d} \cdots, \qquad (25)$$

where the terms beginning with the third are fine and i is the inclusion. This sequence is obviously exact at the second and third terms. Next, if α is a locally defined form on V which represents an element of $\mathfrak{C}_1^p$ and if $d\alpha = 0$, we have by Poincaré's Lemma $\alpha = d\beta$, and if β has the $(0, p-1)$-component $\beta_{0,p-1}$ it follows that $d_{01}\beta_{0,p-1} = 0$, since α has no $(0, p)$-component. Then, we have from Theorem 1, $\beta_{0,p-1} = d_{01}\gamma_{0,p-2}$, and the form $\beta - d\gamma_{0,p-2}$ defines an element of $\mathfrak{C}_1^{p-1}$ such that $\alpha = d(\beta - d\gamma_{0,p-2})$. This implies that the sequence (25) is exact and

$$H^r(V, \Psi(V)) = H^{r+1}(C_1), \qquad (26)$$

follows, which is the interpretation sought.

From Formula (26) and the exact sequence (23) we get the following result.

14 Theorem *For every Riemannian foliated manifold there is an associated exact cohomology sequence*

$$\cdots \to H^{r-1}(V, \Psi(V)) \to H^r(V, R) \xrightarrow{pr} H^r(V, \mathfrak{G}(V))$$
$$\to H^r(V, \Psi(V)) \to \cdots \qquad (27)$$

This sequence could also be obtained directly by noting that d_{10} induces an isomorphism of the sheaves $\mathfrak{G}(V)/R$ and $\Psi(V)$. Indeed, by assigning to the germ of the foliated function f the germ of $d_{10}f$ we get (since $d^2 = 0$) a homomorphism $d' : \mathfrak{G}(V) \to \Psi(V)$ whose kernel is R and which is an epimorphism by Poincaré's Lemma. Then, we have the exact sequence of sheaves

$$0 \to R \xrightarrow{i} \mathfrak{G}(V) \xrightarrow{d'} \Psi(V) \to 0, \qquad (28)$$

and (27) is just its exact cohomology sequence.

2. Theorems of de Rham for Foliated and Complex Manifolds

We also remark that we can consider the spectral sequence of the complex C, as defined in Section 2.5, which is related to the Riemannian foliated structure.

Using the previously introduced obstructions we get the solution of the following extension problem: given a differential form of type $(0, r)$, closed with respect to the operator d_f, find a form of degree r closed with respect to d such that the given form is its $(0, r)$-component. The extension exists if and only if all the obstructions of the given form vanish.

A similar study can be made for an n-dimensional complex manifold V. In this case, we have the double positive complex

$$K = \bigoplus_{p,q} A^{pq}(V)$$

of complex valued differential forms on V, with the differential $d = d_{10} + d_{01} = d_z + d_{\bar{z}}$ and $A^u(K)$ is the complex of forms of type (u, q) with the differential $d_{\bar{z}}$.

With de Rham's and Dolbeault's theorems we have

$$H^r(K) = H^r(V, C),$$
$$H^r(A^u) = H^{ur}(V) = H^r(V, \Omega^u(V)), \qquad (29)$$
$$H^r(A) = H^r(V, \Omega(V)),$$

C being the complex field. The following theorem results.

15 Theorem *For a complex manifold V there are canonical projection homomorphisms*

$$p^r: H^r(V, C) \to H^r(V, \Omega(V))$$

induced by the inclusion of C into $\Omega(V)$, and p^0 is a monomorphism; if

$$H^{p,r-p} = 0, \quad p = 1, \ldots, r \geqslant 1,$$

p^r is a monomorphism; if

$$H^{p,r-p+1}(V) = 0, \quad p = 1, \ldots, r+1 \geqslant 1,$$

p^r is an epimorphism.

16 Corollary *If* $0 < p < n$ *and if* $H^{n-p,n}(V) = \cdots = H^{n,n-p}(V) = 0$, *then* $H^{2n-p}(V, C) = 0$.

The proof is the same as for Corollary 13. We remark that this result is a straightforward consequence of Dolbeault's theory [10]. In Dolbeault's paper one also finds many other results regarding the cohomology of the complex K, including the resolution (25) for the complex case. This gives immediately the analogue of Theorem 14, etc.

The extension problem becomes: *given a form of type* $(0, r)$, *closed with respect to* $d_{\bar{z}}$, *find a closed form of degree r on V, whose* $(0, r)$-*component is the given form*. Again, the solution exists if and only if all the obstructions of the given form vanish.

3 Characteristic Classes of Differentiable Vector Bundles

In Section 3.4 we introduced the characteristic classes of vector bundles and, in particular, the Chern, Pontrjagin, and Euler classes. In the case of real cohomology, de Rham's Theorem shows that these classes can be represented by closed differential forms, which we wish to obtain in this section. This is a very important application of de Rham's Theorem.[†]

Let $\pi: S \to V$ be a finite dimensional differentiable vector bundle with p-dimensional fiber. This condition is always supposed satisfied in the sequel, but it is not assumed that V is finite dimensional. According to Definition 3.4.12, in order to obtain real characteristic classes of S we must attach to S some closed differential forms on V. To do this, let D be a connection on S and let Ω be its curvature form, which is $L(S, S)$-valued. The necessary forms will be defined with the help of Ω. We study these problems following [6].

Consider functions $\mathscr{W}(\lambda_1, \ldots, \lambda_h)$ whose values are real if S is real and complex if S is complex, i.e., they are $R(C)$-valued, and whose arguments are square matrices of order p with entries in $R(C)$. Suppose that these functions have the following invariance property:

$$\mathscr{W}(g\lambda_1 g^{-1}, \ldots, g\lambda_h g^{-1}) = \mathscr{W}(\lambda_1, \ldots, \lambda_h) \qquad (1)$$

[†] For other, more recent, applications see T. Nagano: "Homotopy Invariants in Differential Geometry." *Memoirs Amer. Math. Soc.*, 100 (1970).

3. Characteristic Classes of Differentiable Vector Bundles

for every nonsingular matrix g of order p, and that they are multilinear functions.

Such functions $\mathscr{W}$ are called *invariant functions* on matrices and they can be considered as functions defined on the Lie algebra of the general linear group, invariant under the action of its adjoint group. It follows that one can consider similar functions for any Lie group.

If we use the natural basis of the space of matrices of order p, the function $\mathscr{W}$ may be expressed as

$$\mathscr{W}(\lambda_1, \ldots, \lambda_h) = \sum a_{\alpha_1 \cdots \alpha_h}^{\beta_1 \cdots \beta_h} \lambda_{1\beta_1}^{\alpha_1} \cdots \lambda_{h\beta_h}^{\alpha_h}. \tag{2}$$

This permits one to define the expression $\mathscr{W}(\lambda_1, \ldots, \lambda_h)$ for matrices $\lambda_1, \ldots, \lambda_h$ whose elements are differential forms defined on an open subset of V by replacing the ordinary products of (2) with exterior products.

Let $\mathscr{W}$ be an invariant function on matrices and let $\theta_1, \ldots, \theta_h$ be h $L(S,S)$-valued forms of degrees $d_1, \ldots, d_h$, respectively, on the differentiable manifold V. Then, we get a differential form $\mathscr{W}(\theta_1, \ldots, \theta_h)$ of degree $d_1 + \cdots + d_h$, globally defined on V, with the local coordinate expressions

$$\mathscr{W}(\theta_1, \ldots, \theta_h) = \sum a_{\alpha_1 \cdots \alpha_h}^{\beta_1 \cdots \beta_h} \theta_{1\beta_1}^{\alpha_1} \wedge \cdots \wedge \theta_{h\beta_h}^{\alpha_h}, \tag{3}$$

where the matrices $(\theta_{j\beta}^{\alpha})$ are the local expressions of θ_j with respect to a basis of S (Section 4.4). Formula (3) defines globally a form on V because of the condition (1), and since, if the θ_j are $L(S,S)$-valued forms, the corresponding local matrices satisfy, on the intersection of two trivializing neighborhoods of S, transition relations of the form $\theta'_j = g\theta_j g^{-1}$, where the g are the transition functions of S.

Consider in (1) a matrix g of the form $I + tg'$ where I is the unit matrix and t is a sufficiently small real parameter. Then

$$g^{-1} = I - tg' + \cdots$$

where the dots contain terms in t^u ($u \geq 2$). Replacing these values in (1) and taking only the coefficient of t we get

$$\sum_{i=1}^{h} \mathscr{W}(\lambda_1, \ldots, g'\lambda_i - \lambda_i g', \ldots, \lambda_h) = 0 \tag{4}$$

for every (even singular) matrix g' of order p.

1 Proposition *If $\mathscr{W}$ is an invariant function in h matrix arguments $\theta_1, \ldots, \theta_h$ as in (3), and θ is a matrix of Pfaff forms on V, then*

$$\sum_{i=1}^{h} (-1)^{d_1 + \cdots + d_{i-1}} \mathscr{W}(\theta_1, \ldots, [\theta, \theta_i], \ldots, \theta_h) = 0 \tag{5}$$

In fact, (1) and hence (4) obviously hold if we replace λ_i by θ_i, g' being a matrix of functions. From this we deduce that (5) holds for the case $\theta = \alpha g'$ with an arbitrary Pfaff form α and, hence, also for an arbitrary θ, since we have $\theta = \sum \alpha g'$ and the left-hand side of (5) depends linearly on θ.

Using this result we prove the following proposition.

2 Proposition *For a differential form $\mathscr{W}(\theta_1, \ldots, \theta_h)$ as above, we have*

$$d(\mathscr{W}(\theta_1, \ldots, \theta_h))$$
$$= \sum_{i=1}^{h} (-1)^{d_1 + \cdots + d_{i-1}} \mathscr{W}(\theta_1, \ldots, D\theta_i, \ldots, \theta_h), \tag{6}$$

where $D\theta_i$ is the exterior covariant derivative given by Formula 4.4(31).

It suffices to prove (6) for the respective coordinate expressions. From Formula (3) we have

$$d(\mathscr{W}(\theta_1, \ldots, \theta_h)) = \sum_{i=1}^{h} (-1)^{d_1 + \cdots + d_{i-1}} \mathscr{W}(\theta_1, \ldots, d\theta_i, \ldots, \theta_h)$$

and (6) follows in view of Formula 4.4(32) and Proposition 1.

Now, in the previous calculations we can take $\theta_1 = \cdots = \theta_h = \Omega$, which is the curvature form of a connection on S. It follows that for every invariant function on matrices $\mathscr{W}$ there is an associated differential form of degree $2h$ globally defined on V and whose local expression is

$$\mathscr{W}\Omega = \sum_{i=1}^{h} a_{\alpha_1 \cdots \alpha_h}^{\beta_1 \cdots \beta_h} \Omega_{\beta_1}^{\alpha_1} \wedge \cdots \wedge \Omega_{\beta_h}^{\alpha_h}, \tag{7}$$

where the Ω_β^α are the local curvature forms (Section 4.4).

From Proposition 2, we obtain the following result.

3 Corollary *For every invariant function $\mathscr{W}$, the form $\mathscr{W}\Omega$ is closed.*

This is an immediate consequence of Formula (6) and of the Bianchi identities 4.4(35).

We consider in the sequel only *symmetric* invariant functions, i.e., functions whose values do not change by a permutation of the arguments. These functions make up an algebra if the product is defined by symmetrizing the ordinary product.[†] This is the so-called *Weil algebra* $W(GL)$ of the general linear group. One obtains similar Weil algebras for arbitrary Lie groups G. The following result is of fundamental importance.

4 Theorem (A. Weil.) *The de Rham cohomology class of a form $\mathscr{W}\Omega$, where $\mathscr{W} \in W(GL)$ does not depend on the choice of the connection of S.*

Suppose that $\mathscr{W}\Omega$ is constructed from the connection D and let $\tilde{D}$ be another connection on S. The pair $(D, \tilde{D})$ defines an associated $L(S, S)$-valued Pfaff form η on V by the formula

$$\eta(\xi)(s) = \tilde{\nabla}_\xi s - \nabla_\xi s, \qquad s \in S, \tag{8}$$

and, if we put

$$\nabla^t_\xi s = \nabla_\xi s + t\eta(\xi)(s), \qquad t \in [0, 1],$$

we get, easily, a family of connections on S which gives D for $t = 0$ and $\tilde{D}$ for $t = 1$.

This enables one to write

$$\mathscr{W}\tilde{\Omega} - \mathscr{W}\Omega = \int_0^1 \frac{d}{dt}(\mathscr{W}\Omega_t)\, dt, \tag{9}$$

where the index t denotes the elements related to ∇^t. We calculate the integrand on the right-hand side of (9).

First, from Formula 4.4(18),

$$\Omega_t = \Omega + tD\eta - \tfrac{1}{2}t^2[\eta, \eta], \tag{10}$$

where Ω and η denote the corresponding local matrices.

[†] Recall that the symmetrization of arguments (or indices) is the operation which sends $\alpha(i_1, \ldots, i_h)$ to $(1/h!) \Sigma\, \alpha(\sigma(i_1), \ldots, \sigma(i_h))$, where the σ denote the permutations of the arguments.

If we put

$$\mathscr{W}(A, \underbrace{B, \ldots, B}_{h-1 \text{ times}}) = \mathscr{L}(A, B),$$

we get, by (10),

$$\frac{d}{dt}\mathscr{W}\Omega_t = h\mathscr{L}(D\eta - t[\eta, \eta], \Omega_t). \tag{11}$$

Next, we compute the exterior derivative of the form $\mathscr{L}(\eta, \Omega_t)$. Using (6), we get

$$d\mathscr{L}(\eta, \Omega_t) = \mathscr{L}(D\eta, \Omega_t) - (h-1)\mathscr{W}(\eta, \underbrace{D\Omega_t, \Omega_t, \ldots, \Omega_t}_{h-2 \text{ times}})$$

By means of the Bianchi identity it follows that

$$D\Omega_t = t[\eta, \Omega_t] + t^2[\eta, D\eta],$$

where we have also used 4.4(34).

Applying Proposition 1 for

$$\mathscr{W}(\eta, \underbrace{\Omega_t, \ldots, \Omega_t}_{h-1 \text{ times}})$$

and with η instead of θ, we finally get

$$d\mathscr{L}(\eta, \Omega_t) = \mathscr{L}(D\eta, \Omega_t) - t\mathscr{L}([\eta, \eta], \Omega_t) = \frac{1}{h}\frac{d}{dt}\mathscr{W}\Omega_t.$$

Now, Formula (9) becomes

$$\mathscr{W}\tilde{\Omega} - \mathscr{W}\Omega = hd\left(\int_0^1 \mathscr{L}(\eta, \Omega_t)dt\right),$$

which proves Theorem 4.

As a consequence of the previous theory, we have a procedure for associating real (complex) cohomology classes of the base space with finite dimensional differentiable vector bundles. More precisely, to every element

3. Characteristic Classes of Differentiable Vector Bundles

of the Weil algebra there corresponds such a class. This defines an algebra homomorphism

$$w: W(GL) \to H^*(V, R(C)),$$

which is called *Weil's homomorphism* and can be generalized for arbitrary Lie groups.

It is possible to find all the elements of $W(GL)$ [25], but for our purpose it suffices to consider only some of them (which can be seen to generate the Weil algebra).

It is classical that for a square matrix A of order p we have

$$\det(A - \lambda I) = \sum_0^p (-1)^j \Delta_{p-j} \lambda^j,$$

where

$$\Delta_j = \frac{1}{j!} \sum_{\alpha,\beta} \delta_{\alpha_1 \cdots \alpha_j}^{\beta_1 \cdots \beta_j} a_{\beta_1}^{\alpha_1} \cdots a_{\beta_j}^{\alpha_j} \tag{12}$$

is the sum of the principal minors of order j of the matrix A. It is also known that Δ_j remains invariant if we replace A by gAg^{-1} for a non-degenerate matrix g. Hence, the symmetric functions on matrices

$$\mathscr{W}_j(A_1, \ldots, A_j) = \frac{1}{j!} \sum_{\alpha,\beta} \delta_{\alpha_1 \cdots \alpha_j}^{\beta_1 \cdots \beta_j} a_{\beta_1 (i}^{\alpha_1} \cdots a_{\beta_j j)}^{\alpha_j} \tag{13}$$

are invariant functions, $((1, \ldots, j)$ means the symmetrization of the indices), which can be expressed by the Δ_j as

$$\mathscr{W}_j(A_1, \ldots, A_j) = \frac{1}{j!} \Bigg[\Delta_j(A_1 + \cdots + A_j) - \sum_{h=1}^j \Delta_j(A_1 + \cdots + \hat{A}_h + \cdots + A_j) - \cdots - \sum_{h=1}^j \Delta_j(A_h) \Bigg],$$

where the sign "^" denotes, as usual, the absence of the respective term, and the dots mean sums of Δ_j of sums of $j - 2, \ldots, 2$ matrices A_h.

The functions (13) are just the required elements of the Weil algebra. Consider the case when the bundle S is complex and has complex p-

dimensional fiber, and construct, by the previous theory, the differential forms of degree $2j$ ($j = 0, \ldots, p$)

$$\mathscr{C}_j(S) = \frac{(-1)^j}{(2\pi i)^j j!} \mathscr{W}_j \Omega$$

$$= \frac{(-1)^j}{(2\pi i)^j j!} \sum_{\alpha,\beta} \delta^{\beta_1 \cdots \beta_j}_{\alpha_1 \cdots \alpha_j} \Omega^{\alpha_1}_{\beta_1} \wedge \cdots \wedge \Omega^{\alpha_j}_{\beta_j}, \tag{14}$$

where Ω is the curvature of a connection on S ($\mathscr{C}_0(S) = 1$). These forms are closed and globally defined on V and, hence, by de Rham's Theorem (Remark 1.6) they define some complex cohomology classes $C_j(S) \in H^{2j}(V, C)$ which depend only on the isomorphism class of S.

5   Definition   The cohomology class $C_j(S)$ is called the *jth Chern characteristic class of the bundle S*.

In the sequel, we show that this definition agrees with the definition in Section 3.4.

Consider the element $C(S) \in H^*(V, C)$, the cohomology algebra of the manifold V, defined by

$$C(S) = \sum_{j=0}^{p} C_j(S), \quad C_0(S) = 1, \tag{15}$$

which is the *total Chern class of S*. It is easy to see that this is represented in the sense of de Rham's Theorem by the nonhomogeneous differential form

$$\mathscr{C}(S) = \det\left(I - \frac{1}{2\pi i}\Omega\right), \tag{16}$$

where Ω is the curvature matrix of an arbitrary connection on S and the determinant of a matrix of differential forms is defined by the same expression as the determinant of a numerical matrix, but with the ordinary products replaced by exterior products.

We first prove the following result.

6   Proposition $C(S)$ *is a characteristic class as given by Definition 3.4.1, with the category* Top *replaced by* Diff.

As we have already remarked, the total Chern class depends only on S

3. Characteristic Classes of Differentiable Vector Bundles

and not on the choice of the connection on S (there is a connection since we assumed that V has differentiable partitions of unity). Moreover, $C(S)$ depends only on the isomorphism class of S, since the connection is defined by the matrices (ω_α^β) which have only to satisfy the transition relations on the intersection of two coordinate neighborhoods, and these relations depend only on the transition functions of S. Hence, for two isomorphic bundles we can use the same matrices ω, and we find the same form $\mathscr{C}(S)$.

Now let S and S' be two differentiable complex vector bundles on V and $S \oplus S'$ their Whitney sum. If D and D' are connections on S and on S', respectively, then

$$(D + D')(s)(\xi) = Ds_1(\xi) + D's_2(\xi), \qquad s = s_1 + s_2,$$

defines a connection on $S \oplus S'$ whose local expression is given by the matrix

$$\tilde{\omega} = \begin{pmatrix} \omega & 0 \\ 0 & \omega' \end{pmatrix}$$

where ω is the local matrix of D and ω' that of D'.

The corresponding curvature matrix is

$$\tilde{\Omega} = \begin{pmatrix} \Omega & 0 \\ 0 & \Omega' \end{pmatrix},$$

and one easily gets

$$\det\left(I - \frac{1}{2\pi i}\tilde{\Omega}\right) = \det\left(I - \frac{1}{2\pi i}\Omega\right) \wedge \det\left(I - \frac{1}{2\pi i}\Omega'\right).$$

This relation means that

$$C(S \oplus S') = C(S)C(S') \tag{17}$$

where the operation on the right-hand side is the cup product.

Next, if $f: V \to W$ is a differentiable map and S is a vector bundle on W with the induced bundle f^*S on V, then for the connection D on S with local matrix ω there is an associated connection on f^*S with local matrix $f_*\omega$. Using this connection we immediately get

$$C(f^*S) = f^*C(S). \tag{18}$$

The relations (17) and (18) yield Proposition 6.

7 Proposition *The Chern classes are real cohomology classes of the manifold V.*

This means that $C(S)$ is the image of a real cohomology class under the natural injection $H^*(V, R) \to H^*(V, C)$. To prove this, we consider a Hermitian metric on the bundle S, which exists because V has differentiable partitions of unity, and express this metric locally in *unitary frames* in the fibers of S, i.e., bases e_α such that $(e_\alpha, e_\beta) = \delta_{\alpha\beta}$. We shall use a general Hermitian connection on S (Section 4.6).

Now, Formula 4.6(20) reduces to

$$\omega + \bar{\omega}^t = 0, \tag{19}$$

where ω is the local matrix of the connection, the bar denoting the complex conjugate and t the transpose of the matrix.

From (19) one gets for the curvature matrix

$$\Omega + \bar{\Omega}^t = 0 \tag{20}$$

and, by an easy calculation,

$$\overline{\det\left(I - \frac{1}{2\pi i}\Omega\right)} = \det\left(I - \frac{1}{2\pi i}\Omega\right),$$

which shows that the corresponding differential form is real and proves Proposition 7.

In Section 3.4 we gave an axiomatic definition of the integral Chern classes. Propositions 6 an 7 above show that the Chern classes considered here satisfy the first three axioms of Section 3.4 when integer cohomology is replaced by real cohomology. We prove here that the fourth axiom is also satisfied.

8 Proposition *Let S be a complex line bundle on V. Then*

$$C_1(S) = jc_1(S),$$

3. Characteristic Classes of Differentiable Vector Bundles

where $j: H^2(V, J) \to H^2(V, R)$ and $c_1: H^1(V, C_c^*) \to H^2(V, J)$ is the isomorphism given by Formula 3.4(5).

We use the notation of Formulas 3.4(4) and 3.4(5). Let $g_{ij}: U_i \cap U_j \to C^*$ be the cocycle which defines the isomorphism class of S in $H^1(V, C_c^*)$, with respect to some open covering $\{U_i\}$ of V. This cocycle is the image by the homomorphism e of 3.3(4) of the cochain with sheaf coefficients C_c, defined by the system of functions

$$g' = \left\{\frac{1}{2\pi i} \ln g_{ij}\right\}.$$

The boundary of this cochain is given by the system $(\delta g')_{hjk}$ of functions on $U_i \cap U_j \cap U_k$, where

$$(\delta g')_{hjk} = \frac{1}{2\pi i}(\ln g_{jk} - \ln g_{hk} + \ln g_{hj}). \tag{21}$$

This is a cocycle of J and its cohomology class is just $c_1(S)$. The same functions, viewed as a real cocycle, define the cohomology class $jc_1(S)$.

On the other side, $C_1(S)$ is the cohomology class which corresponds by de Rham's Theorem to the closed differential form $-(1/2\pi i)\Omega$, where Ω is the curvature form of a Hermitian connection on S and is globally defined on V.

To get the desired result, we must calculate the isomorphism of de Rham's Theorem for differential forms of the second degree; it corresponds to $(r^2)^{-1}$ of Formula 2.5(6''). Following the proof of de Rham's abstract theorem of Section 2.5, we start from the resolution 1(3) of R, denote by $K_h = \ker d_h$ ($d_h = d$ applied to forms of degree h) and consider the exact sequence of sheaves

$$0 \to K_h \to \mathscr{A}^h(V) \xrightarrow{d_h} K_{h+1} \to 0.$$

Taking $h = 0, 1$, and writing the corresponding exact cohomology sequences, we have

$$\cdots \to 0 \to H^1(V, K_1) \xrightarrow{\partial} H^2(V, R) \to 0 \to \cdots,$$
$$\cdots \to H^0(V, \mathscr{A}^1(V)) \to H^0(V, K_2) \xrightarrow{\partial'} H^1(V, K_1) \to 0 \to \cdots,$$

where ∂ and ∂' are the respective coboundary homomorphisms.

We see that ∂ is an isomorphism and that $H^1(V, K_1) \approx H^0(V, K_2)/dA^1(V)$, and de Rham's isomorphism is the composition of these two isomorphisms. Then, take $-(1/2\pi i)\Omega \in H^0(V, K_2)$ and note that for the local forms of the connection ω_h on the neighborhood U_h we have $d\omega_h = \Omega_h$. It follows that to the element $-(1/2\pi i)\Omega$ there corresponds in $H^1(V, K_1)$ the cohomology class of the cocycle

$$\eta_{hj} = -\frac{1}{2\pi i}(\omega_j - \omega_h)$$

on $U_h \cap U_j$. But, from the transition relations 4.4(8) of the matrices of a connection, namely,

$$\omega_j = dg_{jh} \cdot g_{jh}^{-1} + g_{jh}\omega_h g_{jh}^{-1},$$

we obtain

$$-2\pi i \eta_{hj} = d \ln g_{jh}$$

in the 1-dimensional case. Using the definition of ∂ we finally get that the corresponding element of $H^2(V, R)$ is represented by the cocycle (21). Q.E.D.

Thus, we have established that the Chern classes of Definition 5 satisfy the axioms given in Section 3.3 for the Chern classes, if the integral cohomology is replaced by the real one, and this justifies Definition 5. Hence, the existence of the real Chern classes defined axiomatically in Section 3.3 is proved for the differentiable case.

One can prove the uniqueness of these classes too. Indeed, one can prove that for every vector bundle $S \to V$ there is a map $f: V_1 \to V$ such that f^*S is completely reducible (Section 3.4) and that $f^*: H^*(V, R) \to H^*(V_1, R)$ is a monomorphism. But then uniqueness is an immediate consequence of Axioms A2, A3, and A4 of Section 3.4. The proof can be found in [24], among other texts.

Obviously, the real images of the integral Chern classes, whose existence and uniqueness was stated in Section 3.4 (proof in [24]) satisfy the axioms of the real Chern classes. From uniqueness it follows that these images are the classes introduced in Definition 5.

Now, let $\pi: S \to V$ be a real differentiable vector bundle and $S^c = S \otimes (V \times C)$ its complexification. Then, S^c is a complex vector bundle, and if D is a connection of S it extends by linearity to a connection D^c on S^c. Locally, the connections D and D^c are represented by the same ma-

3. Characteristic Classes of Differentiable Vector Bundles

trices ω, whose elements are real Pfaff forms. Hence, they also have the same curvature matrix Ω. We shall use this connection in order to represent the Chern classes of S^c.

9   Proposition

$$C_{2h+1}(S^c) = 0.$$

To get this, we use a Euclidean connection of some Riemannian metric on S i.e., a connection such that the covariant derivative of the metric tensor vanishes. By using local orthonormal bases, it follows immediately from condition (a) of Theorem 4.5.2 that the matrices ω and Ω of the connection are skew-symmetric. Hence, every principal minor of odd order of Ω vanishes and the proposition follows.

10   Definition   The cohomology class

$$P_j(S) = (-1)^j C_{2j}(S^c) \in H^{4j}(V, R)$$

is called the *jth real Pontrjagin class* of the bundle S.

From the theory of Chern classes developed above, we get the following result.

11   Theorem   *The class $P_j(S)$ is represented in the de Rham cohomology by the closed differential form*

$$\mathscr{P}_j(S) = \frac{1}{(2\pi^{2j})(2j)!} \sum_{\alpha,\beta} \delta^{\beta_1\cdots\beta_{2j}}_{\alpha_1\cdots\alpha_{2j}} \Omega^{\alpha_1}_{\beta_1} \wedge \cdots \wedge \Omega^{\alpha_{2j}}_{\beta_{2j}}, \tag{22}$$

where Ω is the curvature of an arbitrary connection on S.

The *total Pontrjagin class* is defined as

$$P(S) = \sum_j P_j(S) \in H^*(V, R)$$

and it follows that it is represented by the nonhomogeneous differential form

$$\det\left(I - \frac{1}{2\pi}\Omega\right), \tag{23}$$

where, again, Ω is the curvature of a connection of S.

Remark that the definition given here for the Pontrjagin classes agrees with the definition of the integer Pontrjagin classes in Section 3.4. Hence, the Pontrjagin classes of Definition 10 are the real images of the integral classes of Section 3.4. With the same proof as for Proposition 6, we see that $P(S)$ is a characteristic class in the sense of Definition 3.4.1, though the integral total Pontrjagin class was not so.

In order to introduce another important characteristic class, let us consider a real skew-symmetric matrix of even order $p = 2q$, $A = (a_{ij})$. We give some formulas related to this matrix, following [4].

Introduce the expressions

$$A(k) = \sum_{i,j} \delta^{j_1 \cdots j_{2q}}_{i_1 \cdots i_{2q}} a_{i_1 i_2} \cdots a_{i_{2k-1} i_{2k}} a_{j_1 j_2} \cdots a_{j_{2k-1} j_{2k}}$$
$$\cdot a_{i_{2k+1} j_{2k+1}} \cdots a_{i_{2q} j_{2q}},$$

$$A'(k) = \sum_{i,j,h} \delta^{j_1 \cdots j_{2q-1} h}_{i_1 \cdots i_{2q-1} h} a_{i_1 i_2} \cdots a_{i_{2k-1} i_{2k}} a_{j_1 j_2} \cdots a_{j_{2k-1} j_{2k}}$$
$$\cdot a_{i_{2k+1} j_{2k+1}} \cdots a_{i_{2q-2} j_{2q-2}} a_{i_{2q-1} h} a_{j_{2q-1} h}, \quad 0 \leq k \leq q - 1.$$

If we take in $A(k)$ the terms such that

$$i_{2k} = j_1, \ldots, i_{2k} = j_{2k}$$

we get $A'(k - 1)$ every time. The terms such that

$$i_{2k} = j_{2k+1}, \ldots, i_{2k} = j_{2q}$$

give 0. Hence

$$A(k) = 2k A'(k - 1).$$

Similarly, by taking first the terms with $i_{2k+1} = j_1, \ldots, j_{2k+1}$ we get 0 and then, from the terms with $i_{2k+1} = j_{2k+2}, \ldots, j_{2q}$ we get $A'(k)$, which shows that

$$A(k) = (2q - 2k - 1) A'(k).$$

Combining the two results, we have

$$A(k) = \frac{2k}{2q - 2k + 1} A(k - 1),$$

3. Characteristic Classes of Differentiable Vector Bundles

whence, by iteration

$$A(k) = \frac{2^k k!}{(2q - 2k + 1)(2q - 2k + 3) \cdots (2q - 3)(2q - 1)} A(0). \quad (24)$$

Now, we remark that $A(0) = (2q)! \det (a_{ij})$ and

$$A(q) = (\sum_\sigma \varepsilon(\sigma) a_{i_1 i_2} \cdots a_{i_{2q-1} i_{2q}})^2,$$

where $\sigma = (i_1, \ldots, i_{2q})$ and hence

$$\det (a_{ij}) = \frac{1}{(2^q q!)^2} (\sum_\sigma \varepsilon(\sigma) a_{i_1 i_2} \cdots a_{i_{2q-1} i_{2q}})^2. \quad (25)$$

From the fact that $\det (a_{ij}) = \Delta_p$ is defined by the function $\mathscr{W}_p$ of (13) of the Weil algebra $W(GL)$, we deduce that the paranthesis of the right-hand side in (25) is defined analogously by an element of the Weil algebra $W(GL^+)$ where GL^+ is the group of square matrices with positive determinant.

Suppose that we take only oriented bundles S with even-dimensional ($p = 2q$) fiber (see Section 3.3). Then, we are led to introduce the closed differential form

$$\mathscr{E}(S) = \frac{(-1)^q}{2^{2q}\pi^q q!} \sum_\sigma \varepsilon_{i_1 \cdots i_{2q}} \Omega^{i_1}_{i_2} \wedge \cdots \wedge \Omega^{i_{2q-1}}_{i_{2q}}, \quad (26)$$

where Ω is the curvature matrix of a Euclidean connection on S. Its corresponding cohomology class $E(S)$ does not depend on the choice of the Euclidean connection on S. We must consider only Euclidean connections in order to have, in orthonormal bases, a skew-symmetric matrix Ω.

12 Definition   The class $E(S)$ is called the *Euler class* of the oriented bundle S. If S has an odd-dimensional fiber, we put $E(S) = 0$.

From this definition and Formula (26) we get, just as for the total Chern class, the following proposition.

13 Proposition   *The Euler class is a characteristic class in the sense of Definition 3.4.1. It is defined for oriented differentiable real vector bundles and takes values in $H^p(V, R)$.*

Consequently, we have

$$E(S \oplus S') = E(S)E(S'), \qquad E(f^*S) = f^*E(S), \qquad f: W \to V. \qquad (27)$$

14 Proposition *If the bundle S has its structure group reducible to $GL(q, C)$, then*

$$E(S) = C_q(S), \qquad (28)$$

where the Chern class C_q is calculated for the complex structure of the bundle S.

In fact, if we take a Hermitian connection on S, it gives a real Euclidean connection on the real structure of S, and by comparing the expressions of $\mathscr{C}_q(S)$ and $\mathscr{E}(S)$ we get the result.

The previous result agrees with the definition of the Euler class in Section 3.4 and it shows that $E(S)$ is the real image of the integral Euler class of Section 3.4. Moreover, the Euler class here is extended to all the oriented bundles (this extension was used in connection with Proposition 3.4.2).

In [25], Kobayashi and Nomizu give an axiomatic definition of the real Euler class, taking as axioms the relations (27) and the relation (28) for a complex line bundle, and show as above the existence of the Euler class. They also show that these axioms uniquely define the real Euler class, though this is not true for the integral Euler class.

Following Section 3.4 we also consider the Stiefel–Whitney classes. In fact, it is possible to give a representation of these classes by pairs of differential forms as in the theorem of Allendoerfer–Eells, but this representation is not based on the theory of the Weil algebra. The development of this representation would take us too far afield, and for this reason we only mention Eells' paper [11], where the representation is given. A similar representation of the integral Chern classes is given in [30].

Another application of Weil's theory gives the representative differential form of the real Chern character of a differentiable complex vector bundle S. Following the definition given in Section 3.4 we must consider the formal factorization of the Chern class

$$\det\left(I - \frac{1}{2\pi i}\Omega\right) = \prod_1^p (1 + \gamma_h),$$

which means that the symbols γ_h are the formal eigenvalues of the matrix

3. Characteristic Classes of Differentiable Vector Bundles

$-(1/2\pi i)\Omega$ and, hence, satisfy the same rules for calculating the eigenvalues of a numerical matrix. Then, taking into account that the eiganvalues of the exponential of a matrix are the exponentials of the eigenvalues of the matrix, Formula 3.4(19) shows that the real Chern character, i.e., the real image of the rational Chern character, is represented by the closed nonhomogeneous differential form $\mathscr{C}hS$ given by

$$\mathscr{C}hS = \text{Tr}\left(\exp\frac{-1}{2\pi i}\Omega\right). \tag{29}$$

Hence, the real kth Chern character is given by the differential form

$$\mathscr{C}h_k S = \text{Tr}\frac{(-1)^k}{(2\pi i)^k k!}\Omega^k, \tag{30}$$

where the form of Ω is an exterior power. One sees that these forms correspond to elements of the Weil algebra and, hence, they have the corresponding properties.

The previous results can be applied to the tangent bundle of a finite dimensional differentiable manifold, and we consider again Definition 4.1.11.

15 Definition The Pontrjagin classes of the tangent bundle of a differentiable manifold are called the *Pontrjagin classes of the manifold*. The Euler class of the tangent bundle of an oriented manifold is the *Euler class of the manifold*. The Chern classes and characters of the tangent bundle of an almost complex or complex manifold are the *Chern classes and characters of the manifold*.

We point out a very important consequence of some topological results (for instance, Theorem 7.2, Chapter 17, of [24]).

16 Theorem (Gauss–Bonnet.) *If V is a compact, orientable differentiable manifold, the Euler–Poincaré characteristic $\chi(V)$ is given by the formula*

$$\chi(V) = \int_V \mathscr{E}(T(V)), \tag{31}$$

where the form $\mathscr{E}$ is expressed in terms of a Riemannian connection of the manifold.

If V is almost complex we can use Proposition 14.

The results of this section give important relations between the characteristic classes of a differentiable manifold and the curvature of the connections on this manifold. The consequences of these relations have been intensively studied. As an example, we have the following very simple corollary of Theorem 11 and of (26).

17 Proposition *The Pontrjagin classes of a locally affine manifold vanish. The Euler class of a locally Euclidean manifold vanishes.*

In fact, we have, by hypotheses, a linear (respectively Euclidean) connection with vanishing curvature. We cannot deduce in the same manner that the Euler class of a locally affine manifold vanishes because in this case we must use a Euclidean connection.

We give also another typical example of a consequence of the relations between curvature and characteristic classes.

18 Theorem (Chern [4].) *If the differentiable manifold V has a Riemannian metric of constant sectional curvature k, the Pontrjagin classes $P_j(V)$ vanish for $j > 0$. If, moreover, V is compact and orientable and has even dimension $2q$, $\chi(V)$ has the sign of k^q and vanishes if and only if $k = 0$.*

To prove this, we shall first obtain another expression of the forms $\mathscr{P}_j(S)$ which represent the Pontrjagin classes of V. We shall use the curvature of the Riemannian metric given by the hypotheses of Theorem 18. Moreover, because we work with local expressions, we shall use orthonormal bases and, hence, the matrix (Ω_β^α) will be skew-symmetric.

Then, $\mathscr{P}_j(V)$ is given by a sum of skew-symmetric determinants and we can use for their expression Formula (25). Hence, we introduce the local differential forms

$$\Theta^{(j)}_{\alpha_1 \cdots \alpha_{2j}} = \sum_\beta \delta^{\beta_1 \cdots \beta_{2j}}_{\alpha_1 \cdots \alpha_{2j}} \Omega^{\beta_1}_{\beta_2} \wedge \cdots \wedge \Omega^{\beta_{2j-1}}_{\beta_{2j}}$$

and we shall have

$$\mathscr{P}_j(V) = \frac{1}{(2^j j!)^2 (2j)! (2\pi)^{2j}} \sum_\alpha \Theta^{(j)}_{\alpha_1 \cdots \alpha_{2j}} \wedge \Theta^{(j)}_{\alpha_1 \cdots \alpha_{2j}}. \tag{32}$$

It is classical (see for instance [25]) that for a Riemannian metric of constant sectional curvature k

$$\Omega_\alpha^\beta = -k\omega^\beta \wedge \omega^\alpha \tag{33}$$

where ω^α is the dual cobasis of the given orthonormal basis.

It follows that

$$\Theta_{\alpha_1\cdots\alpha_{2j}}^{(j)} = (-1)^j k^j \omega^{\alpha_1} \wedge \cdots \wedge \omega^{\alpha_{2j}}, \tag{34}$$

and by introducing this in (32) we get the first part of the theorem.

For the second part, we observe that $\mathscr{E}(V)$ given by (26) can be put in the form

$$\mathscr{E}(V) = \frac{(-1)^q}{2^{2q}\pi^q q!} \Theta_{1\cdots 2q}^{(q)} \tag{35}$$

and the desired result follows from (34) and the Gauss–Bonnet theorem.

This theorem was generalized by Thorpe [47] for manifolds which have Riemannian metrics with higher order constant sectional curvature.

Finally, while many results, such as Chern's Theorem above, are known, the inverse problem has not been considered systematically. This problem consists in obtaining results of the following type: if the characteristic classes satisfy certain conditions, then the manifold has connections whose curvature or holonomy group has a determined behavior. Particular results of this type were given, for instance, by Milnor [31].

4 Elliptic Operators. Elliptic Complexes

Returning to the problem of representing cohomology classes by differential forms, the next stage is to look for canonical representative forms. This problem has, in many important cases, an affirmative answer, which is given by the theory of harmonic forms. The later is an important consequence of the theory of elliptic differential operators, a chapter of functional analysis whose complete exposition is not possible here because it would take us too far afield; we shall only sketch the most important results for our purposes. This material will be treated in this section, while the applications of harmonic forms to cohomology will be treated in the next section. Our exposition is an abbreviated version of the complete exposition given by G. Lusztig [29]. In the sequel, all the differentiable manifolds are assumed Riemannian, compact (hence finite dimensional), and oriented, and all the vector bundles on V will be assumed complex finite dimensional with a Hermitian metric. In particular, this is assumed for the linear spaces.

Let $p = (p_1, \ldots, p_h)$ be a system of nonnegative integers. Put $|p| = p_1 + \cdots + p_h$ and introduce the operator

$$D^p = \frac{\partial^{|p|}}{\partial x_1^{p_1} \cdots \partial x_h^{p_h}} \tag{1}$$

acting on the differentiable functions on open subsets of R^h, with values in a complex k-dimensional linear space L which has a fixed basis $(e_1, \ldots, e_k)$ (i.e., $L = C^k$ and we use L here because this is a more convenient notation).

1 Definition A *differential operator* in R^h is an operator of the type

$$D = \sum_{|p| \leq k} a_p(x) D^p, \tag{2}$$

where $a_p(x): L \to M$ is a homomorphism of vector spaces and depends differentiably on x.

Consequently, if $f: R^h \dashrightarrow L$, $Df: R^h \dashrightarrow M$. The operator D given by (2) is said to be of *order* $\leq k$.

2 Definition The differential operator (2) is called *k-elliptic* at the point $x \in R^h$ if, for every $\xi = (\xi_1, \ldots, \xi_h) \in R^h$, $\xi \neq 0$,

$$\sum_{|p|=k} a_p(x) \xi_{p_1} \cdots \xi_{p_h}$$

is an isomorphism $L \approx M$. The operator is k-elliptic on a subset of R^h if it is such at every point of the subset.

For the theory of elliptic operators, it is convenient to consider that the point x of (1) and (2) varies on the h-dimensional torus $T^h = R^h/2\pi J^h$ (the quotient of R^h by the group of translations generated by h independent vectors). Locally, this is the same as before. Then, the function space on which the operator D acts is $C^\infty(T^h, L)$, the space of L-valued differentiable functions on T^h, and the range of D is $C^\infty(T^h, M)$.

It is known that $C^\infty(T^h, L)$ can be made into a topological vector space in such a way that the convergence of a sequence of functions to a limit

4. Elliptic Operators. Elliptic Complexes

means the uniform convergence of the sequence and of all its derivative sequences. In the sequel, $C^\infty(T^h, L)$ always has this topology.

We also consider the space $\mathcal{D}' = \mathcal{D}'(T^h, L)$ of antilinear continuous functionals on $C^\infty(T^h, L)$; these are called *L-valued distributions*[†] on T^h. The space $\mathcal{D}'$, with the strong topology is again a topological vector space and there is a continuous injection $C^\infty \to \mathcal{D}'$, with dense image, which sends the function f to the functional $\int f(f, g)dx$ ($dx = dx^1 \wedge \cdots \wedge dx^h$), where g is the argument and (f, g) is the canonical Hermitian product ($\sum z\bar{z}$) with respect to the basis of L. Hence, we can talk of a function as an element of $\mathcal{D}'$. The action of a differential operator D on $C^\infty(T^h, L)$ extends naturally to $\mathcal{D}'$.

For every $u \in \mathcal{D}'$ one introduces the *Fourier coefficients*

$$\hat{u}(n) = (u(e^{inx} e_1), \ldots, u(e^{inx} e_k)) \in L,$$

where

$$n \in J^h, \qquad nx = \sum_{\alpha=1}^{h} n_\alpha x_\alpha,$$

and then the spaces

$$H^s(T^h, L) = \{u \mid u \in \mathcal{D}' \text{ and } (2\pi)^{-h} \sum_n |\hat{u}(n)|^2 (1 + \|n\|^2)^s < \infty\},$$

where s is an integer, $|v|^2 = (v, v)$, $v \in L$, and $\|n\| = (\sum n_\alpha^2)^{1/2}$. We use the notation

$$|u|_s^2 = (2\pi)^{-h} \sum_n |\hat{u}(n)|^2 (1 + \|n\|^2)^s. \tag{3}$$

Every space H^s has a scalar product given by

$$(u, v)_s = (2\pi)^{-h} \sum_n (\hat{u}(n), \hat{v}(n))(1 + \|n\|^2)^s, \tag{4}$$

and H^s is a Hilbert space with respect to this scalar product.

With this notation one can prove the following fundamental result.

3 Theorem (Friedrichs' inequality.) *Let D be a k-elliptic operator, s and s_0 integers, and $u \in C^\infty(T^h, V)$. Then the following inequality,*

[†] These distributions are, of course, different from the distributions of Section 4.3.

where c is a constant depending on s and s_0, holds

$$|u|_{s+k} \leqslant c(|Du|_s + |u|_{s_0}). \tag{5}$$

4 Definition The differential operator D is called *hyperelliptic* if $Du \in C^\infty(T^h, M)$ for a distribution u implies that $u \in C^\infty(T^h, L)$.

The following result is of fundamental importance.

5 Theorem *Every k-elliptic differential operator is hyperelliptic.*

Now, let us go over to the global version of the previous results.

6 Definition Let V be a differentiable manifold and E, F vector bundles on V (with the conventions announced at the beginning of this section). A linear map $D: C^\infty(E) \to C^\infty(F)$ between the spaces of differentiable global cross sections of the bundles is called a *differential operator of order* $\leqslant k$, if the local coordinate expression of this map is of the form (2).

Here, we write $C^\infty(E)$, instead of $\Gamma(E)$ of the previous chapters, in order to have the same notation as in the local case. The definition obviously has an invariant meaning, since the form of the expression (2) is not changed if we change the local charts.

Let ξ be a tangent covector of V at a point x. It is easy to see that in a coordinate neighborhood U of x there is a homomorphism $A_U: L \to M$, between the fibers of E and F, given by

$$A_U = \sum_{|p|=k} a_p(x)\xi^p, \qquad \xi^p = \xi_{p_1} \cdots \xi_{p_h}, \tag{6}$$

where ξ_α are the coordinates of ξ with respect to natural cobases. Since there is the same transition relation between the operators A_U, $A_{U'}$ of two coordinate neighborhoods as for the corresponding local expressions for the operator D, there is a homomorphism

$$\sigma(D, \xi): E_x \to F_x \tag{7}$$

of the respective local fibers, whose coordinate expression is just A_U. The homomorphism (7) is called the *symbol* of D for the covector ξ at the point x.

4. Elliptic Operators. Elliptic Complexes

7 Definition The operator D of Definition 6 is called *k-elliptic at* x if, for every ξ, $\sigma(D, \xi)$ is an isomorphism, and D is *k-elliptic* if this is true for every $x \in V$.

It follows easily that D is k-elliptic if and only if its local coordinate expressions are k-elliptic.

We now introduce the necessary function spaces. Since V is compact, we can consider a finite trivializing covering for E, $\{U_i\}$, and let $\{\lambda_i\}$ be a subordinate differentiable partition of unity. We can consider the coordinates on U_i to be T^h-valued. A scalar function which is defined on a subset of a space will be automatically extended by the value 0 on the whole space. Define

$$Q: C^\infty(E) \to \bigoplus_i C^\infty(T^h, L),$$

by associating to a cross section s the sum Qs of the coordinate expressions of the functions $\lambda_i s|_{U_i}$. Next, denote by μ_i differentiable functions with the support contained in U_i and which take the value 1 on support of λ_i. Then, we shall define

$$P: \bigoplus_i C^\infty(T^h, L) \to C^\infty(E)$$

by $P(\sum_i g_i) = \sum_i \mu_i g'_i$, where g'_i is the local cross section of E over U_i whose coordinate expression is g_i.

For the operators P and Q one finds

$$PQ = I, \quad (QP)^2 = QP.$$

The first relation shows that Q is an injection and the second, that QP is a continuous *projector*, whence it follows that $C^\infty(E)$ can be identified with a closed subspace of $\bigoplus_i C^\infty(T^h, L)$, which gives it a structure of a topological vector space. One can see that the topology is uniquely defined and this space $C^\infty(E)$ plays the role of the space $C^\infty(T^h, L)$ of the local theory.

Let Ω_V be the canonical line bundle of V (Section 3.4), i.e., the bundle whose transition functions are the Jacobians of the coordinate transformations on V. The space $\mathcal{D}'(E)$ of *E-valued distributions on* V is defined as the space of linear continuous functionals $C^\infty(E^* \otimes \Omega_V)$ with the topology given by an identification with a closed subspace of $\bigoplus_i \mathcal{D}'(T^n, L)$ obtained using the operators P and Q just as for functions. A cross section $s \in C^\infty(E)$ can be considered as an element of $\mathcal{D}'(E)$, i.e., as the functional which sends $f \otimes \omega \in C^\infty(E \otimes \Omega_V)$ to $\int_V f(s)\omega$. The injection $C^\infty(E) \to \mathfrak{D}'(E)$ obtained in this manner is continuous and sends $C^\infty(E)$ into $\mathfrak{D}'(E)$, which allows to extend by continuity the action of differential operators to distributions.

Let $u \in \mathfrak{D}'(E)$. Denote by $u(i)$ the coordinate expression of $\lambda_i u|_{U_i}$ and introduce $\|u\|_s^2 = \sum_i |u(i)|_s^2$ (see Formula (3)). We now define the space

$$H^s(E) = \{u \mid u \in \mathfrak{D}'(E), \|u\|_s^2 < \infty\},$$

where s is an integer, which can again be identified with a closed subspace of the direct sum $\oplus_i H^s(T^h, C^k)$ and, hence, is a Hilbert space.

By using these spaces and the usual techniques of coordinate expressions one obtains the global version of Theorem 3 and 5.

8 Theorem *Let E and F be differentiable vector bundles with compact basis V and let $D: C^\infty(E) \to C^\infty(F)$ be a k-elliptic operator. Then*

(a) *we have the Friedrichs' inequality*

$$\|u\|_{s+k}^2 \leq (\text{const.})(\|Du\|_s^2 + \|u\|_{s_0}^2);$$

(b) *D is hyperelliptic, i.e., if $Du \in C^\infty(F)$ then $u \in C^\infty(E)$.*

The following lemma is required for the proof of Theorem 10.

9 Lemma (L. Schwartz.) *If $T: H \to H'$ is a continuous injection with closed image of two Hilbert spaces H and H', and if $K: H \to H'$ is a compact operator, then $T + K$ has finite dimensional kernel and closed image.*

Let $N = \ker(T + K)$ and e_i ($i = 1, 2, \ldots$) be an infinite orthonormal sequence in N. Then, $Te_i + Ke_i = 0$ and, since K is compact, we can suppose $Ke_i \to f$, whence $Te_i \to -f$, which is impossible. Hence N is finite dimensional, and we have the first part of the lemma. For the second part, let $a \in \overline{(T + K)(H)}$. Then there is a sequence f_n in H such that $(T + K)f_n \to a$. Assuming $Kf_n \to b$, then $Tf_n \to a - b$ and $f_n \to f$ with $Tf = a - b, K(f) = b$. It follows that $a = (T + K)f$, whence $a \in (T + K)(H)$.
Q.E.D.

10 Theorem *If $D: C^\infty(E) \to C^\infty(F)$ is a k-elliptic operator, $\ker D$ is finite dimensional, and $\operatorname{im} D$ is closed.*

We sketch the proof. From the properties of D it follows that the map

4. Elliptic Operators. Elliptic Complexes

$(D \oplus 1)u = (Du + u)$ can be viewed as a continuous injection

$$D \oplus 1: H^{s+k}(E) \to H^s(F) \oplus H^{s+k-1}(E)$$

(one can show that there is an injection $H^t \to H^s$ for $t \geq s$). The Friedrichs inequality for $s_0 = s + k - 1$ leads to the fact that $D \oplus 1$ is a topological isomorphism of $H^{s+k}(E)$ onto its image.

From the decomposition

$$D = D \oplus 0 = D \oplus 1 + 0 \oplus (-1): H^{s+k}(E) \to H^s(F), \qquad (8)$$

we have a sum of a continuous injection with closed image and a compact operator, whence, by Lemma 9, $\dim \ker D < \infty$ and $\operatorname{im} D$ of (8) is closed. But the kernel of the operator (8) coincides with $\ker (D: C^\infty(E) \to C^\infty(F))$ because of the hyperellipticity of D (Theorem 8) and, for the same reason,

$$D(C^\infty(E)) = \bigcap_s D(H^s(E))$$

which is closed as an intersection of closed sets. This proves the theorem.

11 Definition Let $D: C^\infty(E) \to C^\infty(F)$ be a differential operator. The *formal adjoint* of D is the operator $D^*: C^\infty(F^* \otimes \Omega_V) \to C^\infty(E^* \otimes \Omega_V)$, such that

$$\int_V Df \cdot g = \int_V f \cdot D^*g, \qquad f \in C^\infty(E), g \in C^\infty(F^* \otimes \Omega_V), \qquad (9)$$

where $a \cdot (b \otimes \omega)$ means $b(a)\omega$.

If for the integral (9) we use the volume element of the Riemannian metric of V, the adjoint operator can be viewed as $D^*: C^\infty(F^*) \to C^\infty(E^*)$. For the adjoint operator one proves the following result.

12 Proposition *Every differential operator D of order $\leq k$ has a unique formal adjoint D^* with order $\leq k$ whose symbol is $\sigma(D^*, \xi) = (-1)^k \cdot {}^t\sigma(D, \xi)$.*

The transposed operator is induced by the given operator.

In Proposition 12, D^* is considered to be an operator $C^\infty(F^*) \to C^\infty(E^*)$ in the sense of the previous remark. The uniqueness of D^* follows immediately from (9). Next, suppose that the local coordinate expression of D is (2) and consider the global operator D^* defined with the local expressions

$$D^*g = \sum_{|p|\leq k}(-1)^{|p|}D^p({}^t a_p(x)g), \qquad g \in C^\infty(F^*).\tag{10}$$

Let us give a simple formula on the torus. Let f, g be two scalar differentiable functions on T^h. We have

$$\frac{\partial}{\partial x^\alpha}(fg) = \frac{\partial f}{\partial x^\alpha}g + f\frac{\partial g}{\partial x^\alpha},$$

whence, by integration on T^h and by the help of Stokes' formula, we get

$$\int_{T^h}\frac{\partial f}{\partial x^\alpha}g\,dx = -\int_{T^h}f\frac{\partial g}{\partial x^\alpha}dx.$$

Applying this relation, step by step, $|p|$ times we get

$$\int_{T^h}(D^p f)g\,dx = (-1)^{|p|}\int_{T^h}f(D^p g)\,dx.\tag{11}$$

Formula (11) explains the validity of the following calculation:

$$\int_{T^h}a_p(x)D^p f\cdot g\,dx = \int_{T^h}D^p f\cdot {}^t a_p(x)g\,dx = \int_{T^h}(-1)^{|p|}f\cdot D^p({}^t a_p(x)g)\,dx,$$

where $f \in C^\infty(T^h, L)$, $g \in C^\infty(T^h, L^*)$.

By taking such calculations into account it follows that the operator (10) satisfies the condition (9), whence the existence of the formal adjoint.

Finally, the relation between the symbols of D and D^* is an immediate consequence of Formula (10), thereby ending the proof of Proposition 12. For a detailed proof, see [36].

13   Corollary   *If D is k-elliptic, the same is true for D^*.*

Consider the case of a differential operator $D: C^\infty(E) \to C^\infty(E)$. Then, D^* can be identified with an operator $D^*: C^\infty(E) \to C^\infty(E)$, if we go over from E to E^* by lowering indices with the Hermitian metric of E. This is characterized by the relation

4. Elliptic Operators. Elliptic Complexes

$$\int_V (Ds, t)dv = \int_V (s, D^*t)dv, \qquad s, t \in C^\infty(E), \qquad (12)$$

where dv is the volume element of the Riemannian metric of V.

14 Definition The differential operator $D: C^\infty(E) \to C^\infty(E)$ is called *self-adjoint* if $D = D^*$.

15 Theorem *If the operator $D: C^\infty(E) \to C^\infty(E)$ is k-elliptic, and self-adjoint, we have the decomposition into a direct topological sum*

$$C^\infty(E) = \ker D \oplus \operatorname{im} D. \qquad (13)$$

Let $s \in \ker D$ and $t \in \operatorname{im} D$, i.e., $t = Du$. Then

$$\int_V (s, t)dv = \int_V (s, Du)dv = \int_V (Ds, u)dv = 0,$$

which shows that $\ker D$ and $\operatorname{im} D$ are orthogonal subspaces with respect to the scalar product given by $\int_V (s, t)dv = (s, t)$, considered in $L^2(E)$, the space of square summable cross sections of E.

Next, let u be orthogonal to $\ker D \oplus \operatorname{im} D$ in $L^2(E) \subseteq \mathcal{D}'(E)$. Then, since u is orthogonal to $\operatorname{im} D$, we have $(u, Ds) = (Du, s) = 0$ for every $s \in C^\infty(E)$, which implies $Du = 0$ and, from the hyperellipticity of D, $u \in C^\infty(E)$. Hence $u \in \ker D$. But u is also orthogonal to $\ker D$, whence $(u, u) = \int_V |u|^2 \, dv = 0$ and $u = 0$. By a classical theorem, if every vector orthogonal to $\ker D \oplus \operatorname{im} D$ vanishes, then this subspace is dense in $C^\infty(E)$ and, since it is closed by Theorem 10, it is $C^\infty(E)$, which proves Theorem 15.

16 Theorem *If $D: C^\infty(E) \to C^\infty(F)$ is a k-elliptic operator, then*

$$C^\infty(E) = \ker D \oplus \operatorname{im} D^*, \quad C^\infty(F) = \ker D^* \oplus \operatorname{im} D.$$

The proof is the same as that of Theorem 15. From this theorem it follows that $\operatorname{coker} D = \ker D^*$, which shows, because of Theorem 10, that $\operatorname{coker} D$ is also finite dimensional. A differential operator which has finite dimensional kernel and cokernel and closed image is called a *Fredholm operator*. Hence, we have the following result.

17 Corollary *Every elliptic operator is a Fredholm operator.*

In connection with the operator D of theorem 15, we introduce two other important operators: $\mathcal{H}$, the projection of $C^\infty(E)$ onto ker D, given by Formula (13), and the *Green's operator* $\mathcal{G}: C^\infty(E) \to $ im D obtained as follows. The operator $D:$ im $D \to $ im D is a continuous bijection (since $D\alpha_1 = D\alpha_2$ and $\alpha_1, \alpha_2 \in $ im D imply $\alpha_1 - \alpha_2 \in $ ker $D \cap $ im D and, hence, $\alpha_1 = \alpha_2$), whence it is a topological isomorphism and $\mathcal{G}$ is defined as the composition

$$\mathcal{G}: C^\infty(E) \xrightarrow{\text{prim } D} \text{im } D \xrightarrow{D^{-1}} \text{im } D.$$

We then have for every $u \in C^\infty(E)$

$$u = \mathcal{H}u + D\mathcal{G}u. \tag{14}$$

One can see that $\mathcal{G}$ is self-adjoint and that $\mathcal{G}$ and $\mathcal{H}$ extend to the space $\mathfrak{D}'(E)$, giving the decomposition

$$\mathfrak{D}'(E) = \text{im } D \oplus \text{ker } D,$$

where D is considered on $\mathfrak{D}'(E)$. Considering then the spectral theory for the operator $\mathcal{G}: L^2(E) \to L^2(E)$ one gets: *if $D: C^\infty(E) \to C^\infty(E)$ is a self-adjoint elliptic operator, there is in $L^2(E)$ an orthonormal basis consisting of eigenvectors of D.*

We consider cochain complexes with elliptic operators.

18 Definition A *differential complex* is a system $\mathfrak{C}$ consisting of a sequence of differentiable vector bundles $(E_i)_{i \in J}$ on the compact manifold V and a sequence of differential operators $D_i: C^\infty(E_i) \to C^\infty(E_{i+1})$, such that $E_i = 0$ for $i < 0$ and $i > i_0$ and $D_{i+1}D_i = 0$. The complex is k-*elliptic* if each D_i has the order $\leq k$ and the sequence of symbols of order k

$$0 \longrightarrow E_{0,x} \xrightarrow{\sigma(D_0,\xi)} E_{1,x} \xrightarrow{\sigma(D_1,\xi)} \cdots \xrightarrow{\sigma(D_{i_0-1})} E_{i_0,x} \longrightarrow 0 \tag{15}$$

is exact for every nonzero tangent covector to V at an arbitrary point $x \in V$.

For a differential complex $\mathfrak{C}$, the spaces $C^\infty(E_i)$ and the operators D_i form a cochain complex and its cohomology spaces $H^i(\mathfrak{C}) = $ ker $D_i/$im D_{i-1} are called the *cohomology spaces of* $\mathfrak{C}$.

4. Elliptic Operators. Elliptic Complexes

We can extend the usual operations with vector bundles to differential complexes.

Thus, if $\mathfrak{C} = (E_i, D_i)$ and $\mathfrak{C}' = (E_i', D_i')$ are two differential complexes on V, their *direct sum* is defined by $\mathfrak{C} \oplus \mathfrak{C}' = (E_i \oplus E_i', D_i \oplus D_i')$. Obviously, if $\mathfrak{C}$ and $\mathfrak{C}'$ are elliptic, the same is true of $\mathfrak{C} \oplus \mathfrak{C}'$.

The *tensor product* $\mathfrak{C}'' = \mathfrak{C} \otimes \mathfrak{C}' = (E_i'', D_i'')$ is defined by

$$E_i'' = \bigoplus_{j+k=i} (E_j \otimes E_k)$$

and

$$D_i''(s \otimes t) = D_j s \otimes t + (-1)^j s \otimes D_k' t, \qquad s \in C^\infty(E_j), t \in C^\infty(E_k).$$

It is easily seen that this is really a differential complex and the sequence (15) for $\mathfrak{C}''$ is obtained from the corresponding sequences of $\mathfrak{C}$ and $\mathfrak{C}'$ by the same procedure, i.e.,

$$E_{i,x}'' = \bigoplus_{j+k=i} (E_{j,x} \otimes E_{k,x})$$

and

$$\sigma(D_i'', \xi)(s_x \otimes t_x) = \sigma(D_j, \xi) s_x \otimes t_x + (-1)^j s_x \otimes \sigma(D_k', \xi) t_x,$$

where $s_x \in E_{j,x}$, $t_x \in E_{k,x}'$.

This tensor product can be generalized. Suppose that $\mathfrak{C}$ and $\mathfrak{C}'$ have different bases V and V'. Then they induce complexes on $V \times V'$ by the projections of this manifold on V and V', and we shall consider that the tensor product of $\mathfrak{C}$ and $\mathfrak{C}'$ is the tensor product of these induced complexes. The previous formulas for E'', D'', and $\sigma(D'', \cdot)$ hold with the remark that the second argument of σ has to be a pair (ξ, ξ') where ξ is a tangent covector of V and ξ' of V' and at least one of them does not vanish.

If $\mathfrak{C}$ is a differential complex, its *dual* $\mathfrak{C}^*$ is defined by the bundles $E_i' = E_{i_0-i}^* \otimes \Omega_V$ and the operators $D_i' = D_{i_0-i-1}^*$. Considering a fixed Riemannian metric on V, E_i can be identified with $E_{i_0-i}^*$ (see the previous remarks regarding the definition of the adjoint operator). By Formula (10), we see that if $\mathfrak{C}$ is k-elliptic, $\mathfrak{C}^*$ is also k-elliptic.

In the sequel we consider a fixed Riemannian metric on the compact oriented manifold V and Hermitian metrics on the bundles of $\mathfrak{C}$, and using these metrics, we identify E_i with E_i^* as in Definition 14.

We define, for every index i, the operator

$$\Box_i = D_i^* D_i + D_{i-1} D_{i-1}^* : C^\infty(E_i) \to C^\infty(E_i) \qquad (16)$$

(when no confusion arises we shall omit the indices) and call it the *Laplace operator* (*laplacian*) of the complex $\mathfrak{C}$. It is immediate that, if the D_i have orders $\leq k$, the $\Box_i$ have orders $\leq 2k$ and they are self-adjoint. We also have

$$\Box_{i+1} D_i = D_i \Box_i, \qquad \Box_i D_i^* = D_i^* \Box_{i+1}, \qquad (17)$$

i.e., $\Box$ commutes with D and D^*.

19 Proposition *The operators $\Box$ of a k-elliptic differential complex are 2k-elliptic operators.*

To prove this, we remark first, from the definition of the symbol of an operator, that for the composition of two differential operators A and B $\sigma(B \circ A, \xi) = \sigma(B, \xi) \circ \sigma(A, \xi)$.

Hence,

$$\sigma(\Box, \xi) = (-1)^k (\sigma(D, \xi) \sigma(D, \xi)^* + \sigma(D, \xi)^* \sigma(D, \xi))$$

and, for $v \in E_{i,x}$ such that $\sigma(\Box, \xi)v = 0$,

$$(\sigma(\Box, \xi)v, v) = |\sigma(D, \xi)v|^2 + |\sigma(D, \xi)^* v|^2 = 0,$$

which holds if and only if

$$\sigma(D, \xi)v = \sigma(D, \xi)^* v = 0.$$

Then, if $\xi \neq 0$, since the complex is k-elliptic, we can put $v = \sigma(D, \xi)w$. Now, $\sigma(D, \xi)^* v = 0$ implies

$$(\sigma(D, \xi)^* \sigma(D, \xi) w, w) = |\sigma(D, \xi)w|^2 = 0,$$

whence $v = 0$.

This means that if $\xi \neq 0$, $\sigma(\Box, \xi)$ is an injection, hence an isomorphism, since it is an injection of a finite dimensional vector space onto itself. This proves Proposition 19.

Now, we shall establish the fundamental properties of elliptic complexes.

4. Elliptic Operators. Elliptic Complexes

20 Theorem (Kodaira.) *For an elliptic complex $\mathfrak{C}$ there is the following decomposition as a direct topological sum of closed orthogonal subspaces*:

$$C^\infty(E_i) = \ker \square_i \oplus \operatorname{im} D_{i-1} \oplus \operatorname{im} D_i^*. \tag{18}$$

The three subspaces on the right-hand side are closed in view of Theorem 10. Their pairwise orthogonality is shown by the relations

$$(D_{i-1}u, D_i^*v) = (D_i D_{i-1}u, v) = 0,$$
$$(\alpha, D_{i-1}u) = (D_{i-1}^*\alpha, u) = 0,$$
$$(\alpha, D_i^*v) = (D_i\alpha, v) = 0,$$

where $\alpha \in \ker \square_i$. The last two relations follow from the fact that $\square_i \alpha = 0$ implies

$$|D_i\alpha|^2 + |D_{i-1}^*\alpha|^2 = 0,$$

whence $D_i\alpha = D_{i-1}^*\alpha = 0$.

Then, if $u \in C^\infty(E_i)$ and $\mathscr{H}_i, \mathscr{G}_i$ are the operators associated with $\square_i$, we have from Formula (14)

$$u = \mathscr{H}_i u + \square_i \mathscr{G}_i u = \mathscr{H}_i u + D_{i-1}(D_{i-1}^* \mathscr{G}_i u) + D_i^*(D_i \mathscr{G}_i u),$$

which shows that (18) is a decomposition into an algebraic direct sum. But $\mathscr{H}_i$, $D_{i-1}D_{i-1}^*\mathscr{G}_i$, and $D_i^* D_{i-1}\mathscr{G}_i$ being continuous, this is also a topological direct product, thereby ending the proof of the theorem.

We remark that by using the commutation properties (17) one can prove that $\mathscr{H}$ and $\mathscr{G}$ also commute with D and D^*. This leads to the decomposition (18) for $\mathfrak{D}'(E_i)$.

21 Theorem (Hodge.) *For an elliptic complex $\mathfrak{C}$,*

$$H^i(\mathfrak{C}) \approx \ker \square_i, \tag{19}$$

and these spaces are finite dimensional.

Let $u \in C^\infty(E_i)$ and

$$u = \mathscr{H}_i u + D_{i-1}s + D_i^*t$$

be the Kodaira decomposition of u. If $u \in \ker D_i$, we have

$$D_i u = D_i D_i^* t = 0,$$

whence $(D_i D_i^* t, t) = |D_i^* t|^2 = 0$ and $D_i^* t = 0$. It follows that u defines the same cohomology class in $H^i(\mathbb{C})$ as $\mathcal{H}_i u$. If $u = D_{i-1} s$, $\mathcal{H}_i u = 0$ and conversely, hence the correspondence $u \mapsto \mathcal{H}_i u$ defines the isomorphism (19). A similar result can be obtained by replacing $C^\infty(E_i)$ with $\mathfrak{D}'(E_i)$. The last part of Hodge's theorem is a corollary of Theorem 10.

Related to this theorem we consider the following class of objects.

22 Definition A cross section $\alpha \in C^\infty(E_i)$ which is in $\ker \square_i$ is called a *harmonic cross section*.

Hence α is harmonic if and only if

$$\square_i \alpha = 0 \qquad (20)$$

or, as was shown in the proof of Theorem 20, if and only if

$$D_i \alpha = 0, \qquad D_{i-1}^* \alpha = 0. \qquad (21)$$

Theorem 21 shows that $H^i(\mathbb{C})$ is isomorphic to the space of harmonic cross sections of E_i.

We have seen that a cocycle u defines the same cohomology class as the corresponding harmonic cross section $\mathcal{H} u$. Since the difference of two different harmonic cross sections cannot be D_{i-1}-exact, it follows that $\mathcal{H} u$ is the unique harmonic cross section which is cohomologous to u. Hence the isomorphism (19) can be viewed as a representation of the cohomology of $\mathbb{C}$ by uniquely determined cross sections of $C^\infty(E)$, namely, the harmonic cross sections. This means that we have found canonical representatives of the cohomology classes.

23 Theorem (a) *If $\mathbb{C}$ and $\mathbb{C}'$ are k-elliptic complexes on V, $H^i(\mathbb{C} \oplus \mathbb{C}')$ is isomorphic to $H^i(\mathbb{C}) \oplus H^i(\mathbb{C}')$.* (b) *If $\mathbb{C}$ and $\mathbb{C}'$ are k-elliptic complexes on V and V', respectively, then we have Künneth's formula*

$$H^i(\mathbb{C} \otimes \mathbb{C}') = \bigoplus_{j+k=i} H^j(\mathbb{C}) \otimes H^k(\mathbb{C}') \qquad (22)$$

and $\mathbb{C} \otimes \mathbb{C}'$ is also k-elliptic. (c) *If $\mathbb{C}$ is a k-elliptic complex we have the*

4. Elliptic Operators. Elliptic Complexes

Poincaré's duality

$$H^{i_0-i}(\mathbb{C}^*) = (H^i(\mathbb{C}))^*. \tag{23}$$

Part (a) is a straightforward consequence of Theorem 21. For (b), consider in $L^2(E_j)$ the orthonormal basis e_α consisting of eigenvectors of $\square_\mathbb{C}$ and such that $e_1, \ldots, e_a \in \ker \square_\mathbb{C}$ and, similarly, consider in $L^2(E_k')$ the orthonormal basis f_β where f_β are eigenvectors of $\square_{\mathbb{C}'}$ and

$$f_1, \ldots, f_b \in \ker \square_{\mathbb{C}'}.$$

Then, $\{e_\alpha \otimes f_\beta\}$ is an orthonormal basis of $L^2(E_j \otimes E_k')$.
From the definition of $\square$, we immediately obtain

$$\square_{\mathbb{C} \otimes \mathbb{C}'} = \square_\mathbb{C} \otimes 1 + 1 \otimes \square_{\mathbb{C}'}, \tag{24}$$

whence

$$\square_{\mathbb{C} \otimes \mathbb{C}'}(e_\alpha \otimes f_\beta) = 0$$

for $\alpha \leq a$ and $\beta \leq b$.

Consider $g \in \ker \square_{\mathbb{C} \otimes \mathbb{C}'}$ orthogonal to $e_\alpha \otimes f_\beta$ ($\alpha \leq a, \beta \leq b$). It follows that

$$g = \sum_{\substack{\alpha > a \\ \text{or} \\ \beta > b}} a_{\alpha\beta} e_\alpha \otimes f_\beta$$

and applying $\square_{\mathbb{C} \otimes \mathbb{C}'}$ we get, using (24),

$$0 = \sum_{\substack{\alpha > a \\ \text{or} \\ \beta > b}} (\lambda_\alpha + \mu_\beta) a_{\alpha\beta} e_\alpha \otimes f_\beta,$$

where λ_α and μ_β are the eigenvalues of e_α and f_β.

Hence, $(\lambda_\alpha + \mu_\beta)a_{\alpha\beta} = 0$ ($\alpha > a$ or $\beta > b$), and since in the first case $\lambda_\alpha > 0$, and in the second case $\mu_\beta > 0$, and in the remaining cases $\lambda_\alpha, \mu_\beta \geq 0$ we get $a_{\alpha\beta} = 0$, i.e., $g = 0$.

It follows that $e_\alpha \otimes f_\beta$ ($\alpha \leq a, \beta \leq b$) is a basis of $\ker \square_{\mathbb{C} \otimes \mathbb{C}'}$, from which

$$\ker (\square_{\mathbb{C} \otimes \mathbb{C}'})_i = \bigoplus_{j+k=i} [(\ker \square_\mathbb{C})_j \otimes (\ker \square_{\mathbb{C}'})_k]. \tag{25}$$

Applying this result for the sequences of symbols of $\mathfrak{C}$ and $\mathfrak{C}'$ which are elliptic complexes over a point, we find that $\mathfrak{C} \otimes \mathfrak{C}'$ is a k-elliptic complex. But then, because of Theorem 21, formula (25) is just (22), and assertion (b) is proved.

Finally, for (c), we remark that the map

$$H^i(\mathfrak{C}) \otimes H^{i_0-i}(\mathfrak{C}^*) \to C,$$

defined by integration on the manifold V, corresponds, in view of Stokes' theorem, to the scalar product induced by $L^2(E)$ on ker $\Box_i$, from which the desired result follows by a classical theorem of Riesz and Theorem 21 above. Recall that for complete proofs of the results of this section we refer the reader to [29].

5 Cohomology and Harmonic Forms

In this section we shall apply the theory of elliptic complexes to the study of cohomology with sheaf coefficients for the cases of Sections 1 and 2. The conventions made in Section 4 for the differentiable manifolds and vector bundles under consideration are followed in the sequel.

By the de Rham theorem of Section 1, the complex cohomology of an n-dimensional differentiable manifold V is isomorphic to the cohomology of the differential complex

$$\mathfrak{R} = (\Lambda^p T^{*c}(V), d), \qquad p = 0, \ldots, n, \tag{1}$$

i.e., of the complex consisting of the bundles of covariant skew-symmetric tensors on V and by the exterior derivative d which, obviously, is a differential operator of order 1, acting on differential forms, cross sections of the bundles considered.

Here, V is a compact, oriented, Riemannian manifold and we give the bundle $\Lambda^p T^{*c}(V)$ the Hermitian product

$$(a, b) = \frac{1}{p!} a_{i_1 \cdots i_p} \bar{b}^{i_1 \cdots i_p}, \tag{2}$$

5. Cohomology and Harmonic Forms

where $a \ldots$ and $b \ldots$ are the components of the tensors a and b.

1 Proposition *The differential complex (1) is 1-elliptic.*

Consider the local expression of a differential form of degree p:

$$\alpha = \frac{1}{p!} a_{i_1 \cdots i_p} dx^{i_1} \wedge \cdots \wedge dx^{i_p}. \tag{3}$$

Then, (Section 4.3)

$$(d\alpha)_{i_1 \cdots i_{p+1}} = \frac{1}{p!} \delta^{jj_1 \cdots j_p}_{i_1 \cdots i_{p+1}} \frac{\partial a_{j_1 \cdots j_p}}{\partial x^j}, \tag{4}$$

which is the local expression of the operator d.

Now, if ξ is a tangent covector of V, we easily obtain

$$\sigma(d, \xi)\alpha(x) = (\xi \wedge \alpha)(x)$$

and the sequence of symbols of the complex (1) is

$$0 \longrightarrow \Lambda^0 T_x^{*c} \xrightarrow{\xi \wedge} \Lambda^1 T_x^{*c} \xrightarrow{\xi \wedge} \cdots \longrightarrow 0.$$

This sequence is exact since $\xi \wedge (\xi \wedge \alpha) = 0$ and, if $\xi \wedge \alpha = 0$, we see, by expressing α with a cobasis containing ξ, that $\alpha = \xi \wedge \beta$. Q.E.D.

We wish to calculate the adjoint operator d^*, denoted here by δ, and the laplacian $\Box_\Re$, denoted by Δ. To do this, we use the Levi–Civita connection of the compact oriented Riemannian manifold V. Expressing in (4) the partial derivatives with the help of the covariant derivatives we immediately get

$$(d\alpha)_{i_1 \cdots i_{p+1}} = \frac{1}{p!} \delta^{jj_1 \cdots j_p}_{i_1 \cdots i_{p+1}} a_{j_1 \cdots j_p/j}. \tag{5}$$

Introduce the important auxiliary operator

$$*: A^p(V) \to A^{n-p}(V),$$

given by the formula

$$(*\alpha)_{j_1 \cdots j_{n-p}} = \frac{1}{p!} \eta_{i_1 \cdots i_p j_1 \cdots j_{n-p}} a^{i_1 \cdots i_p}, \tag{6}$$

where $\eta \cdots$ are the components of the form

$$dv = \frac{1}{n!}\eta_{i_1\cdots i_n} dx^{i_1} \wedge \cdots \wedge dx^{i_n},$$

dv being the volume element of the metric of V. It is known [28] that this form is determined by $\eta_{1\cdots n} = \sqrt{g}$, $g = \det(g_{ij})$.

By a straightforward computation one gets

$$**\alpha = (-1)^{p(n-p)}\alpha, \tag{7}$$

whence $*^{-1}$ exists and is given by

$$*^{-1} = (-1)^{p(n-p)}*, \tag{8}$$

where p is the degree of the form on which $*^{-1}$ acts.

According to the general theory of Section 4, we must consider the *scalar product* of forms on V given by integrating $(a, b)dv$ on V, with (a, b) of (2). But, using the operator $(*)$ we get, again by straightforward calculation, that for the forms α, β with components as in Formula (2),

$$\alpha \wedge *\beta = (a, b)dv, \tag{9}$$

from which

$$\alpha \wedge *\beta = \beta \wedge *\alpha \tag{10}$$

and the scalar product is given by

$$(\alpha, \beta) = \int_V \alpha \wedge *\beta \tag{11}$$

2 Proposition *The adjoint of the operator d is*

$$\delta = (-1)^p *^{-1} d* \tag{12}$$

where p is the degree of the form on which δ acts.

In fact, $\delta: A^p(V) \to A^{p-1}(V)$ and, for $\deg \alpha = p - 1$ and $\deg \beta = p$, we have $(\alpha, \delta\beta) = \int_V \alpha \wedge *\delta\beta = (-1)^p \int_V \alpha \wedge d*\beta = \int_V d\alpha \wedge *\beta = (d\alpha, \beta)$, where the third equality sign follows by integrating the relation

5. Cohomology and Harmonic Forms

$$d(\alpha \wedge *\beta) = d\alpha \wedge *\beta + (-1)^{p-1}\alpha \wedge d*\beta$$

using Stokes' Theorem, 4.3(21).

Now, apply the operator δ to the form (3) and compute the components of the resulting form. We shall use the well known formula of Riemannian geometry [28] (which follows by a straightforward calculation from the expression 4.5(8) of the Christoffel symbols)

$$\eta_{i_1 \cdots i_n/h} = 0 \tag{13}$$

and the formula

$$\delta^{h i_1 \cdots i_p}_{j_1 \cdots j_{p+1}} = \sum_{\alpha=1}^{p+1} (-1)^{\alpha-1} \delta^h_{j_\alpha} \delta^{i_1 \cdots i_p}_{j_1 \cdots \hat{j}_\alpha \cdots j_{p+1}}, \tag{14}$$

obtained from

$$\delta^{s_1 \cdots s_p}_{t_1 \cdots t_p} = \det(\delta^{s_i}_{t_j})$$

by developing the determinant of the left-hand side of (14) with respect to the first row.

We get

$$(d*\alpha)_{h_1 \cdots h_{n-p+1}} = \frac{1}{p!} \sum_{\alpha=1}^{n-p+1} (-1)^{\alpha-1} \eta_{i_1 \cdots i_p h_1 \cdots \widehat{h_\alpha} \cdots h_{n-p+1}} a^{i_1 \cdots i_p}_{/h_\alpha}$$

Next

$$(*d*\alpha)_{s_1 \cdots s_{p-1}} = \frac{1}{p!(n-p)!} \eta^l_{k_1 \cdots k_{n-p} s_1 \cdots s_{p-1}} \eta^{i_1 \cdots i_p k_1 \cdots k_{n-p}} \cdot a_{i_1 \cdots i_p/l},$$

from which

$$(\delta\alpha)_{s_1 \cdots s_{p-1}} = -a^l_{s_1 \cdots s_{p-1}/l}. \tag{15}$$

To calculate Δ we start with the definition

$$\Delta = d\delta + \delta d \tag{16}$$

and use Formulas (5) and (15) for d and δ. The calculation has only technical interest and it uses the commutation formulas 4.4(28) for mixed covariant derivatives, which introduce in the expression of Δ the curvature tensor of V. The result of this calculation is

$$(\Delta\alpha)_{i_1\cdots i_p} = -g^{ij}a_{i_1\cdots i_p/ij}$$
$$+ \sum_{\rho=1}^{p} a_{i_1\cdots i_{\rho-1}ji_{\rho+1}\cdots i_p} R^j_{i_\rho}$$
$$+ \frac{1}{2}\sum_{\sigma=1}^{p}\sum_{\rho=1}^{p} a_{i_1\cdots i_{\rho-1}ji_{\rho+1}\cdots i_{\sigma-1}ii_{\sigma+1}\cdots i_p} R^{ij}_{i_\rho i_\sigma},$$

where R^i_{jkh} is the curvature tensor and R_{ij} is the Ricci tensor of the Riemannian manifold V.

A differential form α is called *harmonic* if $\Delta\alpha = 0$. From Section 4 we know that this is equivalent to the system $d\alpha = 0$, $\delta\alpha = 0$. We denote by $\mathcal{H}^p_c(V)$ the space of complex valued harmonic forms of degree p on V. From Theorems 4.20, 4.21, and 4.23 we obtain the following results.

3   Theorem   *For a compact oriented Riemannian manifold V^n we have a decomposition as a direct topological sum of closed orthogonal subspaces, namely,*

$$A^p(V) = \mathcal{H}^p_c(V) \oplus \operatorname{im} d \oplus \operatorname{im} \delta.$$

The orthogonality is given with respect to the scalar product (11) and the decomposition of a form is given by

$$\alpha = \mathcal{H}\alpha + \Delta\mathcal{G}\alpha,$$

where $\mathcal{G}$ and $\mathcal{H}$ are the Green's and the harmonic projector operators associated with Δ from the general theory of Section 4.

4   Theorem   *For a compact oriented Riemannian manifold V^n*

$$H^p(V, C) = \mathcal{H}^p_c(V).$$

Moreover, these spaces are finite dimensional.

Though the harmonic forms above are complex valued, it is possible to give a similar result for the real cohomology of V. This is an immediate

5. Cohomology and Harmonic Forms

consequence of Theorem 3. Let α be a real valued form whose decomposition is

$$\alpha = \lambda + d\mu + \delta v, \quad \Delta\lambda = 0.$$

Taking complex conjugates and using the fact that d and δ are real operators, as is seen from Formulas (5) and (15), we get

$$\alpha = \bar{\lambda} + d\bar{\mu} + \delta\bar{v}.$$

Δ is also a real operator, hence $\Delta\lambda = 0$ implies $\Delta\bar{\lambda} = 0$, so, by the uniqueness of the decomposition of a form, $\lambda = \bar{\lambda}$. In other words, if α is a real-valued form, $\mathcal{H}\alpha$ is again real-valued. It follows, as in the proof of Theorem 4.21, that every real closed form is cohomologous to a unique real harmonic form, which gives the following result.

5 Theorem   *For a manifold V, as in Theorem 4,*

$$H^p(V, R) = \mathcal{H}^p(V),$$

where the right-hand side is the space of real harmonic forms of degree p on V.

In our case this is the solution of the problem announced at the beginning of Section 4: the cohomology classes of V can be canonically represented by harmonic forms.

Moreover, from Theorem 4.23, we have the following important results.

6 Theorem

(a) *If V and V' are two compact oriented Riemannian manifolds, then*

$$H^p(V \oplus V', R) = H^p(V, R) \oplus H^p(V', R).$$

(b) *Under the same hypotheses we have Künneth's formula*

$$H^p(V \times V', R) = \bigoplus_{i+j=p} H^i(V, R) \otimes H^j(V', R).$$

(c) *For a manifold V as above we have Poincaré's duality*

$$H^p(V, R) = H^{n-p}(V, R).$$

As before, we can consider the case of the real cohomology. For (a) we apply the assertion (a) of Theorem 4.23 on $V \oplus V'$, for $\mathfrak{R}$ and $\mathfrak{R}'$, and then induce back these complexes on V and V' by the injections of these manifolds into the direct sum. Assertion (b) follows directly from the corresponding part of Theorem 4.23. Assertion (c) can be obtained from (c) of Theorem 4.23 by showing that the complex $\mathfrak{R}$ is self-dual (i.e., it can be identified with its dual complex). But, it is simpler to remark that (c) follows from Theorem 5 and from the fact that the operators $*$ and Δ commute, which can easily be established by a calculation. Hence $*$ induces an isomorphism of $\mathscr{H}^p(V)$ and $\mathscr{H}^{n-p}(V)$. Q.E.D.

Theorem 5 has many important consequences. Recalling that the curvature tensor enters in Formula (17), we see that the cohomology of the manifold V is related to its curvature. Such relations were intensively studied and complete results can be found in [16]. Here, we give only one important result which is typical.

To obtain this result, we shall calculate $(\Delta\alpha, \alpha)$.

Let us obtain first an auxiliary result. For

$$\varphi = a_{i_1\cdots i_p} a^{i_1\cdots i_p} = p!(a,a),$$

$$\delta d\varphi = -2(g^{ih}a^{i_1\cdots i_p}a_{i_1\cdots i_p/ih} + g^{ih}a^{i_1\cdots i_p}{}_{/i}\, a_{i_1\cdots i_p/h}).$$

Since $(\delta d\varphi, 1) = (d\varphi, d1) = 0$, we get

$$-\int_V g^{ih}a^{i_1\cdots i_p} a_{i_1\cdots i_p/ih}\, dv = \int_V A(\alpha)\, dv, \tag{18}$$

where

$$A(\alpha) = g^{ih}a^{i_1\cdots i_p}{}_{/i}\, a_{i_1\cdots i_p/h}.$$

Now, by (17),

$$(\Delta\alpha, \alpha) = \int_V A(\alpha)\, dv + p\int_V F^p(\alpha)\, dv, \tag{19}$$

where

$$F^p(\alpha) = R_{ij}a^{ii_1\cdots i_p}a^j{}_{j_2\cdots i_p}$$

$$+ \frac{p-1}{2} R_{ijkl}a^{iji_3\cdots i_p}a^{kl}{}_{i_3\cdots i_p}. \tag{20}$$

5. Cohomology and Harmonic Forms

The first term of the right member of (19) is positive if $\alpha \neq 0$, and since for an harmonic form α, $(\Delta\alpha, \alpha) = 0$, we derive

7 Theorem (Bochner–Lichnerowicz.) *If on a compact orientable Riemannian manifold V^n the form $F^p(\alpha)$ ($p > 0$), which is a quadratic form in the components of α, is positive definite, then $H^p(V, R) = 0$.*

In fact, under our hypotheses, the only harmonic form of degree p on V is 0.

8 Corollary *If V is as in Theorem 7 and if it has positive constant sectional curvature, then $H^p(V, R) = 0$ for $0 < p < n$.*

In fact, by hypothesis, we easily find that $F^p(\alpha)$ is positive definite for $0 < p < n$.

We can also find the spaces $H^0(V, R)$ and $H^n(V, R)$ for every connected compact orientable Riemannian manifold V; these two spaces are isomorphic to R. Indeed, every harmonic function on V is constant and, for H^n, the form dv is harmonic, and every harmonic form of degree n is of the type λdv, $\lambda = $ const. (or we can use Poincaré duality).

Hence, for the manifolds which are connected and satisfy the hypotheses of Corollary 8, we determined all the real cohomology spaces. Since the n-dimensional sphere is such a manifold, all the manifolds which have the same cohomology spaces are called *cohomology spheres* or *rational homology spheres*.

A deep analysis of the structure of the space of harmonic forms has been worked out for Kählerian manifolds, from which one derives important topological (cohomological) consequences. For this subject we refer the reader to [5] or [16].

The previous theory extends without difficulty to the cohomology of V with coefficients in sheaves of germs of locally constant cross sections of locally constant vector bundles on V. Another remark is that the method used to prove Theorem 7 above can be used for many other questions of global differential geometry. For instance, Bochner and others used this method to prove the nonexistence of certain types of vector fields on a manifold (under suitable conditions) such as Killing vector fields, infinitesimal conformal transformations, etc. For such results, we again refer the reader to [16].

We now consider the case of a compact oriented Riemannian foliated manifold V. The notation will be that of Section 4.5. Theorem 2.2 leads, in this case, to the study of the differential complex of differential forms of type (p, q) with the operator d_f, which, after Formula 4.5(37), is a differential operator of the first order.

Let ξ be a tangent covector to V at the point x. We associate with this covector the form $\zeta = \xi_u \theta^u$ (the indices are also as in Section 4.5). Just as before for the operator d, Formula 4.5(37) shows that the symbol of d_f is given by

$$\sigma(d_f, \xi)\alpha(x) = (\zeta \wedge \alpha)(x)$$

and, since we can have $\zeta = 0$ for $\xi \neq 0$ ($\xi_u = 0$ but $\xi_a \neq 0$) the symbol sequence of the considered complex is no longer exact. Hence this complex is not elliptic. However, the previous methods can be used in part, and will give weaker but interesting results. We consider real-valued differential forms.

We observe that the operator $*$ defined by (6) is *homogeneous*, i.e., it sends a form of type (p, q) to a form of the type $(n - p, m - q)$ (recall that dim $V = n + m$ and that the codimension of the foliation is n).

Introducing the operator

$$\delta_f = (-1)^{p+q} *^{-1} d_f *, \qquad (21)$$

which is of type $(0, -1)$ one gets, just as for the relation $\delta = d^*$, the formula

$$(d_f \alpha, \beta) = (\alpha, \delta_f \beta)$$

(Stokes' formula plays an assential role here), which means that δ_f is the adjoint d_f^* of the operator d_f. It is immediate that δ_f is the part of type $(0, -1)$ of the operator δ.

To obtain the expressions of d_f and δ_f, it is convenient to use the second connection of V. Using 4.5(37), 4.5(38), and 4.5(39),

$$(d_f \alpha)_{a_1 \cdots a_p u_1 \cdots u_{q+1}} = \frac{(-1)^p}{q!} \delta^{vv_1 \cdots v_q}_{u_1 \cdots u_{q+1}} a_{a_1 \cdots a_p v_1 \cdots v_q / v}. \qquad (22)$$

Then, supposing for simplicity that V is a Reinhart space (Definition 4.5.13), we get the auxiliary result

$$\eta_{1 \cdots n + m/v} = \sqrt{\det(h_{uv})} \frac{\partial \sqrt{\det(h_{ab})}}{\partial x^v} = 0$$

5. Cohomology and Harmonic Forms

(h_{ab} and h_{uv} are the components of the metric tensor) and, next, with the same procedure as for δ, we obtain

$$(\delta_f \alpha)_{a_1 \cdots a_p v_1 \cdots v_{q-1}} = \frac{(-1)^{p-1}}{(n-p)!} \tilde{h}^{uv} \alpha_{a_1 \cdots a_p v v_1 \cdots v_{q-1}/u}. \qquad (23)$$

Now, with the definition

$$\Delta_f = d_f \delta_f + \delta_f d_f \qquad (24)$$

and using the commutation rules of the covariant derivatives, we get, again in the case of a Reinhart space,

$$(\Delta_f \alpha)_{a_1 \cdots a_p u_1 \cdots u_q}$$
$$= -\frac{1}{(n-p)!} \tilde{h}^{uv} \alpha_{a_1 \cdots a_p u_1 \cdots u_q/vu} + \frac{1}{(n-p)!} \sum_{h=1}^{q} R^t_{u_h} \alpha_{a_1 \cdots a_p u_1 \cdots t \cdots u_q}$$
$$+ \frac{1}{2(n-p)!} \sum_{s=1}^{q} \sum_{h=1}^{q} R^{vu}{}_{u_s u_h} \alpha_{a_1 \cdots a_p u_1 \cdots u \cdots v \cdots u_q}. \qquad (25)$$

9 Definition A differential form α on V is called a *foliated harmonic form* if $\Delta_f \alpha = 0$ or, equivalently, $d_f \alpha = 0$ and $\delta_f \alpha = 0$.

We denote by $\mathcal{K}^{pq}(V)$ the space of foliated harmonic forms of type (p,q). Denote also by $\mathcal{H}^{pq}(V)$ the space of harmonic forms of type (p,q) on V. From the definitions of d and δ it follows immediately that $\mathcal{H}^{pq}(V)$ is a subspace of $\mathcal{K}^{pq}(V)$.

If $\alpha \in \mathcal{K}^{pq}(V)$, it is d_f-closed and, in view of Theorem 2.2, it defines a cohomology class of $H^{pq}(V)$. This class is 0 if and only if $\alpha = 0$, since from $\alpha = d_f \beta$ it follows that $(\alpha, \alpha) = (\alpha, d_f \beta) = (\delta_f \alpha, \beta) = 0$. Hence the space $\mathcal{K}^{pq}(V)$ is isomorphic to a subspace of $H^{pq}(V)$, and we have the following result.

10 Theorem *For a compact oriented Riemannian foliated manifold V, the space $\mathcal{K}^{pq}(V)$ of foliated harmonic forms is, up to an isomorphism, a subspace of the cohomology space $H^{pq}(V)$ and $\mathcal{H}^{pq}(V)$ is a subspace of $\mathcal{K}^{pq}(V)$.*

Hence in this case we can represent canonically only a part of the cohomology classes. This makes it useful to give a new definition.

11 Definition If the manifold V is as in Theorem 10 and if $H^{pq}(V) \approx \mathcal{H}^{pq}(V)$, then it is called a *cohomologically regular manifold*.
We give, following [49], some consequences of theorem 10.

12 Proposition *If V is a compact oriented Riemannian foliated manifold which has a foliated transverse volume element, then* $\dim H^m(V, \mathfrak{G}(V)) \geq 1$.
Consider on V a form of type $(0, m)$,

$$\tau = \sqrt{\det (h_{uv})}\theta^{n+1} \wedge \cdots \wedge \theta^{n+m}.$$

For it, we find

$$*\tau = (-1)^{nm}\sqrt{\det (h_{ab})}\, dx^1 \wedge \cdots \wedge dx^n.$$

We can always make a conformal transformation of V such that, with respect to the new metric, $\det (h_{ab}) = 1$. Supposing that this transformation is already performed, we get $d*\tau = 0$, whence $\delta_f \tau = 0$. But we also have $d_f \tau = 0$, whence τ is a nonzero element in $\mathcal{H}^{0m}(V)$ which, because of Theorem 10, proves the proposition.

13 Proposition *If V is a compact oriented Reinhart space and if R_{ij} is the Ricci tensor of the second connection of V, then if the quadratic form $R_{uv}\xi^u\xi^v$ is positive definite, there are no nonzero foliated harmonic forms of type $(p, 1)$ on V.*

The proof follows that of Theorem 7. The result can be generalized to forms of type (p, q) and one gets a result analogous to Theorem 7.

Using Theorem 10, we can also obtain results relating the real cohomology and the $\mathfrak{G}(V)$-cohomology of the manifold.

14 Proposition *Let V be a compact oriented Riemannian foliated manifold and let ρ_{ij} be the Ricci tensor of the first (Levi–Civita) connection of V. If* (a) $\rho_{au} = 0$, (b) $\rho_{ab}\zeta^a\zeta^b$ *is a positive definite quadratic form, then* $\dim H^{01}(V) \geq b^1$, *where b^1 is the first Betti number of V.*

In view of Theorem 10, it suffices to show that the hypotheses imply that any harmonic form of degree 1 on V has type $(0, 1)$.
Let

$$\varphi = \varphi_a\, dx^a + \varphi_u \theta^u$$

5. Cohomology and Harmonic Forms

be a form of degree 1. The component of type (1, 0) of $\Delta\varphi$ is given by

$$(\Delta\varphi)_c = -\tilde{h}^{ab}\varphi_{c;ab} - \tilde{h}^{uv}\varphi_{c;uv} + \rho_c^a \varphi_a,$$

where ";" denotes the covariant derivative with respect to the first connection of V.

If φ is harmonic, $(\Delta\varphi)_c = 0$ and we have

$$\int_V (\Delta\varphi)_c \varphi^c \, dv = 0,$$

from which

$$\int_V \rho_{ac}\varphi^a \varphi^c dv - \int_V \tilde{h}^{ab}\varphi_{c;ab}\varphi^c dv - \int_V \tilde{h}^{uv}\varphi_{c;uv}\varphi^c dv = 0.$$

Now, consider the function $\lambda = \varphi_c \varphi^c$. Observing that $(\delta d\lambda, 1) = (d\lambda, d1) = 0$,

$$\int_V (\tilde{h}^{ab}\lambda_{;ab} + \tilde{h}^{uv}\lambda_{;uv})dv = 0.$$

Expressing the covariant derivatives and introducing them into the previous relation we get

$$\int_V \rho_{ab}\varphi^a \varphi^b dv + \int_V (\tilde{h}^{ab}\varphi_{c;a}\varphi_{;b}^c + \tilde{h}^{uv}\varphi_{c;u}\varphi_{;v}^c)dv = 0,$$

from which, by our hypotheses, $\varphi^a = 0$ (the second integral is non-negative).   Q.E.D.

Such results can be made more precise by applying Theorem 2.11. From that theorem and from Propositions 12 and 14 one gets, for instance, the following result.

15   Proposition   *If V satisfies the hypotheses of Proposition 12 and if $H^{m-p+1}(V, \mathfrak{S}^p(V)) = 0$ for $p = 1, \ldots, m+1$, then $b^m \geq 1$ (b^m is the mth Betti number of V).*

This follows from Proposition 12 and from the fact that p^m is, as shown by Theorem 2.11, a surjection.

16 Proposition *If to the hypotheses of Proposition 14 one adds* $H^{11}(V) = 0$ *and* $H^{20}(V) = 0$, *then* $b^1 = \dim H^{01}(V)$.

This follows from Proposition 14 and the fact that p^1 is a surjection (see Theorem 2.11).

We consider applications of the theory of elliptic complexes to the case of complex manifolds.

Let V be a complex compact manifold and S an analytic vector bundle on V; we shall use the notation of Theorem 2.8 of Dolbeault–Serre. If α is an S-valued form on V of type (p, q), it has the local coordinate expression

$$\alpha^1 = \frac{1}{p!\,q!} a^s_{i_1\cdots i_p j_1 \cdots j_q}\, dz^{i_1} \wedge \cdots \wedge dz^{i_p} \wedge d\bar{z}^{j_1} \wedge \cdots \wedge d\bar{z}^{j_q}, \quad (26)$$

where s is the index in the fiber of S. Then

$$(d_{\bar{z}}\alpha)^s_{i_1\cdots i_p l_1 \cdots l_{q+1}} = \frac{1}{q!} \delta^{k j_1 \cdots j_q}_{l_1 \cdots l_{q+1}} \frac{\partial a^s_{i_1\cdots i_p j_1 \cdots j_q}}{\partial \bar{z}^k}. \quad (27)$$

Now, let ξ be a real tangent covector to V, i.e., $\xi = \xi_h dz^h + \bar{\xi}_h d\bar{z}^h$ calculated at the point of contact of ξ. Put $\zeta = \bar{\xi}_h d\bar{z}^h$. Just as for foliated manifolds we deduce

$$\sigma(d_{\bar{z}}, \xi)\alpha(x) = (\zeta \wedge \alpha)(x),$$

but now ζ and ξ vanish simultaneously and we see as for the exterior derivative d that the S-valued differential forms on V and the operator $d_{\bar{z}}$ define a differential elliptic complex. According to Theorem 2.8, the cohomology spaces of this complex are isomorphic with $H^{pq}(V, S)$. In the sequel, we indicate the principal consequences of these results.

Let us obtain first the adjoint and the laplacian of $d_{\bar{z}}$, if Hermitian metrics on V and S are given.

The volume element of V is obviously a form of type (n, n), and one gets by (6) that the operator $*$ sends forms of type (p, q) to forms of type $(n - q, n - p)$.

The scalar product of S-valued forms is given by

$$(\alpha, \beta) = \frac{1}{p!\,q!} \int_V a^s_{i_1\cdots i_p j_1 \cdots j_q} \bar{b}_s^{i_1\cdots i_p j_1 \cdots j_q}\, dv$$

5. Cohomology and Harmonic Forms

which can be put in the form

$$(\alpha, \beta) = \int_V \alpha^t \bar{A} \wedge *\bar{\beta}, \tag{28}$$

where α and β are the matrices (a^s) and (b^s) of exterior forms. A is the local matrix of the Hermitian metric of S.

Consider on S the connection given by proposition 1.6.10 and let θ be the local matrix of this connection, which consists of Pfaff forms of type $(1, 0)$. The exterior covariant derivative of a form α is given by

$$D\alpha = d\alpha + \theta \wedge \alpha = D'\alpha + d_{\bar{z}}\alpha,$$

where the operator D' is defined by

$$D'\alpha = d_z\alpha + \theta \wedge \alpha \tag{29}$$

and is of type $(1, 0)$.

Using this operator we introduce

$$d_{\bar{z}}^* = -*D'* \tag{30}$$

and prove that this is just the adjoint of $d_{\bar{z}}$. A calculation where we take into account that the connection θ is Hermitian and hence (Section 4.6)

$$dA = \theta'A + A\bar{\theta}$$

gives

$$d(\alpha^t \bar{A} \wedge *\bar{\beta}) = (d_{\bar{z}}\alpha)^t \bar{A} \wedge *\bar{\beta} - \alpha^t \bar{A} \wedge \overline{*d_{\bar{z}}^*\beta},$$

from which, by integration and using Stokes' formula,

$$(d_{\bar{z}}\alpha, \beta) = (\alpha, d_{\bar{z}}^*\beta),$$

which is the required relation.

The corresponding laplacian is given by

$$\square_{d_{\bar{z}}} = d_{\bar{z}}d_{\bar{z}}^* + d_{\bar{z}}^*d_{\bar{z}}, \tag{31}$$

whose kernel consists of the S-valued $d_{\bar{z}}$-harmonic forms on V, denoted by $\mathcal{H}^{pq}(V, S)$.

Hence, the principal results are as follows.

17   Theorem

$$H^{pq}(V, S) = \mathcal{H}^{pq}(V, S).$$

This is an immediate consequence of the theorems of Dolbeault–Serre and Hodge.

18   Theorem   (Serre's duality theorem.)

$$H^{pq}(V, S) = H^{n-p, n-q}(V, S^*).$$

This follows from part (c) of Theorem 4.23. Indeed, by considering coordinate expressions, we see that $(A^{p,n-q}(V, S))^*$ can be identified with $A^{n-p,q}(V, S^*)$, and $d_{\bar{z}}^*$ can be identified up to a constant factor with $d_{\bar{z}}$ acting on S^*-valued forms. This proves the theorem. One can see that the isomorphism of this theorem is given by the operator

$$\tilde{*}: A^{pq}(V, S) \to A^{n-p, n-q}(V, S^*)$$

given by

$$\tilde{*}\alpha = *\bar{A}\bar{\alpha}$$

and restricted to harmonic forms.

For a theorem of the Bochner–Lichnerowicz type see, for instance Nakano's theorem in [5].

This ends the program which we set for our book.

REFERENCES

1. Allendoerfer, C. B., and Eells, J. Jr. On the cohomology of smooth manifolds. *Comment. Math. Helv.*, **32**, 1958, pp. 165–179.
2. Atiyah, M. F. *K*-Theory. Harvard Univ. Press, Cambridge, Mass., 1965.
3. Bucur, I. *Algebră omologică*. Ed. did. și ped., București, Romania, 1965.
4. Chern, S. S. On curvature and characteristic classes of a Riemannian manifold. *Abh. Math. Sem. Univ. Hamburg*, **20**, 1955, pp. 117–126.
5. Chern, S. S. Complex Manifolds. Univ. of Chicago, Mimeographed Notes, 1956.
6. Chern, S. S. *Complex Manifolds Without Potential Theory*. D. Van Nostrand Comp., Princeton, New Jersey, 1967.
7. Cohn, P. M. *Universal Algebra*. Harper and Row Publishers, London, 1965.
8. Craioveanu, M. Contributii la studiul unor structuri geometrice pe varietăti infinit dimensionale. Thesis, Univ. of Iași, Romania, 1970.
9. Dieudonné, J. *Foundations of Modern Analysis*. Academic Pres, New York, 1960.
10. Dolbeault, P. Formes différentielles et cohomologie sur une variété analytique complexe. *Ann. Math.*, **64**, 1956, pp. 83–130.
11. Eells, J., Jr. A generalization of the Gauss–Bonnet theorem. *Trans. Amer. Math. Soc.*, **92**, 1959, pp. 142–153.
12. Eells, J., Jr. On the geometry of function spaces. *Symp. Int. Top. Alg.*, Mexico 1958, pp. 303–308.
13. Gheorghiev, Gh., Miron, R., and Papuc, D. I. *Geometrie analitică și diferențială*, I, II. Ed. did. și ped., București, 1968–1969.
14. Gheorghiev, Gh., and Oproiu, V. Geometrie diferentială, I. Univ. Iași, Mimeographed Notes, 1971.
15. Godement, R. *Topologie algébrique et théorie des faisceaux*. Hermann, Paris, 1958.
16. Goldberg, S. I. *Curvature and Homology*. Academic Press, New York, 1962.
17. Griffiths, P. A. On a theorem of Chern. *Illinois J. Math.*, **6**, 1962, pp. 468–479.
18. Grothendieck, A. Sur quelques points d'algèbre homologique. *Tôhoku Math. J.* **2** (9), 1957, pp. 119–221.
19. Hangan, T. Asupra cohomologiei spațiilor cu curbură constantă. *An. Univ. Bucure;ti (Mat.–Mec.)*, 1964, pp. 111–120.

20. Hattori, A. Spectral sequence in the de Rham cohomology of fiber bundles. *J Fac. Sci. Univ. Tokyo, Sec. I*, **VIII**, 1960, pp. 289–331.
21. Helgason, S. *Differential Geometry and Symmetric Spaces.* Academic Press, New York, 1962.
22. Henderson, D. W. Infinite-dimensional manifolds on open subsets of Hilbert space. *Topology*, **8**, 1970, pp. 25–34.
23. Hirzebruch, F. *Topological Methods in Algebraic Geometry.* Springer, Berlin, 1966.
24. Husemoller, D. *Fiber Bundles.* McGraw-Hill, New York, 1966.
25. Kobayashi, S. and Nomizu, K. *Foundations of Differential Geometry*, I, II. Interscience, New York, 1963, 1969.
26. Kodaira, K., and Spencer, D. C. Multifoliate structures. *Ann. Math.*, **74**, 1961, pp. 52–100.
27. Lang, S. *Introduction to Differentiable Manifolds.* Columbia Univ. Press, New York, 1962.
28. Lichnerowicz, A. Théorie globale des connexions et des groupes d'holonomie. Ed. Cremonese, Rome, 1955.
29. Lusztig, Gh. Coomologia complexelor eliptice. *St. şi cerc. mat.*, **21**, 1969, pp. 33–84.
30. Mihai, A. O reprezentare a claselor Chern intregi cu ajutorul formelor cu singularităţi. *St. şi cerc. mat.*, **19**, 1967, pp. 1273–1284.
31. Milnor, J. On the existence of a connection with curvature zero. *Comment. Math. Helv.*, **32**, 1958, pp. 215–223.
32. Miron, R., and Cruceanu, V. Sur les couples de connexions compatibles avec les structures presque complexes. *An. şt. Univ. Iaşi*, **XIII**, 1967, pp. 79–88.
33. Mitchell, B. *Theory of Categories.* Academic Press, New York, 1965.
34. Newlander, A., and Nirenberg, L. Complex analytic coordinates in almost complex manifolds. *Ann. Math.*, **65**, 1957, pp. 391–404.
35. Palais, R. S. *Lectures on the Differential Topology of Infinite Dimensional Manifolds. Math.*, **322**, Brandeis Univ., 1964–65.
36. Palais, R. S. *Seminar on the Atiyah–Singer Index Theorem.* Princeton Univ. Press, Princeton, New Jersey, 1965.
37. Papuc, D. I., and Albu, A. C. *Elemente de geometrie diferenţială globală.* Tip. Univ. Timişoara, 1970.
38. Reinhart, B. L. Foliated manifolds with bundle-like metrics. *Ann. Math.*, **69**, 1959, pp. 119–132.
39. De Rham, G. *Variétés différentiables.* Hermann, Paris, 1955.
40. Schubert, H. *Kategorien.* I, II. Springer-Verlag, Berlin, 1970.
41. Serre, J. P. Faisceaux algébriques cohérents. *Ann. Math.*, **61**, 1955, pp. 197–278.
42. Spanier, E. H. *Algebraic Topology.* McGraw-Hill, New York, 1966.
43. Spencer, D. C. De Rham theorems and Neumann decompositions associated with partial differential equations. (Structures feuilletées.) Coll. C.N.R.S., Grenoble, 1963, pp. 1–20.
44. Steenrod, N.E. *The Topology of Fiber Bundles.* Princeton Univ. Press, Princeton, New Jersey, 1951.
45. Sternberg, S. *Lectures on Differential Geometry.* Prentice-Hall, Inc., Englewood Cliffs, New Jersey, 1964.
46. Teleman, C. Connections and bundles. I, II. *Proc. Koninkl. Akad. van Wetenshappen Amsterdam, Ser. A*, **72**, 1969, pp. 89–112.

References

47. Thorpe, J. A. Sectional curvatures and characteristic classes. *Ann. Math.*, **80**, 1964, pp. 429–443.
48. Vaisman, I. Almost-multifoliate Riemannian manifolds. *An. şt. Univ. Iaşi*, **XVI**, 1970, pp. 97–103.
49. Vaisman, I. Variétés riemanniennes feuilletées. *Czechosl. Math. J.*, **21**, 1971, pp. 46–75.
50. Vaisman, I. Sur une classe de complexes de conchaines. *Math. Ann.*, **194**, 1971, pp. 35–42.
51. Vrănceanu, Gh. *Lecţii de geometrie diferenţială*. IV, Ed. Acad. R. S. România, Bucureşti, 1968.
52. Weil, A. Sur un théorème de de Rham. *Comment. Math. Helv.*, **26**, 1952, pp. 119–145.

INDEX

A
Atlas, 36, 37

B
Betti numbers, 85
Bianchi identities, 174, 178
Bijection, 10
Bochner–Lichnerowicz theorem, 269
Boundaries, 30, 160
Bundle, 95
    analytic, 200
    associated, 106
    completely reducible, 127
    differentiable, 143
    dual, 124
    fiber, 95
    foliated, 216
    induced, 97
    line, 125
    locally constant, 204
    normed, 120
    orientable, 123
    principal, 103
    quotient, 97
    quotient vector, 120
    with structure group, 102
    tangent, 144
    vector, 113

C
$\mathbb{C}$-group, 18
$\mathbb{C}$-cogroup, 18
Category(ies), 5
    Abelian, 23
    additive, 22
    dual, 7
    equivalence of, 34
    isomorphism of, 34
    local, 9
    pointed, 18
    regular, 20
    small, 6
    strengthened, 35
Cauchy–Riemann conditions, 190
Chain, 30
    singular, 160
Character, 132, 144
Chart, local, 37
Chern character, 132, 245
Chern classes, 128, 236, 245
Chern theorem, 246
Class(es), 1
    characteristic, 126, 144
Coboundary, 31
Cocycle, 31
Cohomology, 32, 67, 78
    exact sequence, 32, 68, 82

non-Abelian, 90, 91
Complex, 30
   acyclic, 31
   differential, 256
   double (positive), 76
   double semipositive, 76
   filtered, 71
   k-elliptic, 256
Connection, 100
   almost complex, 195
   Euclidean, 177
   Hermitian, 197, 199, 200
   on a manifold, 167
   on a vector bundle, 165
Coordinate, 37
   expressions, 39
Cross section, 49
   harmonic, 260
Cup product, 88, 89
Curvature, 169
   forms, 169
   of a Riemannian metric, 177
   sectional, 178, 199
   tensor, 169, 177
Curve, 149
   tangent of, 149
Cycle, 29, 30, 161

D

Derivative, 27, 136
   absolute, 165
   covariant, 165
   exterior, 157
   covariant, 173
   foliated, 184
Diffeomorphism, 141
Differential of a map, 142
Distribution, 158
   involutive, 158
Dolbeault theorem, 221
Dolbeault–Serre theorem, 221

E

Epimorphism, 11
Euler class, 130, 243, 245
Euler–Poincaré characteristic, 85, 86
Exact sequence, 21, 22
   short, 23
   splitting of, 23
   of vector bundles, 118
Exterior algebra, 156

F

Fiber, 93
   coordinates, 95
   local, 49
Filtration, 72
   regular, 74
Form
   analytic, 194
   closed, 157
   differential, 155
   exact, 157
   foliated, 185
   harmonic, 266
   foliated, 271
   holomorphic, 194
   S-valued, 159
Fredholm operator, 256
Friedrichs' inequality, 249, 252
Frobenius' theorem, 158
Function, 2
   analytic, 190
   differentiable, 136, 139
   foliated, 183
Functor, 25
   adjoint, 33
   continuous, 115
   contravariant, 25
   covariant, 25
   representable, 33

G

Gauss–Bonnet theorem, 245
Germ, 9, 52
Grothendieck–Dolbeault lemma, 217
Groupoid, 11

H

Hodge theorem, 259
Homology, 29
   exact sequence, 29, 32
   functors, 29, 32
   singular, 160
Homomorphism, 11
   connecting, 29, 32

Index

Homotopy of complexes, 66
Hypercategory, 6
Hyperpseudocategory, 6

I

Imbedding, 142
Immersion, 41, 142
Inductive directed family, 16
Inductive family, 13
Inductive limit, 15
    of morphisms, 16
Injection, 10
    strengthened, 35
Integral, 163, 165
Isometry, 179
Isomorphism, 11
    functorial, 33

K

K-theory, 133
Killing vector field, 179
Kodaira theorem, 259
Künneth formula, 260, 267

L

Laplacian, 258
Leray theorem, 87
Levi-Cività connection, 177
Lie derivative, 153, 154
Lie (Banach) group, 144
    complex, 200
Lift, 100
    horizontal, 168
Lifting function, 100
Lipshitz-Killing curvature, 179

M

Manifold ($\mathscr{P}$), 38
    almost complex, 191
    complex, 188
    differentiable, 140
    foliated, 180
    Hermitian, 195
    Kählerian, 196
    orientable, 164
    Riemannian, 175
    foliated, 184
Map(ping), 2
    differentiable, 141
    foliated, 183
    holomorphic, 189
    partial, 3
Module
    bigraded, 76
    differential, 24
    graded, 24
Monomorphism, 11
Morphism, 3
    coimage of, 19
    cokernel of, 19
    corange of, 13
    functorial, 33
    image of, 19
    inverse of, 11
    kernel of, 18
    range of, 13
    regular, 117, 142

N

Natural basis, 150
Natural equivalence, 33
Natural transformation, 33
Nerve, 61
Nijenhuis tensor, 192

O

Object, 4
    copointed, 18
    final, 12
    initial, 12
    pointed, 18
    quotient, 13
    zero, 12
Operator,
    adjoint, 253
    differential, 248, 250
    hypoelliptic, 250
Operator
    k-elliptic, 248, 251
    self-adjoint, 255
    symbol of, 250

P

$\mathscr{P}$-manifold, 38
$\mathscr{P}$-mapping, 40
$\mathscr{P}$-submanifold, 41

Parallel translation, 168
Poincaré lemma, 201
Pontrjagin classes, 130, 241, 245
Presheaf(ves), 43
    canonical, 53
    exact sequence of, 46
    homomorphism of, 44
    image of, 47
    module, 48
    restriction of, 47
    simple (constant), 47
Projective family, 17
Projective limit, 17
Pseudocategory, 4
    Cantorian, 5
    dual, 7
    topological, 5

R

Reduction of the structure group, 110
Reinhart space, 186
Resolution, 69, 82
    fine, 85
de Rham cohomology, 157
de Rham complex, 157
de Rham theorem, 70, 82, 203
Ricci tensor, 171, 179

S

Section, 44, 49
Serre theorem, 276
Set, 7
    universal, 3
Sheaf(ves), 48
    $\mathfrak{C}$-valued, 54
    coherent, 59
    constant, 60
    direct image of, 54
    fine, 84
    of germs, 52
    homomorphism, 51
    induced, 51
    inverse image of, 51
    module, 58
    of sets, 50
    trivial extension of, 54
Simplex, 159
    boundary of, 160
    singular, 159
    standard, 159
Space
    classifying, 112
    projected, 49
    trivial projected, 93
Spectral sequence, 72
Stalk, 51
Stiefel–Whitney classes, 131
Stokes' formula, 164, 165
Subbundle, 97
Subcategory, 8
    full, 8
Submanifold, 142
Submersion, 41, 142
Subobject, 13
    strengthened, 35
Surjection, 10
    strengthened, 35
System of coefficients, 78

T

Tangent object, 40
Tensor, 124
    field, 151
    on a manifold, 151
    product, 26, 124
Total source, 12
Total target, 12
Torsion forms, 169
Torsion tensor, 170
Transition functions, 37, 39, 94
Trivialization, 93

V

Vector, 150
    contravariant, 150
    covariant, 150
Vector field(s), 147
    bracket of, 148
    foliated, 184

W

Weil algebra, 233
Weil homomorphism, 235
Weil theorem, 233
Whitney sum, 116
Whitney theorem, 119

A CATALOG OF SELECTED
DOVER BOOKS
IN SCIENCE AND MATHEMATICS

CATALOG OF DOVER BOOKS

Mathematics-Bestsellers

HANDBOOK OF MATHEMATICAL FUNCTIONS: with Formulas, Graphs, and Mathematical Tables, Edited by Milton Abramowitz and Irene A. Stegun. A classic resource for working with special functions, standard trig, and exponential logarithmic definitions and extensions, it features 29 sets of tables, some to as high as 20 places. 1046pp. 8 x 10 1/2. 0-486-61272-4

ABSTRACT AND CONCRETE CATEGORIES: The Joy of Cats, Jiri Adamek, Horst Herrlich, and George E. Strecker. This up-to-date introductory treatment employs category theory to explore the theory of structures. Its unique approach stresses concrete categories and presents a systematic view of factorization structures. Numerous examples. 1990 edition, updated 2004. 528pp. 6 1/8 x 9 1/4. 0-486-46934-4

MATHEMATICS: Its Content, Methods and Meaning, A. D. Aleksandrov, A. N. Kolmogorov, and M. A. Lavrent'ev. Major survey offers comprehensive, coherent discussions of analytic geometry, algebra, differential equations, calculus of variations, functions of a complex variable, prime numbers, linear and non-Euclidean geometry, topology, functional analysis, more. 1963 edition. 1120pp. 5 3/8 x 8 1/2. 0-486-40916-3

INTRODUCTION TO VECTORS AND TENSORS: Second Edition--Two Volumes Bound as One, Ray M. Bowen and C.-C. Wang. Convenient single-volume compilation of two texts offers both introduction and in-depth survey. Geared toward engineering and science students rather than mathematicians, it focuses on physics and engineering applications. 1976 edition. 560pp. 6 1/2 x 9 1/4. 0-486-46914-X

AN INTRODUCTION TO ORTHOGONAL POLYNOMIALS, Theodore S. Chihara. Concise introduction covers general elementary theory, including the representation theorem and distribution functions, continued fractions and chain sequences, the recurrence formula, special functions, and some specific systems. 1978 edition. 272pp. 5 3/8 x 8 1/2.
0-486-47929-3

ADVANCED MATHEMATICS FOR ENGINEERS AND SCIENTISTS, Paul DuChateau. This primary text and supplemental reference focuses on linear algebra, calculus, and ordinary differential equations. Additional topics include partial differential equations and approximation methods. Includes solved problems. 1992 edition. 400pp. 7 1/2 x 9 1/4. 0-486-47930-7

PARTIAL DIFFERENTIAL EQUATIONS FOR SCIENTISTS AND ENGINEERS, Stanley J. Farlow. Practical text shows how to formulate and solve partial differential equations. Coverage of diffusion-type problems, hyperbolic-type problems, elliptic-type problems, numerical and approximate methods. Solution guide available upon request. 1982 edition. 414pp. 6 1/8 x 9 1/4. 0-486-67620-X

VARIATIONAL PRINCIPLES AND FREE-BOUNDARY PROBLEMS, Avner Friedman. Advanced graduate-level text examines variational methods in partial differential equations and illustrates their applications to free-boundary problems. Features detailed statements of standard theory of elliptic and parabolic operators. 1982 edition. 720pp. 6 1/8 x 9 1/4. 0-486-47853-X

LINEAR ANALYSIS AND REPRESENTATION THEORY, Steven A. Gaal. Unified treatment covers topics from the theory of operators and operator algebras on Hilbert spaces; integration and representation theory for topological groups; and the theory of Lie algebras, Lie groups, and transform groups. 1973 edition. 704pp. 6 1/8 x 9 1/4.
0-486-47851-3

Browse over 9,000 books at www.doverpublications.com

CATALOG OF DOVER BOOKS

A SURVEY OF INDUSTRIAL MATHEMATICS, Charles R. MacCluer. Students learn how to solve problems they'll encounter in their professional lives with this concise single-volume treatment. It employs MATLAB and other strategies to explore typical industrial problems. 2000 edition. 384pp. 5 3/8 x 8 1/2.   0-486-47702-9

NUMBER SYSTEMS AND THE FOUNDATIONS OF ANALYSIS, Elliott Mendelson. Geared toward undergraduate and beginning graduate students, this study explores natural numbers, integers, rational numbers, real numbers, and complex numbers. Numerous exercises and appendixes supplement the text. 1973 edition. 368pp. 5 3/8 x 8 1/2.   0-486-45792-3

A FIRST LOOK AT NUMERICAL FUNCTIONAL ANALYSIS, W. W. Sawyer. Text by renowned educator shows how problems in numerical analysis lead to concepts of functional analysis. Topics include Banach and Hilbert spaces, contraction mappings, convergence, differentiation and integration, and Euclidean space. 1978 edition. 208pp. 5 3/8 x 8 1/2.   0-486-47882-3

FRACTALS, CHAOS, POWER LAWS: Minutes from an Infinite Paradise, Manfred Schroeder. A fascinating exploration of the connections between chaos theory, physics, biology, and mathematics, this book abounds in award-winning computer graphics, optical illusions, and games that clarify memorable insights into self-similarity. 1992 edition. 448pp. 6 1/8 x 9 1/4.   0-486-47204-3

SET THEORY AND THE CONTINUUM PROBLEM, Raymond M. Smullyan and Melvin Fitting. A lucid, elegant, and complete survey of set theory, this three-part treatment explores axiomatic set theory, the consistency of the continuum hypothesis, and forcing and independence results. 1996 edition. 336pp. 6 x 9.   0-486-47484-4

DYNAMICAL SYSTEMS, Shlomo Sternberg. A pioneer in the field of dynamical systems discusses one-dimensional dynamics, differential equations, random walks, iterated function systems, symbolic dynamics, and Markov chains. Supplementary materials include PowerPoint slides and MATLAB exercises. 2010 edition. 272pp. 6 1/8 x 9 1/4.   0-486-47705-3

ORDINARY DIFFERENTIAL EQUATIONS, Morris Tenenbaum and Harry Pollard. Skillfully organized introductory text examines origin of differential equations, then defines basic terms and outlines general solution of a differential equation. Explores integrating factors; dilution and accretion problems; Laplace Transforms; Newton's Interpolation Formulas, more. 818pp. 5 3/8 x 8 1/2.   0-486-64940-7

MATROID THEORY, D. J. A. Welsh. Text by a noted expert describes standard examples and investigation results, using elementary proofs to develop basic matroid properties before advancing to a more sophisticated treatment. Includes numerous exercises. 1976 edition. 448pp. 5 3/8 x 8 1/2.   0-486-47439-9

THE CONCEPT OF A RIEMANN SURFACE, Hermann Weyl. This classic on the general history of functions combines function theory and geometry, forming the basis of the modern approach to analysis, geometry, and topology. 1955 edition. 208pp. 5 3/8 x 8 1/2.   0-486-47004-0

THE LAPLACE TRANSFORM, David Vernon Widder. This volume focuses on the Laplace and Stieltjes transforms, offering a highly theoretical treatment. Topics include fundamental formulas, the moment problem, monotonic functions, and Tauberian theorems. 1941 edition. 416pp. 5 3/8 x 8 1/2.   0-486-47755-X

CATALOG OF DOVER BOOKS

Mathematics–Probability and Statistics

BASIC PROBABILITY THEORY, Robert B. Ash. This text emphasizes the probabilistic way of thinking, rather than measure-theoretic concepts. Geared toward advanced undergraduates and graduate students, it features solutions to some of the problems. 1970 edition. 352pp. 5 3/8 x 8 1/2. 0-486-46628-0

PRINCIPLES OF STATISTICS, M. G. Bulmer. Concise description of classical statistics, from basic dice probabilities to modern regression analysis. Equal stress on theory and applications. Moderate difficulty; only basic calculus required. Includes problems with answers. 252pp. 5 5/8 x 8 1/4. 0-486-63760-3

OUTLINE OF BASIC STATISTICS: Dictionary and Formulas, John E. Freund and Frank J. Williams. Handy guide includes a 70-page outline of essential statistical formulas covering grouped and ungrouped data, finite populations, probability, and more, plus over 1,000 clear, concise definitions of statistical terms. 1966 edition. 208pp. 5 3/8 x 8 1/2. 0-486-47769-X

GOOD THINKING: The Foundations of Probability and Its Applications, Irving J. Good. This in-depth treatment of probability theory by a famous British statistician explores Keynesian principles and surveys such topics as Bayesian rationality, corroboration, hypothesis testing, and mathematical tools for induction and simplicity. 1983 edition. 352pp. 5 3/8 x 8 1/2. 0-486-47438-0

INTRODUCTION TO PROBABILITY THEORY WITH CONTEMPORARY APPLICATIONS, Lester L. Helms. Extensive discussions and clear examples, written in plain language, expose students to the rules and methods of probability. Exercises foster problem-solving skills, and all problems feature step-by-step solutions. 1997 edition. 368pp. 6 1/2 x 9 1/4. 0-486-47418-6

CHANCE, LUCK, AND STATISTICS, Horace C. Levinson. In simple, non-technical language, this volume explores the fundamentals governing chance and applies them to sports, government, and business. "Clear and lively ... remarkably accurate." – *Scientific Monthly*. 384pp. 5 3/8 x 8 1/2. 0-486-41997-5

FIFTY CHALLENGING PROBLEMS IN PROBABILITY WITH SOLUTIONS, Frederick Mosteller. Remarkable puzzlers, graded in difficulty, illustrate elementary and advanced aspects of probability. These problems were selected for originality, general interest, or because they demonstrate valuable techniques. Also includes detailed solutions. 88pp. 5 3/8 x 8 1/2. 0-486-65355-2

EXPERIMENTAL STATISTICS, Mary Gibbons Natrella. A handbook for those seeking engineering information and quantitative data for designing, developing, constructing, and testing equipment. Covers the planning of experiments, the analyzing of extreme-value data; and more. 1966 edition. Index. Includes 52 figures and 76 tables. 560pp. 8 3/8 x 11. 0-486-43937-2

STOCHASTIC MODELING: Analysis and Simulation, Barry L. Nelson. Coherent introduction to techniques also offers a guide to the mathematical, numerical, and simulation tools of systems analysis. Includes formulation of models, analysis, and interpretation of results. 1995 edition. 336pp. 6 1/8 x 9 1/4. 0-486-47770-3

INTRODUCTION TO BIOSTATISTICS: Second Edition, Robert R. Sokal and F. James Rohlf. Suitable for undergraduates with a minimal background in mathematics, this introduction ranges from descriptive statistics to fundamental distributions and the testing of hypotheses. Includes numerous worked-out problems and examples. 1987 edition. 384pp. 6 1/8 x 9 1/4. 0-486-46961-1

Browse over 9,000 books at www.doverpublications.com

CATALOG OF DOVER BOOKS

Astronomy

CHARIOTS FOR APOLLO: The NASA History of Manned Lunar Spacecraft to 1969, Courtney G. Brooks, James M. Grimwood, and Loyd S. Swenson, Jr. This illustrated history by a trio of experts is the definitive reference on the Apollo spacecraft and lunar modules. It traces the vehicles' design, development, and operation in space. More than 100 photographs and illustrations. 576pp. 6 3/4 x 9 1/4. 0-486-46756-2

EXPLORING THE MOON THROUGH BINOCULARS AND SMALL TELESCOPES, Ernest H. Cherrington, Jr. Informative, profusely illustrated guide to locating and identifying craters, rills, seas, mountains, other lunar features. Newly revised and updated with special section of new photos. Over 100 photos and diagrams. 240pp. 8 1/4 x 11. 0-486-24491-1

WHERE NO MAN HAS GONE BEFORE: A History of NASA's Apollo Lunar Expeditions, William David Compton. Introduction by Paul Dickson. This official NASA history traces behind-the-scenes conflicts and cooperation between scientists and engineers. The first half concerns preparations for the Moon landings, and the second half documents the flights that followed Apollo 11. 1989 edition. 432pp. 7 x 10.
0-486-47888-2

APOLLO EXPEDITIONS TO THE MOON: The NASA History, Edited by Edgar M. Cortright. Official NASA publication marks the 40th anniversary of the first lunar landing and features essays by project participants recalling engineering and administrative challenges. Accessible, jargon-free accounts, highlighted by numerous illustrations. 336pp. 8 3/8 x 10 7/8. 0-486-47175-6

ON MARS: Exploration of the Red Planet, 1958-1978--The NASA History, Edward Clinton Ezell and Linda Neuman Ezell. NASA's official history chronicles the start of our explorations of our planetary neighbor. It recounts cooperation among government, industry, and academia, and it features dozens of photos from Viking cameras. 560pp. 6 3/4 x 9 1/4. 0-486-46757-0

ARISTARCHUS OF SAMOS: The Ancient Copernicus, Sir Thomas Heath. Heath's history of astronomy ranges from Homer and Hesiod to Aristarchus and includes quotes from numerous thinkers, compilers, and scholasticists from Thales and Anaximander through Pythagoras, Plato, Aristotle, and Heraclides. 34 figures. 448pp. 5 3/8 x 8 1/2.
0-486-43886-4

AN INTRODUCTION TO CELESTIAL MECHANICS, Forest Ray Moulton. Classic text still unsurpassed in presentation of fundamental principles. Covers rectilinear motion, central forces, problems of two and three bodies, much more. Includes over 200 problems, some with answers. 437pp. 5 3/8 x 8 1/2. 0-486-64687-4

BEYOND THE ATMOSPHERE: Early Years of Space Science, Homer E. Newell. This exciting survey is the work of a top NASA administrator who chronicles technological advances, the relationship of space science to general science, and the space program's social, political, and economic contexts. 528pp. 6 3/4 x 9 1/4.
0-486-47464-X

STAR LORE: Myths, Legends, and Facts, William Tyler Olcott. Captivating retellings of the origins and histories of ancient star groups include Pegasus, Ursa Major, Pleiades, signs of the zodiac, and other constellations. "Classic." – *Sky & Telescope.* 58 illustrations. 544pp. 5 3/8 x 8 1/2. 0-486-43581-4

A COMPLETE MANUAL OF AMATEUR ASTRONOMY: Tools and Techniques for Astronomical Observations, P. Clay Sherrod with Thomas L. Koed. Concise, highly readable book discusses the selection, set-up, and maintenance of a telescope; amateur studies of the sun; lunar topography and occultations; and more. 124 figures. 26 halftones. 37 tables. 335pp. 6 1/2 x 9 1/4. 0-486-42820-6

Browse over 9,000 books at www.doverpublications.com

Chemistry

MOLECULAR COLLISION THEORY, M. S. Child. This high-level monograph offers an analytical treatment of classical scattering by a central force, quantum scattering by a central force, elastic scattering phase shifts, and semi-classical elastic scattering. 1974 edition. 310pp. 5 3/8 x 8 1/2. 0-486-69437-2

HANDBOOK OF COMPUTATIONAL QUANTUM CHEMISTRY, David B. Cook. This comprehensive text provides upper-level undergraduates and graduate students with an accessible introduction to the implementation of quantum ideas in molecular modeling, exploring practical applications alongside theoretical explanations. 1998 edition. 832pp. 5 3/8 x 8 1/2. 0-486-44307-8

RADIOACTIVE SUBSTANCES, Marie Curie. The celebrated scientist's thesis, which directly preceded her 1903 Nobel Prize, discusses establishing atomic character of radioactivity; extraction from pitchblende of polonium and radium; isolation of pure radium chloride; more. 96pp. 5 3/8 x 8 1/2. 0-486-42550-9

CHEMICAL MAGIC, Leonard A. Ford. Classic guide provides intriguing entertainment while elucidating sound scientific principles, with more than 100 unusual stunts: cold fire, dust explosions, a nylon rope trick, a disappearing beaker, much more. 128pp. 5 3/8 x 8 1/2. 0-486-67628-5

ALCHEMY, E. J. Holmyard. Classic study by noted authority covers 2,000 years of alchemical history: religious, mystical overtones; apparatus; signs, symbols, and secret terms; advent of scientific method, much more. Illustrated. 320pp. 5 3/8 x 8 1/2. 0-486-26298-7

CHEMICAL KINETICS AND REACTION DYNAMICS, Paul L. Houston. This text teaches the principles underlying modern chemical kinetics in a clear, direct fashion, using several examples to enhance basic understanding. Solutions to selected problems. 2001 edition. 352pp. 8 3/8 x 11. 0-486-45334-0

PROBLEMS AND SOLUTIONS IN QUANTUM CHEMISTRY AND PHYSICS, Charles S. Johnson and Lee G. Pedersen. Unusually varied problems, with detailed solutions, cover of quantum mechanics, wave mechanics, angular momentum, molecular spectroscopy, scattering theory, more. 280 problems, plus 139 supplementary exercises. 430pp. 6 1/2 x 9 1/4. 0-486-65236-X

ELEMENTS OF CHEMISTRY, Antoine Lavoisier. Monumental classic by the founder of modern chemistry features first explicit statement of law of conservation of matter in chemical change, and more. Facsimile reprint of original (1790) Kerr translation. 539pp. 5 3/8 x 8 1/2. 0-486-64624-6

MAGNETISM AND TRANSITION METAL COMPLEXES, F. E. Mabbs and D. J. Machin. A detailed view of the calculation methods involved in the magnetic properties of transition metal complexes, this volume offers sufficient background for original work in the field. 1973 edition. 240pp. 5 3/8 x 8 1/2. 0-486-46284-6

GENERAL CHEMISTRY, Linus Pauling. Revised third edition of classic first-year text by Nobel laureate. Atomic and molecular structure, quantum mechanics, statistical mechanics, thermodynamics correlated with descriptive chemistry. Problems. 992pp. 5 3/8 x 8 1/2. 0-486-65622-5

ELECTROLYTE SOLUTIONS: Second Revised Edition, R. A. Robinson and R. H. Stokes. Classic text deals primarily with measurement, interpretation of conductance, chemical potential, and diffusion in electrolyte solutions. Detailed theoretical interpretations, plus extensive tables of thermodynamic and transport properties. 1970 edition. 590pp. 5 3/8 x 8 1/2. 0-486-42225-9